ANTISEMITISM

THE FELIX POSEN BIBLIOGRAPHIC PROJECT ON ANTISEMITISM

Academic Advisor: *Otto Dov Kulka*

Technical Editors: *Sara Grosvald, Sylviane Stampfer*

Abstracting Staff:
Marian Assaf, Yisrael (Elliot) Cohen, Ilana Dana, Ruth Engelberg, Mirjam Factor, Lily Fogel (z"l), Robert E. Kaplan, Daniel Romanovsky, Tamar Stern, Jiří Tobíšek, Ágnes Vázsonyi, Hanna Volovici, Renate Wolfson, Manuel Zkorenblut

We thank the Fondation pour la mémoire de la Shoah for their support of this project

The Vidal Sassoon International Center
for the Study of Antisemitism
The Hebrew University of Jerusalem

ANTISEMITISM

An Annotated Bibliography

Volume 17
2001

Edited by
Susan Sarah Cohen

K · G · Saur München 2004

Bibliographic information published by Die Deutsche Bibliothek
Die Deutsche Bibliothek lists this publication in the Deutsche Nationalbibliografie;
detailed bibliographic data is available in the Internet at
http://dnb.ddb.de

∞
Gedruckt auf säurefreiem Papier / Printed on acid free paper

Printed in Germany

Printed/Bound by Strauss GmbH, Mörlenbach
ISBN 3-598-23716-2

CONTENTS

INTRODUCTION

The Vidal Sassoon International Center for the Study of Antisemitism, established by an international group of scholars in 1982, is an interdisciplinary research center which engages in research on antisemitism throughout the ages. The Felix Posen Bibliographic Project of the Center comprises a current on-line data base accessible through the Israel University Inter-Library Network and printed volumes.

The Bibliographic Project has a unique position in the world of scholarship for several reasons. First, it is truly comprehensive, i.e. it lists books and articles published throughout the world on the subject of antisemitism. This is made possible through the very special connection which the Project's research team has developed with the Jewish National and University Library located on the Givat Ram campus of the Hebrew University of Jerusalem. The National Library, itself a priceless asset for Israel, the Diaspora and world-wide Judaica scholarship, is committed to acquiring the majority of works published around the world dealing with Jews and Judaism. The Felix Posen Bibliographic Project and its highly trained staff of abstractors have immediate access to the library's acquisitions which greatly facilitates its endeavors. These holdings include works from a diverse range of disciplines – history, psychology, sociology, anthropology, literature and art.

The second unique aspect has been the cumulative and consistent service that the Felix Posen Project has provided for researchers in the field since it first began publication in 1984. Other bibliographies are much more limited in scope and rarely offer more than a one-time publication. This project has been continuous, ongoing and in 2004 is celebrating its twentieth anniversary.

Third, the Bibliographic Project is not confined to a particular language or country. One can find entries on works in all European languages, in Hebrew, Yiddish, and other languages where relevant. Thus, its range and sweep is impressive and uncompromisingly global.

Fourth, there is the special quality of our experienced staff of abstractors, who have immigrated to Israel from many countries such as the United States, Russia, Romania, Poland, Hungary, Switzerland and Argentina. Their ability to read works in their original languages and their knowledge of Jewish history makes it possible for them to produce reliable abstracts of high quality.

Fifth, the abstracts themselves are substantial and are provided for each item in the bibliography. This is a service that is unique in the world when it comes to the study of antisemitism – and of immense benefit to professional scholars, students, lay people, community leaders and others who seek a truly dependable guide to the constantly expanding ocean of newly published materials on antisemitism. Since no one person can be expected to master all the relevant publications or languages, our Project becomes literally indispensable for any individual interested both in the broad picture and in the concrete details of this particular subject matter. Elsewhere, the reader will find only bibliographies with no abstracts or at best a very brief one. The Bibliographic Project, on the other hand, offers readers a real window into the essence of what is being published today across the globe. One can hardly exaggerate the significance of this research asset.

Since October 2002, when I became the head of the Vidal Sassoon International Center for the Study of Antisemitism I have come to appreciate more than ever the dedicated staff associated with the Felix Posen Project. At the same time, I have had to rethink and redefine some of the purposes of the bibliography, especially where it concerns borderline cases. Hence, the present edition has been streamlined and somewhat condensed. There have been some changes in the abstracting of publications relating to post-war reactions to the Shoah - e.g. Holocaust commemoration, memory and meaning, study and teaching; these are still being listed, although in smaller numbers.

On the other hand, the bibliography continues to abstract all the publications dealing with antisemitism in the Holocaust period (1933-1945). This treatment is remarkably comprehensive, including not only works on antisemitic ideology, policy and attitudes but also materials dealing with the wider Jewish experience during the Shoah.

We have continued the earlier bibliographic policy regarding the Middle East. Hence we do not include works in the present volume whose main interest is the political conflict between Israel, the Palestinians and the Arab states. Those books and articles listed here, relate strictly to antisemitism in the Arab world. Nevertheless, we will have to take account of the fact that the number of items concerning Arab and Muslim Judeophobia has greatly expanded since 2000. Therefore, I have already committed the Sassoon Center to adopt a much broader coverage of this issue, one that has already had violent overspill effects in Europe, Asia and the American continent. The Bibliographic Project will also be looking to cover relevant works about non-Arab Muslim antisemitism which has been somewhat neglected in the past.

Another future challenge is how our research project will grapple bibliographically with the issue of anti-Zionism. Hitherto, the policy has been to list only the literature that specifically describes anti-Zionism as a form of antisemitism; particularly those works that deal with explicitly antisemitic accusations directed at Zionism or Israel. Clearly, discussions of themes that relate to the "world Jewish conspiracy," "Zionist collaboration with the Nazis" or the equation of Israel/Zionism with racism and Nazism, do warrant an abstract. But what about the less obvious manifestations of contemporary Judeophobia? It seems to me that we may have to widen our research to analyse newer and more subtle examples of antisemitism which deny any moral legitimacy to Israel, which apply double standards to its behavior or attribute to it imaginary crimes. Then, there is also the controversial issue of antisemitism concealing itself under the mask of "anti-racism," "human rights" and protests against globalization ("Americanization"). Such complex discourses will offer a difficult challenge to bibliographers in the coming years.

In discussing these methodological issues, it is important to clarify a common misconception about the Bibliographic Project. This bibliography has never sought to cover antisemitic publications per se. The Felix Posen Project is a bibliography that includes the great bulk of the relevant literature about antisemitism over the past two millennia.

We are living in a time of sweeping and dramatic changes which inevitably affect our research project. The subject of antisemitism has never seemed more urgent and timely in recent decades than it is today. Our project has therefore acquired a special significance at the turn of the new millennium.

Professor Robert S. Wistrich
Head of the Vidal Sassoon International Center for the Study of Antisemitism

ACKNOWLEDGMENTS

Our special thanks to Mr. Felix Posen for his continuing and generous support, to the staff of the Jewish National and University Library for their cooperation and to the Fondation pour la mémoire de la Shoah for helping us to ensure the continuation of this enterprise.

PREFACE

The present edition of the bibliography includes a selection of works about antisemitism – books, dissertations, masters' theses, and articles from periodicals and collections – published in 2001, with supplementary material from 1998-2000 not published in the previous volumes. The bibliography does not include newspaper articles, reviews, audio-visual material, and works of fiction, nor does it cover antisemitic publications (except for critical editions published for scholastic purposes, which include a scholarly introduction – e.g. the diaries of Joseph Goebbels, the speeches of Adolf Hitler).

For the purpose of this bibliography, antisemitism is defined as antagonism toward Jews and Judaism as expressed in writings (e.g. the New Testament, polemical literature, works of fiction), in the visual arts (e.g. art, caricatures, films), and in actions (e.g. massacres and pogroms, discriminatory legislation, the Holocaust). Works are included which deal with philosemitism and Jewish responses to antisemitism. In borderline cases (e.g. dealing with Christian-Jewish relations, anti-Israel and anti-Zionism, works of fiction and films), books and articles are listed when the author of the work relates to the subject of antisemitism.

The bibliography lists all the publications dealing with antisemitism in the Holocaust period (1933-1945). It also lists publications on postwar reactions to the Holocaust – e.g. Holocaust commemoration, memory and meaning, study and teaching; it does not deal with the psychological effects of the Holocaust on survivors and their descendants.

In regard to Soviet Jewry, works are included that deal with Soviet ideology and policy toward the Jews but not with the Jewish experience as such.

In regard to anti-Zionism, works are included that describe anti-Zionism as antisemitism, particularly those that discuss antisemitic accusations vaunted against Zionism and Israel, such as the "world Jewish conspiracy," Zionist collaboration with Nazis, the "Zionism equals racism" formula, and the comparison of Israelis with Nazis.

In regard to the Arab-Israeli conflict, works are included that deal with antisemitism in the Arab world but not with the political conflict between the State of Israel and its Arab neighbors.

The listings were compiled mainly from the holdings of the Jewish National and University Library in Jerusalem, which receives the majority of works on Jewish subjects published throughout the world. The works listed come from a diverse range of disciplines – history, psychology, sociology, anthropology, literature, art.

The bibliography is multilingual, containing works in the European languages, Hebrew, and Yiddish, and in other languages as they come to our attention. An English translation of the title is given in brackets for languages other than French, German, Italian, and Spanish.

The format of each entry follows the rules of the Chicago Manual of Style, with slight modification.

Each item is listed only once – a work dealing with two or more of the chronological or geographical subdivisions is listed in the appropriate "General" section. The detailed subject index at the end of the volume compensates for the "general" classifications.

The geographical chapter divisions are according to present-day boundaries, but exceptions have been made in the case of the former Czechoslovakia, the USSR, and Yugoslavia, which are still given as headings for items dealing with those countries in general.

In the subject index, names of rulers (kings, queens, emperors) are given in the original language (e.g. Enrique, Heinrich, Wilhelm). Names of places are divided – countries and capital cities are given in English, while villages, towns, and cities are given in the original language. Geographical locations are listed according to present-day boundaries (e.g. items on Breslau will be entered under Wrocław and in the chapter for Poland). Cross-references are given in such cases.

Authors of memoirs and autobiographies are listed in both the author and subject indexes.

Items which are a translation or a revised version of a work published previously are accompanied by a note stating where and when the original version appeared. If it was published in 1984 or after (when this bibliography began), the abstract for that item will be found in the appropriate volume of the bibliography. If it was not listed earlier, the abstract will accompany the later edition or translation.

The annotation expresses the view of the author(s) of the work described and not those of the annotator. The length of the annotation does not indicate the importance of a particular work.

ACKNOWLEDGMENTS

The Vidal Sassoon International Center for the Study of Antisemitism wishes to thank the administration and the staff of the Jewish National and University Library for their support and cooperation, without which this volume could not have been produced.

Jerusalem S.S.C.
June 2004 — Tammuz 5764

LIST OF PERIODICALS AND COLLECTIONS

Abgestempelt: Judenfeindliche Postkarten, eds. Helmut Gold, Georg Heuberger. Heidelberg: Umschau Braus, 1999 (no. 859)
Acta Judaica Slovaca. Bratislava: Edicia Judaica Slovaca
American Behavioral Scientist. Thousand Oaks, CA: Sage
American Jewish History. Waltham, MA: American Jewish Historical Society
Anglica Wratislaviensia. Wrocław: Wyd. Uniwersytetu Wrocławskiego
Annales Universitatis Mariae Curie-Skłodowska. Sectio F: Historia. Lublin: Uniwersytet Marii Curie-Skłodowskiej
Anti-Judaism and the Fourth Gospel: Papers of the Leuven Colloquium, 2000, eds. R. Bieringer et al. Assen: Royal Van Gorcum, 2001 (no. 47)
Antisemitismo en Uruguay: Raíces, discursos, imágenes (1870-1940). Montevideo: Trilce, 2000 (no. 175)
Antisemitismus - die deutsche Normalitaet: Geschichte und Wirkungsweise des Vernichtungswahns. Freiburg: ça ira, 2001 (no. 32)
Antisemitizmus na konci 20. storočia: Zborník referatov z konferencie (Nitra, 15.-17. maj 2000), ed. Pavol Mešt'an. Bratislava: Judaica Slovaca, 2001 (no. 716)
Die Araber und die Shoa: Ueber die Schwierigkeit dieser Konjunktion, ed. Goetz Nordbruch. Trier: Kulturverein AphorismA, 2000 (no. 705)
Arcadia: Internationale Zeitschrift fuer Literaturwissenschaft. Berlin: Walter de Gruyter
Arche. Minsk: Redaktsyinaya kalehiya "Arche"
L'Arche. Paris: Fonds Social Juifs Unifié
Archiv fuer Kulturgeschichte. Koeln: Boehlau
Aschkenas: Zeitschrift fuer Geschichte und Kultur der Juden. Wien: Boehlau
Das August-Gottschalk-Haus in Esens: Gedenkstaette und Ausstellung zur neueren Geschichte der ostfriesischen Juden. Esens: Oekumenischer Arbeitskreis Juden und Christen, 2000 (no. 466)
Australian Jewish Historical Society Journal. Sydney and Melbourne
Australian Journal of Jewish Studies. Sydney: Australian Association for Jewish Studies
Avotaynu: The International Review of Jewish Genealogy. Teaneck, NJ: Avotaynu
Belaruski Histarychny Ahlyad. Mensk: Belaruski Histarychny Ahlyad
Bernard Lazare, anarchiste et nationaliste juif, ed. Philippe Oriol. Paris: Honoré Champion, 1999 (no. 196)
Best Contemporary Jewish Writing, ed. Michael Lerner. San Francisco: Jossey-Bass, 2001 (no. 731)
Bijdragen tot de Eigentijdse Herinnering see Les Cahiers de la Mémoire Contemporaine
Biography: An Interdisciplinary Quarterly. Honolulu: University of Hawai`i Press

Bishvil ha-Zikaron. Jerusalem: Yad Vashem
Biuletyn Żydowskiego Instytutu Historycznego. Warszawa: Żydowski Instytut Historyczny
The Bombing of Auschwitz: Should the Allies Have Attempted It?, eds. Michael J. Neufeld, Michael Berenbaum. New York: St. Martin's Press, in association with the United States Holocaust Memorial Museum, 2000 (no. 252)
Buergerliche Frauenbewegung und Antisemitismus, eds. Mechthild Bereswill, Leonie Wagner. Tuebingen: Diskord, 1998 (no. 150)
Bund: 100 lat historii, 1897-1997, eds. Feliks Tych, Juergen Hensel. Warszawa: Żydowski Instytut Historyczny, 2000 (no. 581)
Cahier International sur le Témoignage Audiovisuel: Etudes sur le témoignage audiovisuel des victimes des crimes et génocides nazis. Bruxelles: Centre d'Etudes et de Documentation - Fondation Auschwitz
Les Cahiers de la Mémoire Contemporaine. Bruxelles: Fondation de la Mémoire Contemporaine
Cahiers de l'IRP. Lausanne: Institut Romand de Pastorale
Cahiers du Monde Russe. Paris: Ecole des Hautes Etudes en Sciences Sociales
Cardozo Law Review. New York: Yeshiva University, Benjamin N. Cardozo School of Law
Cattolici, ebrei ed evangelici nella guerra: Vita religiosa e società, 1939-1945, eds. Bartolo Gariglio, Riccardo Marchis. Milano: FrancoAngeli, 1999 (no. 510)
CCAR Journal: A Reform Jewish Quarterly. New York: Central Conference of American Rabbis
Central European History. Leiden: Brill
Christian-Jewish Relations through the Centuries, eds. Stanley E. Porter, Brook W.R. Pearson. Sheffield: Sheffield University Press, 2000 (no. 22)
Chroniques du désastre: Témoignages sur la Shoah dans les ghettos polonais, ed. Nathan Weinstock. Genève: Metropolis, 1999 (no. 583)
Chulyot: Dapim le-mehkar be-sifrut yiddish ve-zikotehah le-sifrut ha-ivrit. Haifa: Haifa University; Tel-Aviv: Tel-Aviv University
Church History. Red Bank, NJ: American Society of Church History
Commentary. New York: American Jewish Committee
Le Comunità ebraiche a Modena e a Carpi, dal medioevo all'età contemporanea, eds. Franco Bonilauri, Vincenza Maugeri. Firenze: Giuntina, 1999 (no. 37)
Contemporary Literature. Madison: University of Wisconsin Press
Contemporary Portrayals of Auschwitz: Philosophical Challenges, eds. Alan Rosenberg et al. Amherst, NY: Humanity Books, 2000 (no. 736)
Cross Currents. New York: Association for Religion and Intellectual Life
La cuestión judía: Seminario y... - un seminario, debates, análisis y opiniones sobre el tema. Buenos Aires: Ediciones Cuadernos Marxistas, 2001 (no. 144)
Dachauer Hefte: Studien und Dokumente zur Geschichte der nationalsozialistischen Konzentrationslager. Dachau: Dachauer Hefte
Debatte: Review of Contemporary German Affairs. Basingstoke, Hants: Carfax
Debreceni Szemle. Debrecen: Debreceni Szemle

Defusim shel hantsahah: Asufat ma'amarim, eds. Matitiahu Mayzel, Ilana Shamir. Tel-Aviv: Ministry of Defence, 2000 (no. 709)
Deutsche Vierteljahrsschrift fuer Literaturwissenschaft und Geistesgeschichte. Stuttgart: J.B. Metzler
Deutschland Archiv: Zeitschrift fuer das vereinigte Deutschland. Leverkusen: Leske + Budrich
Dia-Logos: Almanakh. Moskva: "Istina i Zhizn"
Dialectical Anthropology. Dordrecht: Kluwer
Dimensions: A Journal of Holocaust Studies. New York: International Center for Holocaust Studies, ADL
Dor gerush Sefarad: Kovetz ma'amarim, eds. Yom Tov Assis, Yosef Kaplan. Jerusalem: Zalman Shazar Center for Jewish History, 1999 (no. 57)
Drugoi narod, Vol. 1, ed. Ye.S. Chekanovkina. Lugansk: Znaniye, 2000 (no. 68)
Duchowość żydowska w Polsce: Materiały z międzynarodowej konferencji dedykowanej pamięci Profesora Chone Shmeruka, Kraków, 1999, ed. Michał Galas. Kraków: Księgarnia Akademicka, 2000 (no. 40)
East European Jewish Affairs. London: Frank Cass
East European Quarterly. Boulder, CO: University of Colorado
ELH - English Literary History. Baltimore: Johns Hopkins University Press
Emile Zola, "J'accuse...!" Réactions nationales et internationales, ed. Karl Zieger. [Valenciennes]: Presses Universitaires de Valenciennes, 1999 (no. 180)
Encyclopedia of the Holocaust, eds. Robert Rozett, Shmuel Spector. Jerusalem: Yad Vashem; New York: Facts on File, 2000 (no. 10)
Esprit. Paris: Esprit
Etudes Balkaniques. Sofia: Académie Bulgare des Sciences
European Judaism. Oxford: Berghahn Books; London: Leo Baeck College
The European Legacy. Basingstoke, Hants: Carfax
Evangelischer Arbeitskreis Kirche und Israel in Hessen und Nassau: Materialdienst. Heppenheim
Exil: Forschung, Erkenntnisse, Ergebnisse 1933-1945. Maintal, Germany: Edita Koch
Exilforschung. Muenchen: Text + Kritik
Flucht oder Tod: Erinnerungen und Briefe pommerscher Juden ueber die Zeit vor und nach 1945, ed. Wolfgang Wilhelmus. Rostock: Ingo Koch Verlag, 2001 (no. 586)
Fluechtlingspolitik und Fluchthilfe, eds. Susanne Heim et al. Berlin: Verlag der Buchlaeden Schwarze Risse, Rote Strasse, 1999 (no. 244)
"...das Fluestern eines leisen Wehens...": Beitraege zu Kultur und Lebenswelt europaeischer Juden: Festschrift fuer Utz Jeggle, ed. Freddy Raphael. Konstanz: UVK, 2001 (no. 35)
Frauen erinnern: Widerstand - Verfolgung - Exil, 1933-1945, eds. Inge Hansen-Schaberg, Beate Schmeichel-Falkenberg. Berlin: Weidler Buchverlag, 2000 (no. 775)
Freiburger Rundbrief: Zeitschrift fuer christlich-juedische Begegnung. Freiburg i.Br.: Freiburger Rundbrief

The French Review. Carbondale, IL: American Association of Teachers of French
Fritz Bauer Institut: Jahrbuch zur Geschichte und Wirkung des Holocaust, 2001 see "Gerichtstag halten ueber uns selbst..."
Furchtbare Antisemiten, ehrbare Antizionisten: Ueber Israel und die linksdeutsche Ideologie. Freiburg i.Br.: ça ira, 2000 (no. 773)
Le Genre Humain. Paris: Seuil
"Gerichtstag halten ueber uns selbst...": Geschichte und Wirkung des ersten Frankfurter Auschwitz-Prozesses, ed. Irmtrud Wojak. Frankfurt: Campus, 2001 (nos. 663, 747)
German History: The Journal of the German History Society. London: Arnold
German Life and Letters. Oxford: Blackwell
German Studies Review. Northfield, MN: German Studies Association at Carleton College
Geschichte in Wissenschaft und Unterricht: Zeitschrift des Verbandes der Geschichtslehrer Deutschlands. Seelze: Erhard Friedrich Verlag
Geschichte und Kultur der Juden in Schwaben II: Neuere Forschungen und Zeitzeugenberichte, ed. Peter Fassl. Stuttgart: Thorbecke, 2000 (no. 436)
Gesher: Journal of Jewish Affairs. Jerusalem: World Jewish Congress
Gesher: Ktav-et le-inyanim yehudiyim. Jerusalem: World Jewish Congress
The "Goldhagen Effect": History, Memory, Nazism - Facing the German Past, ed. Geoff Eley. Ann Arbor: University of Michigan Press, 2000 (no. 725)
Hagigei Giv`ah. Giv`at Washington, Israel: Giv`at Washington Religious Educational College
La haine de soi: Difficiles identités, eds. Esther Benbassa, Jean-Christophe Attias. Bruxelles: Editions Complexe, 2000 (no. 118)
Die Hamburger Kauffrau Glikl: Juedische Existenz in der Fruehen Neuzeit, ed. Monika Richarz. Hamburg: Christians, 2001 (nos. 86, 94)
Handbuch zur Geschichte der Juden in Europa, Vol. 1-2, eds. Elke-Vera Kotowski et al. Darmstadt: Primus, 2001 (no. 18)
Harvard Theological Review. Cambridge, MA: Harvard University
Hass, Verfolgung und Toleranz: Beitraege zum Schicksal der Juden von der Reformation bis in die Gegenwart, eds. Thomas Sirges, Kurt Erich Schoendorf. Frankfurt a.M.: Peter Lang, 2000 (no. 25)
Hat Israel noch eine Chance? Palaestina in der neuen Weltordnung, ed. Hermann L. Gremliza. Hamburg: Konkret, 2001 (no. 751)
Hebrew Language and Jewish Studies, eds. Aharon Ben-David, Isaac Gluska. Netanya: Association for Society and Culture, 2001 (no. 164)
Heine-Jahrbuch. Stuttgart: Verlag J.B. Metzler
Hinukh le-tarbut be-hevrah rav-tarbutit, ed. Miriam Barlev. Jerusalem: Hebrew University, School of Education, 2000 (no. 720)
Hinukh ve-historia: Heksherim tarbutiim u-politiim, eds. Rivka Feldhay, Immanuel Etkes. Jerusalem: Zalman Shazar Center for Jewish History, 1999 (nos. 709, 826)
L'Histoire. Paris: Société d'Editions Scientifiques
Historical Journal of Film, Radio and Television. Basingstoke, Hants: Carfax

Historical Reflections/Réflexions Historiques. Alfred, NY: Historical Reflections
Historische Zeitschrift. Muenchen: R. Oldenbourg
Historisches Jahrbuch. Freiburg i.Br.: Karl Alber
History of Political Economy. Durham, NC: Duke University Press
History Today. London: History Today
Hochschule Ost: Leipziger Beitraege zu Hochschule & Wissenschaft. Leipzig: Arbeitskreis Hochschulpolitische Oeffentlichkeit beim StuRa der Universitaet Leipzig. 8, 1-2 see Juedische Intellektuelle in der DDR
Holocaust and Genocide Studies. Cary, NC: Oxford University Press
The Holocaust and the Text: Speaking the Unspeakable, eds. Andrew Leak, George Paizis. Basingstoke, Hants: Macmillan Press, 2000 (no. 879)
Holocaust Füzetek. Budapest: Magyar Auschwitz Alapítvány
A Holocaust Reader: Responses to the Nazi Extermination, ed. Michael L. Morgan. New York: Oxford University Press, 2001 (no. 652)
Human Rights Quarterly. Baltimore: Johns Hopkins University Press
Hundert Jahre Zionismus - Befreiung oder Unterdrueckung? Beitraege der Gegentagung zum Herzl-Jubilaeum. Koeln: ISP, 1998 (no. 154)
In God's Name: Genocide and Religion in the Twentieth Century, eds. Omer Bartov, Phyllis Mack. New York: Berghahn Books, 2001 (no. 265)
Indice: Revista de ciencias sociales. Buenos Aires: DAIA, Centro de Estudios Sociales
Indonesia. Ithaca, NY: Cornell University
Inquisición y sociedad, ed. Ángel de Prado Moura. Valladolid: Universidad de Valladolid, 1999 (nos. 56, 80)
International Journal of Psychoanalysis. London: Institute of Psychoanalysis
International Journal on Audio-Visual Testimony see Cahier International sur le Témoignage Audiovisuel
International Review of Social History: Supplement. Cambridge: Cambridge University Press, for the International Institute of Social History
Internationale Kirchliche Zeitschrift. Bern: Staempfli
Internationales Josephus-Kolloquium, Amsterdam 2000, ed. Juergen U. Kalms. Muenster: LIT, 2001 (no. 46)
Isis: An International Review Devoted to the History of Science and Its Cultural Influences. Chicago: University of Chicago Press
Israel als Gegenueber: Vom Alten Orient bis in die Gegenwart: Studien zur Geschichte eines wechselvollen Zusammenlebens, ed. Folker Siegert. Goettingen: Vandenhoeck & Ruprecht, 2000 (no. 24)
Israel as Center Stage: A Setting for Social and Religious Enactments, eds. A. Paul Hare, Gideon M. Kressel. Westport, CT: Bergin & Garvey, 2001 (no. 824)
Israel Studies. Bloomington: Indiana University Press
Istoki: Vestnik Narodnovo Universiteta Yevreiskoi Kultury v Vostochnoi Ukraine. Kharkov: "Yevreiskii Mir"
Istoriya yevreiskovo naroda: Materialy shestoi yezhegodnoi mezhdunarodnoi mezhdistsiplinarnoi konferentsii po iudaike, eds. Rashid M. Kaplanov et al. Moskva: Sefer, 1999 (no. 16)

Jahrbuch des Freien Deutschen Hochstifts. Tuebingen: Max Niemeyer
Jahrbuch fuer Antisemitismusforschung. Frankfurt a.M.: Campus
Jahrbuch fuer Biblische Theologie. Neukirchen-Vluyn: Neukirchener Verlag
Jewish Bible Quarterly. Jerusalem: Jewish Bible Association
Jewish Exile in India, 1933-1945, eds. Anil Bhatti, Johannes H. Voigt. New Delhi: Manohar, in association with Max Mueller Bhavan, 1999 (no. 635)
Jewish Frontier: A Labor Zionist Journal. New York: Labor Zionist Letters
Jewish Social Studies. Bloomington: Indiana University Press
Jewish Studies. Jerusalem: World Union of Jewish Studies
Jewish Themes in English and Polish Culture, eds. Irena Janicka-Świderska et al. Łódź: Wydawn. Uniwersytetu Łódzkiego, 2000 (no. 866)
Jews and Slavs, Vol. 6, eds. Wolf Moskovich et al. Jerusalem: Center for Slavic Languages and Literatures, Hebrew University; Ljubljana: Scientific Research Center of the Slovenian Academy of Sciences and Arts, 1999 (nos. 117, 901)
The Jews and the Expansion of Europe to the West, 1450 to 1800, eds. Paolo Bernardini, Norman Fiering. New York: Berghahn Books, 2001 (no. 84)
Jews in Eastern Europe. Jerusalem: Hebrew University, Centre for Research and Documentation of East European Jewry
Jews in Silesia, eds. Marcin Wodzinski, Janusz Spyra. Cracow: Księgarnia Akademicka; [Wrocław]: University of Wrocław, Research Centre for the Culture and Languages of Polish Jews; [Český Těšín]: Muzeum Těšínska, 2001 (no. 361)
Journal of Church and State. Waco, TX: Baylor University
Journal of Contemporary History. London: Sage
Journal of Ecumenical Studies. Philadelphia: Temple University
Journal of European Studies: Literature and Ideas from the Renaissance to the Present. Chalfont St Giles, Bucks: Alpha Academic
Journal of Genocide Research. Basingstoke, Hants: Carfax
The Journal of Holocaust Education. London: Frank Cass
The Journal of Indo-Judaic Studies. Saskatoon: University of Saskatchewan
Journal of Jewish Studies. Oxford: Oxford Centre for Hebrew and Jewish Studies
The Journal of Medieval and Early Modern Studies. Durham, NC: Duke University Press
The Journal of Modern History. Chicago: University of Chicago Press
The Journal of Psychohistory. New York: Association for Psychohistory
Journal of Psychology and Judaism. New York: Human Sciences Press
Journal of the American Academy of Religion. Cary, NC: Oxford University Press
Journal of the Australian Jewish Historical Society see Australian Jewish Historical Society Journal
Journalism & Mass Communication Quarterly. Columbia, SC: Association for Education in Journalism and Mass Communication
JPCS - Journal for the Psychoanalysis of Culture & Society. Gardiner, NY: Critical Press

Judaica: Beitraege zum Verstaendnis des juedischen Schicksals in Vergangenheit und Gegenwart. Basel: Stiftung fuer Kirche und Judentum
Judaica Bohemiae. Praha: Státní Židovské Muzeum
Judaica Latinoamericana: Estudios historico-sociales. Jerusalem: Magnes, Universidad Hebrea
Judaism. New York: American Jewish Congress
Juden, Buerger, Deutsche: Zur Geschichte von Vielfalt und Differenz, 1800-1933, eds. Andreas Gotzmann et al. Tuebingen: Mohr Siebeck, 2001 (no. 152)
Juden in Berlin, eds. Andreas Nachama et al. Berlin: Henschel, 2001 (no. 34)
Das Juedische Echo: Zeitschrift fuer Kultur und Politik. Wien: Vereinigung juedischer Hochschueler Oesterreichs; Juedischer Akademiker Oesterreichs
Juedische Intellektuelle in der DDR: Politische Strukturen und Biographien, ed. Georg Schuppener. Leipzig: Arbeitskreis Hochschulpolitische Oeffentlichkeit beim StuRa der Universitaet Leipzig, 1999 (no. 679)
Juedisches Kinderleben im Spiegel juedischer Kinderbuecher, Vol. 1, eds. Helge Ulrike Hyams et al. Oldenburg: Bibliotheks- und Informationssystem der Universitaet Oldenburg, 2001 (no. 447)
Justice. Tel-Aviv: International Association of Jewish Lawyers and Jurists
Kholokost, soprotivleniye, vozrozhdeniye: Yevreiskii narod v gody Vtoroi mirovoi voiny i poslevoyennyi period 1939-1948, eds. Ilya A. Altman, Pinchas Agmon. Moskva: Mezhregionalnyi fond "Kholokost," 2000 (no. 260)
Kirche und Israel: Neukirchener theologische Zeitschrift. Neukirchen-Vluyn: Neukirchener Verlag des Erziehungsvereins
Kirchliche Zeitgeschichte: Internationale Zeitschrift fuer Theologie und Geschichtswissenschaft. Goettingen: Vandenhoeck & Ruprecht
Komponisten in Theresienstadt, ed. Cornelis Witthoefft. 2nd, rev. ed. Hamburg: Initiative Hans Krása, 2001 (no. 922)
Kóściół katolicki w Polsce a walka z antysemityzmem: Wymiana doświadczeń amerykańskich i polskich, ed. Bohdan W. Oppenheim. Kraków: Fundacja Judaica-Centrum Kultury Żydowskiej, 2000 (no. 793)
Kroke - Kazimierz - Kraków: Mehkarim be-toldot yehudei Kraków, ed. Elchanan Reiner. Tel-Aviv: Center for the History of Polish Jewry, Diaspora Research Institute, Tel-Aviv University, 2001 (no. 97)
Kwartalnik Historii Żydów. Warszawa: Żydowski Instytut Historyczny
Landjudentum im deutschen Suedwesten waehrend der Fruehen Neuzeit, eds. Rolf Kiessling, Sabine Ullmann. Berlin: Akademie Verlag, 1999 (no. 98)
Leo Baeck Institute Year Book. London: Leo Baeck Institute
Liberalism, Anti-Semitism, and Democracy: Essays in Honour of Peter Pulzer, eds. Henning Tewes, Jonathan Wright. Oxford: Oxford University Press, 2001 (no. 138)
Lietuvos žydų zudynių byla: Dokumentų ir straipsnių rinkinys, ed. Alfonsas Eidintas. Vilnius: Vaga, 2001 (no. 518)
Lithuania - Crime and Punishment. Tel-Aviv: Association of the Lithuanian Jews in Israel

Lo Nishkach! Shnaton mukdash le-shoat yehudei Yavan u-le-hantsahatah. Tel-Aviv: Irgun Dor ha-Hemshekh le-Hantsahat Shoat Yehudei Yavan
Losy żydowskie: Świadectwo żywych, Vol. II, ed. Marian Turski. Warszawa: Stowarzyszenie Żydów Kombatantów i Poszkodowanych w II Wojnie Światowej, 1999 (no. 350)
Maguén - Escudo. Caracas: Asociación Israelita de Venezuela, Centro de Estudios Sefardíes
Mahzor Wolfsberg, 5705: Mahaneh le-avodat kfiyah, Germania, 1944, eds. Bella Gutterman, Naomi Morgenstern. Jerusalem: Yad Vashem, 2000 (no. 555)
The Majlis: Interreligious Encounters in Medieval Islam, eds. Hava Lazarus-Yafeh et al. Wiesbaden: Harrassowitz, 1999 (nos. 64, 79)
Massuah: Kovetz shnati le-toda`at ha-shoah ve-ha-gvurah. Tel-Aviv: Massuah
Materialy naukovoyi konferentsiyi, Kharkiv 16 hrudnya 1998, ed. Larisa Volovik. Kharkiv: Kharkivskyi Prosvitnytskyi Tsentr "Holokost," 1999 (no. 619)
Me'ah She`arim: Iyunim be-olamam ha-ruhani shel Yisrael bi-yemei ha-beinayim, le-zekher Yitzhak Twersky, eds. Ezra Fleischer et al. Jerusalem: Magnes Press, Hebrew University, 2001 (no. 63)
La memoria della legislazione e della persecuzione antiebraica nella storia dell'Italia repubblicana. Milano: FrancoAngeli, 1999 (no. 685)
Memory and History in Christianity and Judaism, ed. Michael Alan Signer. Notre Dame, IN: University of Notre Dame Press, 2001 (no. 842)
The Memory of the Holocaust in the 21st Century: The Challenge for Education, ed. Stephanie McMahon-Kaye. Jerusalem: Yad Vashem, 2001 (no. 250)
Merkur: Deutsche Zeitschrift fuer europaeisches Denken. Stuttgart: Klett-Cotta
Mezhdu vostokom i zapadom: Yevrei v russkoi i yevropeiskoi kulture, eds. I. Belobrovtseva et al. Tallinn: Tallinskii Pedagogicheskii Universitet, 2000 (nos. 226, 230)
MFS - Modern Fiction Studies. Baltimore: Johns Hopkins University Press
Middle East Quarterly. Piscataway, NJ: Transaction
Middle Eastern Studies. London: Frank Cass
Midstream: A Monthly Jewish Review. New York: Theodor Herzl Foundation
Mishpat ve-historia, eds. Daniel Gutwein, Menachem Mautner. Jerusalem: Zalman Shazar Center for Jewish History, 1999 (no. 15)
Mishpokha. Vitebsk: Yevreiskii kulturnyi tsentr "Mishpokha"
Mitteilungen des Instituts fuer Oesterreichische Geschichtsforschung. Wien: Oldenbourg
Mittellateinisches Jahrbuch. Stuttgart: Anton Hiersemann Verlag
Modern Austrian Literature. Riverside, CA: University of California, International Arthur Schnitzler Research Association
Modern Judaism. Baltimore: Johns Hopkins University Press
Moritz Daniel Oppenheim: Die Entdeckung des juedischen Selbstbewusstseins in der Kunst: Jewish Identity in 19th Century Art, eds. Georg Heuberger, Anton Merk. Frankfurt a.M.: Wienand Verlag, 1999 (no. 443)
Múlt és Jövő. Budapest: Múlt és Jövő
The Musical Quarterly. Cary, NC: Oxford University Press

Nativ: Ktav et le-mahshavah medinit, hevrah ve-tarbut. Sha`arei Tikvah, Israel: Nativ
Nel nome della razza: Il razzismo nella storia d'Italia, 1870-1945, ed. Alberto Burgio. Bologna: Il Mulino, 1999 (no. 122)
I nemici sono gli "altri": Convegno sull'Olocausto, eds. Laura Fontana, Giorgio Giovagnoli. Firenze: Giuntina, 1999 (no. 165)
Neterpimost v Rossii: Starye i novye fobii, eds. Galina S. Vitkovskaya, Alexei V. Malashenko. Moskva: Moskovskii Tsentr Karnegi, 1999 (no. 801)
Neue Rundschau. Frankfurt a.M.: S. Fischer
Neuer Anbruch: Zur deutsch-juedischen Geschichte und Kultur, eds. Michael Brocke et al. Berlin: Metropol, 2001 (nos. 99, 450)
Neuer Antisemitismus - alte Vorurteile?, ed. Christina Tuor-Kurth. Stuttgart: W. Kohlhammer, 2001 (no. 764)
"Niemand zeugt fuer den Zeugen": Erinnerungskultur und historische Verantwortung nach der Shoah, ed. Ulrich Baer. Frankfurt a.M.: Suhrkamp, 2000 (no. 648)
Nietzsche-Studien. Berlin: Walter de Gruyter
Nit vi shoff tzu der shechitah: Zichroynos fun anteil-nemers in aktivn vidershtand kegn di natzis in Kovner geto (1941-1944), ed. Grigorijus Smoliakovas. Tel-Aviv: I.L. Peretz Pub. House, 2001 (no. 514)
Nordisk Judaistik: Scandinavian Jewish Studies. Lund: Nordisk Judaistik
La Nouvelle Revue Française. Paris: Gallimard
Nouvelle Revue Théologique. Bruxelles: Nouvelle Revue Théologique
NS-Verbrechen und der militaerische Widerstand gegen Hitler. Darmstadt: Primus, 2000 (no. 351)
El Olivo: Documentación y estudios para el diálogo entre judíos y cristianos. Madrid: Centro de Estudios Judeo-Cristianos
The Papon Affair: Memory and Justice on Trial, ed. Richard J. Golsan. New York: Routledge, 2000 (no. 769)
Partisan Review. Boston: Partisan Review at Boston University
Patterns of Prejudice. London: Routledge
Peace Review. Basingstoke, Hants: Carfax
Perspectives. Jerusalem: Magnes Press; Hebrew University
Philosophy & Social Criticism. London: Sage
Plegarias de Rosh Hashana, Wolfsberg, 5705: Campo de trabajo, Alemania, 1944, eds. Bella Gutterman, Naomi Morgenstern. Jerusalén: Yad Vashem, 2001 (no. 556)
The Plunder of Jewish Property during the Holocaust: Confronting European History, ed. Avi Beker. Basingstoke, Hants: Palgrave, 2001 (no. 269)
Poetics Today: International Journal for Theory and Analysis of Literature and Communication. Durham, NC: Duke University Press
Polacy i sąsiedzi - dystanse i przenikanie kultur, Vol. II, ed. Roman Wapiński. Gdańsk: Wydawnictwo "Stepan design," 2001 (no. 169)
Polen, Deutsche und Juden in Lodz 1820-1939: Eine schwierige Nachbarschaft, ed. Juergen Hensel. Osnabrueck: Fibre, 1999 (nos. 550, 561)
Polin: Studies in Polish Jewry. Oxford: Littman Library of Jewish Civilization

Prepodavaniye temy Kholokosta v XXI veke, ed. Ilya A. Altman. Moskva: Fond "Kholokost," 2000 (no. 261)
Probing the Depths of German Antisemitism: German Society and the Persecution of the Jews, 1933-1941, ed. David Bankier. New York: Berghahn Books; Jerusalem: Yad Vashem; Leo Baeck Institute, 2000 (no. 428)
Quadrant. Balmain, NSW: Quadrant Magazine Co.
Queen's Quarterly: A Canadian Review. Kingston, Ont.: Queen's University
Rachunek sumienia: Kościół polski wobec antysemityzmu, 1989-1999, ed. Bohdan W. Oppenheim. Kraków: Wydawn. WAM, 1999 (no. 794)
Radical History Review. Cambridge: Cambridge University Press
Raíces: Revista judía de cultura. Madrid: Sefarad Ediciones
Rambam: Tidsskrift for jødisk kultur og forskning. København: C.A. Reitzels Forlag
Raritan. New Brunswick, NJ: Rutgers University
La Rassegna Mensile di Israel. Roma: Unione delle Comunità Israelitiche Italiane
Réflexions Historiques see Historical Reflections
Revue des Etudes Juives. Paris: Peeters
Revue d'Histoire de la Shoah - Le Monde Juif. Paris: Centre de Documentation Juive Contemporaine
Rhode Island Jewish Historical Notes. Providence, RI: Rhode Island Jewish Historical Association
Richard Wagner und die Juden, eds. Dieter Borchmeyer et al. Stuttgart: J.B. Metzler, 2000 (no. 121)
Romance Notes. Chapel Hill, NC: University of North Carolina
The Russian Review: An American Quarterly Devoted to Russia Past and Present. Columbus, OH: Ohio State University Press
Sachor: Zeitschrift fuer Antisemitismusforschung, juedische Geschichte und Gegenwart. Essen: Klartext Verlag
Scepticism and Hope: Sixteen Contemporary Slovak Essays, ed. Miro Kollar. Bratislava: Kalligram, 1999 (no. 803)
Schriften der Erich-Muehsam-Gesellschaft. Luebeck: Erich-Muehsam-Gesellschaft
Sefarad: Revista de estudios hebraicos, sefardíes y de Oriente Próximo. Madrid: Consejo Superior de Investigaciones Cientificas
Sefárdica. Buenos Aires: Centro de Investigación y Difusión de la Cultura Sefardí
Shakespeare Jahrbuch. Bochum: Kamp
Shedding Light on the Darkness: A Guide to Teaching the Holocaust, eds. Nancy A. Lauckner, Miriam Jokiniemi. New York: Berghahn Books, 2000 (no. 831)
Shofar: An Interdisciplinary Journal of Jewish Studies. Lincoln: University of Nebraska Press
Shvut: Studies in Russian and East European Jewish History and Culture. Tel-Aviv: Tel-Aviv University, Diaspora Research Institute; Beer-Sheva: Ben-Gurion University of the Negev
The Six-Day War and World Jewry, ed. Eli Lederhendler. Bethesda, MD: University Press of Maryland, 2000 (no. 688)

Society. New Brunswick, NJ: Transaction
The Solomon Goldman Lectures, Vol. VII, ed. Dean Phillip Bell. Chicago: Spertus Institute of Jewish Studies Press, 1999 (no. 83)
Spanien und die Sepharden: Geschichte, Kultur, Literatur, eds. Norbert Rehrmann, Andreas Koechert. Tuebingen: Max Niemeyer, 1999 (no. 45)
Die Stadt ohne Juden, eds. Guntram Geser, Armin Loader. Wien: Filmarchiv Austria, 2000 (no. 367)
Le storie estreme e la storia: I racconti della Shoah, eds. Alessandra Deoriti et al. Chiaravalle: Orecchio di Van Gogh, 1999 (no. 651)
Storiografia: Rivista annuale di storia. Pisa: Istituti Editoriali e Poligrafici Internazionali
Studia Judaica. Cluj-Napoca: EFES
Studia nad Faszyzmem i Zbrodniami Hitlerowskimi. Wrocław: Wydawnictwo Uniwersytetu Wrocławskiego
Studia Patristica. Leuven: Peeters
Studies in Philosophy and Education. Dordrecht: Kluwer Academic
Tanulmányok a Holokausztról, I, ed. Randolph L. Braham. Budapest: Balassi Kiadó, 2001 (no. 160)
Tarbiz: Riv`on le-mada`ei ha-yahadut. Jerusalem: Magnes Press, Hebrew University, Institute of Jewish Studies
Teaching History. London: Historical Association
Teaching the Holocaust: Educational Dimensions, Principles and Practice, ed. Ian Davies. London: Continuum, 2000 (no. 724)
Television Quarterly. New York: National Academy of Television Arts and Sciences
Les Temps Modernes. Paris: Les Temps Modernes
Text + Kritik: Zeitschrift fuer Literatur. Muenchen: Text + Kritik
Les textes judéophobes et judéophiles dans l'Europe chrétienne à l'époque moderne, ed. Daniel Tollet. Paris: Presses Universitaires de France, 2000 (no. 111)
Theoretical Interpretations of the Holocaust, ed. Dan Stone. Amsterdam: Rodopi, 2001 (no. 654)
Theresienstaedter Studien und Dokumente. Prag: Stiftung Theresienstaedter Initiative (Nadace Terezinska iniciativa) im Verlag Academia
Tirosh. Moskva: "Sefer"
The Torah u-Madda Journal. New York: Rabbi Isaac Elchanan Theological Seminary, Yeshiva University
Totengebet: 60 Jahre Beginn des Zweiten Weltkriegs und der Schoah in Polen, ed. Erhard Roy Wiehn. Konstanz: Hartung-Gorre, 1999 (no. 584)
Transylvanian Review. Cluj-Napoca: Centrul de Studii Transilvane; Fundaţia Culturală Română (no. 588)
Tribuene: Zeitschrift zum Verstaendnis des Judentums. Frankfurt a.M.: Tribuene
Troubled Souls: Conversos, Crypto-Jews, and Other Confused Jewish Intellectuals from the Fourteenth through the Eighteenth Century, eds. Charles Meyers, Norman Simms. Hamilton, N.Z.: Outrigger, 2001 (no. 103)

Verschwiegenes Wortspiel: Kommentare zu den Werken Ilse Aichingers: Akten des Internationalen Colloquiums vom 27. bis 28. April 1998 an der Vrije Universiteit Brussel, ed. Heidy Margrit Mueller. Bielefeld: Aisthesis, 1999 (no. 861)
Verschwoerungstheorien: Anthropologische Konstanten - historische Varianten, eds. Ute Caumanns, Mathias Niendorf. Osnabrueck: fibre, 2001 (no. 126)
Vertreibung juedischer Kuenstler und Wissenschaftler aus Duesseldorf 1933-1945, eds. Kurt Duewell et al. Duesseldorf: Droste, 1998 (no. 435)
Vestnik Yevreiskovo Universiteta. Moskva: Mosty Kultury; Ierusalim: Gesharim
Victorian Literature and Culture. New York: Cambridge University Press
Vingtième Siècle: Revue d'histoire. Paris: Presses de Sciences Po
Virginia Quarterly Review. Charlottesville: University of Virginia
Visual Culture and the Holocaust, ed. Barbie Zelizer. New Brunswick, NJ: Rutgers University Press, 2001 (no. 925)
"Warum immer Vergangenheit?" Leben und Werk Jenny Alonis, ed. Hartmut Steinecke. Muenster: Ardey-Verlag, 1999 (no. 473)
Wiener Jahrbuch fuer juedische Geschichte, Kultur & Museumswesen. Wien: Verlag Christian Brandstaetter
Wiener Slawistischer Almanach. Wien: Gesellschaft zur Foerderung slawistischer Studien
Więź. Warszawa: Więź
"Wir lebten wie sie...": Juedische Lebensgeschichten aus Tirol und Vorarlberg, ed. Thomas Albrich. Innsbruck: Haymon, 1999 (no. 259)
Women's Studies International Forum. New York: Pergamon
World Congress of Jewish Studies, 12th (1997): Proceedings. Division E: Contemporary Jewish Society. Jerusalem: World Union of Jewish Studies, 2001 (no. 131)
XXe Siècle see Vingtième Siècle
Yad Vashem: Kovetz mehkarim. Jerusalem: Yad Vashem
Yad Vashem Studies. Jerusalem: Yad Vashem
Yahadut America ha-latinit: Kovetz ma'amarim li-khvod Haim Avni, eds. Margalit Bejarano et al. Jerusalem: Amilat – Research Association for Latin American Jewry; Institute of Contemporary Jewry, Hebrew University, 2001 (no. 166)
Yahadut Hofshit. Jerusalem: Techila - the Israeli Secular Movement for Humanistic Judaism
Yakar le'Mordecai: Jubilee Volume in Honor of Rabbi Mordecai Waxman: Essays on Jewish Thought, American Judaism, and Jewish-Christian Relations, ed. Tseviyah Ben-Yosef Ginor. Hoboken, NJ: Ktav, 1999 (no. 13)
The Yale Journal of Criticism. New Haven, CT: Yale University Press
Yalkut Moreshet. Tel-Aviv: Moreshet
Yehudei Brit ha-Mo`atzot be-Ma`avar. Jerusalem: Hebrew University
Yevrei Belarusi: Istoriya i kultura: Sbornik statei. Minsk: Chetyre Chetverti

Yevreiskaya kultura i kulturnye kontakty: Materialy shestoi yezhegodnoi mezhdunarodnoi mezhdistsiplinarnoi konferentsii po iudaike, eds. Rashid M. Kaplanov et al. Moskva: Sefer, 1999 (no. 41)
Yevreiskaya mysl skvoz veka: Sbornik nauchnykh trudov po iudaike, yevreiskoi istorii i kulture, Vol. 3, ed. Lyubov E. Chernova. Dnepropetrovsk: Dnepropetrovskii Institut Sotsialnykh i Obshchinnykh Rabotnikov, 1999 (no. 173)
Yevreiskii mir: Sbornik 1944 goda. Ierusalim: Gesharim; Moskva: Mosty kultury, 2001 (no. 27)
Zeitschrift fuer Antikes Christentum. Berlin: Walter de Gruyter
Zeitschrift fuer Geschichtswissenschaft. Berlin: Metropol
Zeitschrift fuer Literaturwissenschaft und Linguistik: Eine Zeitschrift der Universitaet Gesamthochschule Siegen. Stuttgart: Verlag J.B. Metzler
Zeitschrift fuer Neues Testament. Tuebingen: Francke Verlag
Zeitschrift fuer Theologie und Kirche. Tuebingen: J.C.B. Mohr (Paul Siebeck)
Židé a Morava: Sborník příspěvků přednesených na konferenci konané 11. listopadu 1998 v Kroměříži, ed. Petr Pálka. Kroměříž: Muzeum Kroměřížska, 1999 (no. 190)
Židovské spolky v českých zemích v letech 1918-1948. Praha: Sefer, 2001 (no. 407)
Zmanim: Riv`on le-historia. Tel-Aviv: Tel-Aviv University
Znak. Kraków: Społeczny Instytut Wydawniczy Znak
Zwei Entwuerfe zum Holocaust-Denkmal in Berlin, ed. Matthias Reichelt. Nuernberg: Verlag fuer Moderne Kunst, 2001 (no. 904)
Zwischenwelt: Zeitschrift fuer Kultur des Exils und des Widerstands. Wien: Theodor Kramer Gesellschaft
Żydzi i komunizm, eds. Michał Bilewicz, Bogna Pawlisz. Warszawa: Stowarzyszenie "Jidełe," 2000 (no. 120)

BIBLIOGRAPHIES AND REFERENCE WORKS

0001 Adamushka, Uladzimir I. et al., eds.: *Spravochnik o mestakh prinuditelnovo soderzhaniya grazhdanskovo naseleniya na okkupirovannoi territorii Belarusi 1941-1944 = Handbuch der Haftstaetten fuer Zivilbevoelkerung auf dem besetzten Territorium von Belarus 1941-1944*. Minsk: Natsionalnyi Arkhiv Respubliki Belarus, 2001. 157 pp. In Russian (pp. 1-76) and German (pp. 77-157).

A directory with four columns per page, giving the name of the place of detention, the category (ghetto, labor camp, killing site, etc.), a short description, and reference to archival documents from which this information was derived, all of them Belorussian archives. At some of these sites, the massacre of Belorussian and foreign Jews was carried out. *DR*

0002 Barnavi, Elie; Friedlaender, Saul: *Les Juifs et le XXe siècle: Dictionnaire critique*. Paris: Calmann-Lévy, 2000. 815 pp.

A dictionary comprising essays on events, personalities, doctrines, and movements in 20th-century Jewish history. The following deal with antisemitism and the Holocaust: Friedlaender, Saul: Antisémitisme(s) moderne(s) (19-35); Dubois, Marcel Jacques: Eglise et les Juifs (l') (77-93); Friedlaender, Saul: Solution finale (origine de la) (231-246); Young, James Edward: Auschwitz: Concept et "lieu de mémoire" (273-283); Traverso, Enzo: Primo Levi (621-628); Le Rider, Jacques: Otto Weininger (713-719); Bartal, Israel: 1881 (733-743); Diner, Dan: 1933 (759-765). *HV*

0003 Braham, Randolph L.: *The Holocaust in Hungary: A Selected and Annotated Bibliography, 1984-2000*. New York: Rosenthal Institute for Holocaust Studies, Graduate Center, City University of New York; Boulder, CO: Social Science Monographs, 2001. xii, 252 pp. A sequel to the author's "The Hungarian Jewish Catastrophe: A Selected and Annotated Bibliography" (2nd, rev. ed., 1984).

A multilingual bibliography, with 1,500 items classified under 43 sections, divided by genre (e.g. diaries, interviews, fictional accounts) and by subject (e.g. antisemitism; the Christian Churches and leaders; intellectuals and cultural life; anti-Jewish legislation; concentration and detention camps; rescue). *YC*

0004 Chabord, Marie-Thérèse; Pouëssel, Jean, eds.: *Inventaire des archives du Commissariat général aux questions juives et du Service de restitution des biens des victimes des lois et mesures de spoliation. Sous-série AJ38.* Paris: Centre Historique des Archives Nationales, 1999. xxxi, 325 pp. On title page: 1998.

The introduction (pp. xiii-xxxi) gives an overview of the history, purpose, organization, and functioning of the Commissariat général aux questions juives, created by the Vichy government in March 1941 with the purpose of executing its anti-Jewish policy. The Commissariat was in charge of registering Jewish persons and keeping files of and administering confiscated Jewish property. After the liberation, the Service de restitution was given the task of returning the property to its owners. The inventory of the archives (designated as AJ38 in the Archives nationales) is organized according to subjects and geographical divisions, the largest section being on Aryanization (pp. 63-240). The entries generally contain a brief note on the contents of the file. *EO*

0005 Gerasimova, Inna P.; Papernaya, S.M., eds.: *Istoriya Kholokosta na territorii Belarusi: Bibliograficheskii ukazatel* [The History of the Holocaust on the Territory of Belarus: Bibliographical Guide]. Vitebsk: UPP "Vitebskaya oblastnaya tipografiya," 2001. 104 pp.

A bibliographical list of 916 books and articles in Russian and Belorussian published between 1990-August 2001. *DR*

0006 *Guide européen des sources d'archives sur la Shoah = European Guide of Archival Sources on the Shoah.* Paris: Centre de Documentation Juive Contemporaine, 2000. 191 pp.

A guidebook to archival sources on the Shoah in 23 European countries, Israel, and the U.S. Each country compiled its own listings; the descriptions are given in English, French, or German. Some of the listings present only addresses and access information (telephone, fax, e-mail, person to contact); others give more information. *SSC*

0007 Kruglov, Aleksandr I.: *Katastrofa ukrainskovo yevreistva 1941-1944 gg.: Entsiklopedicheskii spravochnik* [The Catastrophe of Ukrainian Jewry 1941-1944: Encyclopedic Directory]. Kharkov: Karavella, 2001. 376 pp.

Presents information on the Holocaust in more than 2,000 cities, towns, and villages in Ukraine (according to present-day borders). Each entry gives an estimate of the number of Jews (either local or brought from other places) who were killed there, the date of the events, and in some cases a detailed description of the genocide in that town. *DR*

0008 Laqueur, Walter Zeev; Baumel, Judith Tydor, eds.: *The Holocaust Encyclopedia.* New Haven: Yale University Press, 2001. xxxix, 765 pp.

A one-volume encyclopedia of the Holocaust, including information collected in the last decade due to the opening of archives in the 1990s and the widening range of topics related to Holocaust study (e.g. Jewish cultural activities during the Holocaust, non-Jewish victims, religion and theology, and Holocaust denial). This work focuses on issues rather than personalities and the geography of the mass murder (although these are included). The entries were written by more than 100 authors from 11 countries. Pp. xiii-xix contain an introduction by Laqueur. Pp. xxi-xxxvii present a chronology of the Holocaust. Pp. 721-735 contain a bibliographical essay by Robert Rozett. *DR*

0009 Liebersohn, Harry; Schneider, Dorothee: *"My Life in Germany before and after January 30, 1933": A Guide to a Manuscript Collection at Houghton Library, Harvard University*. Philadelphia: American Philosophical Society, 2001. 130 pp. (Transactions of the American Philosophical Society, 91, 3).

Pt. 1 (pp. 1-30) contains the editors' introduction. The collection consists of 263 files containing the memoirs of ca. 230 people who lived in Germany or Austria during the 1930s. Most of the entries were written during the period August 1939-April 1940 in response to an essay contest sponsored by three Harvard professors (one of them was Gordon Allport). The memoirists were Jews, part-Jews, and Christians. The introduction examines various themes in the memoirs, such as identity conflict, persecution and flight, professional occupation, women as memoirists. Pt. 2 (pp. 31-124) contains the catalogue; each entry gives basic biographical information on the memoirist and a synopsis of the content of the memoir. *SSC*

0010 Rozett, Robert; Spector, Shmuel, eds.: *Encyclopedia of the Holocaust*. Jerusalem: Yad Vashem; New York: Facts on File, 2000. viii, 528 pp.

A one-volume Holocaust encyclopedia for students and teachers. Pp. 101-496 contain the encyclopedic entries, including people, places, institutions, concepts, events, etc. Pp. 1-100 contain essays on the Jews in European culture, antisemitism, and the Holocaust. Partial contents: Spector, Shmuel: The Jews of Europe between the Two World Wars [Includes discussion of anti-Jewish legislation in various countries.] (5-17); Fraenkel, Daniel: Nazi Ideology and Its Roots (18-30); Weiss, Aharon: The Destruction of European Jewry, 1933-1945 (45-55); Rozett, Robert: On Being a Jew in the Holocaust (56-71); Silberklang, David: The Allies and the Holocaust [Appeared in "Yad Vashem Studies" 24 (1994).] (72-84); Cesarani, David: The Aftermath of the Holocaust [Discusses, inter alia, reparations, war crimes trials, and postwar reactions to the Holocaust in various countries.] (85-100). *YC*

0011 Spector, Shmuel, ed.: *The Encyclopedia of Jewish Life before and during the Holocaust. Vol. 1-3*. Trans.: Fred Skolnick. Jerusalem: Yad Vashem, 2001. 1770 pp.

An abridged translation of the encyclopedic entries (without the introductory essays) in the multi-volume series "Pinkas Hakehillot: Encyclopaedia of Jewish Communities from Their Foundation till after the Holocaust" (Jerusalem: Yad Vashem: 1969-1999) – the ten published volumes and volumes not yet published. *SSC*

ANTISEMITISM THROUGHOUT THE AGES

General

0012 Baumgarten, Albert I.: Marcel Simon's "Verus Israel" as a Contribution to Jewish History. *Harvard Theological Review* 92, 4 (Oct 1999) 465-478.

Simon's theological work "Verus Israel" (1948) was an attempt, impelled by the prewar upsurge of racism and the Holocaust, to reexamine the nature of the relationship between ancient Christianity and Judaism. Simon suggested that ancient Christian anti-Judaism grew out of a competition between the two groups for the souls of possible proselytes; it was not a conflict between two truths. Simon tried to overcome Christian theological prejudice against Judaism (which was depicted as a "legalistic" and "ritualistic" religion); yet, he is tainted sometimes with the very anti-Judaism that he claimed to oppose. His theory that the Jews of late antiquity were engaged in missionary activities was criticized by modern historians. With all that, Simon's "conflict theory" is an important tool for understanding the relations between early Christians and Jews. *DR*

0013 Ben-Yosef Ginor, Tseviyah, ed.: *Yakar le'Mordecai: Jubilee Volume in Honor of Rabbi Mordecai Waxman: Essays on Jewish Thought, American Judaism, and Jewish-Christian Relations*. Hoboken, NJ: Ktav; Great Neck, NY: Temple Israel of Great Neck, 1999. x, 394 pp.

Partial contents: Taitz, Emily: Jewish-Christian Relations in the Middle Ages: The Underside of a Shared Culture (189-201); Ascher, Abraham: Prime Minister P.A. Stolypin and His "Jewish" Adviser [Appeared in the "Journal of Contemporary History" 30 (1995).] (203-220); Wigoder, Geoffrey: Jewish-Christian Relations in the Light of the Holocaust (221-232); Riegner, Gerhart M.: Thirty Years of "Nostra Aetate": Developments and Lessons (233-243); Fisher, Eugene Joseph: Update on Catholic Education and Jews and Judaism (245-261); Keeler, William: New Moments in Catholic-Jewish Relations [A Catholic Cardinal discusses developments since the Second Vatican Council.] (263-272); Perlman, Mark: The Opening of American Academia to Jews in the Post-World War II Period (327-341). *YC*

0014 Carroll, James: *Constantine's Sword: The Church and the Jews: A History*. Boston: Houghton Mifflin, 2001. xii, 756 pp.

Presents a history of Catholic antisemitism from the ancient period to the present, from a personal point of view. Takes as a point of departure the symbol of the cross, seen as representing Jesus' suffering, which the Church has betrayed by scapegoating the Jews. This theme is introduced by discussing the recent issue of the convent and the cross at Auschwitz. The next chapters are: "New Testament Origins of Jew Hatred," "Constantine, Augustine, and the Jews," "From Crusades to Conversionism," "The Inquisition: Enter Racism," "Emancipation, Revolution, and a New Fear of Jews," and "The Church and Hitler." The last part, "A Call for Vatican III," details further steps that the Church should take in order to shed elements of antisemitism still present in its theology, liturgy, etc. *YC*

0015

גוטויין, דניאל; מאוטנר, מנחם, עורכים: משפט והיסטוריה. ירושלים: מרכז זלמן שזר לתולדות ישראל, 1999. 513 עמ'.

[Gutwein, Daniel; Mautner, Menachem, eds.: *Law and History*. Jerusalem: Zalman Shazar Center for Jewish History, 1999. 513 pp.]

תוכן חלקי: ביבלניק, פנחס: משפטי האינקוויזיציה נגד המתייהדים במכסיקו (1659-1642) (145-127); ויסטריך, רוברט סלומון: אמת וצדק בפרשת דרייפוס: לזאר, זולא וקלמנסו (272-247); אהרונסון, שלמה: ישראל קסטנר, OSS ו"תיאוריית ראש החץ" בנירנברג (338-305); שלף, ליאון: בין רשימת שינדלר לבין הרכבת של קסטנר: על מציאות היסטורית, מיתוסים תקשורתיים ואמת משפטית (353-339); ויץ, יחיעם: בין קתרזיס לקרב סכינים: "משפט אייכמן" ו"משפט קסטנר" והשפעתם על החברה הישראלית (421-395); מעוז, אשר: שיפוט היסטורי: משפט קסטנר וועדת החקירה בעניין רצח ארלוזורוב (471-441).

A collection based on papers presented at the annual conference of the Historical Society of Israel and the Zalman Shazar Center in 1995. Partial contents: [Bibelnik, Pinhas: Inquisitorial Trials against Judaizers in Mexico (1642-1659) (127-145); Wistrich, Robert Solomon: Truth and Justice in the Dreyfus Affair: Lazare, Zola and Clemenceau (247-272); Aronson, Shlomo: Israel Kasztner, the OSS and the "Arrowhead Theory" in Nuremberg on the roles of Rudolf Kasztner, Charles Dwork, Robert Jackson, Jacob Robinson, and Franz Neumann in the trials.] (305-338); Sheleff, Leon: From Schindler's List to Kasztner's Train: On Historical Reality, Media Myths and Judicial Truth (339-353); Weitz, Yehiam: Between Catharsis and Knife-Fight: The "Eichmann Trial" and the "Kasztner Trial" and Their Influence on Israeli Society [Appeared in English in "Israel Studies" 1 (1996).] (395-421); Maoz, Asher: Historical Judgment: The Kasztner Trial and the Investigating Committee of the Arlozorov Murder (441-471).] *RE*

0016 Kaplanov, Rashid M.; Mochalova, Viktoriya V.; Chulkova, L.A., eds.: *Istoriya yevreiskovo naroda: Materialy shestoi yezhegodnoi mezhdunarodnoi mezhdistsiplinarnoi konferentsii po iudaike* [History of the Jewish People: Materials from the Sixth Annual International Interdisciplinary Conference on Jewish Studies]. Moskva: Sefer, 1999. 304 pp. Part 2 of the conference proceedings. See also part 3: "Yevreiskaya kultura i kulturnye kontakty" (1999).

Papers delivered at the conference held in Moscow. Partial contents: Zaretskii, Yurii: Anusim, svyataya inkvizitsiya i avtobiografiya Terezy Avilskoi [Conversos, the Holy Inquisition, and the Autobiography of St. Teresa of Avila] (56-76); Kozyrev, Oleg: Antiyevreiskiye narodnicheskiye proklamatsii v Ukraine (70-ye – per. pol. 80-kh gg. XIX v.) [Anti-Jewish Leaflets of the Narodniks in Ukraine (1870s-first half of the 1880s)] (98-110); Gaidai, Olga: Pravovoye polozheniye yevreyev v sfere narodnovo obrazovaniya v XIX-nach. XX st. [Legislation on the Jews in the Sphere of People's Education in the 19th-Early 20th Century] (127-130); Budnitskii, Oleg V.: "Yevreiskii vopros" v Grazhdanskoi voine (istoriograficheskiye zametki) [The "Jewish Question" in the Civil War (Historiographical Notes). On White Army pogroms in Russia in 1918-20.] (144-173); Milyakova, Lidiya V.: K istorii yevreiskoi obshchiny goroda Lida i rasstrelov v aprele 1919 g. [Toward the History of the Jewish Community in the Town of Lida and of the Shootings of April 1919] (189-197); Yelisavetskii, Ster Ya.: Polveka zabveniya [Half a Century of Oblivion. On the participation of Jews in the partisan warfare in Ukraine during World War II.] (260-268); Batvinnik, Marat B.: Issledovaniye chelovecheskovo faktora v izuchenii genotsida yevreyev Belarusi (materialy o genotside v Borisove) [Examination of the Human Factor in Research on the Genocide of Jews in Belarus (Materials on the Genocide in Borisov)] (269-273); Yelenskaya, Irina E.: Novye istochniki po istorii Kholokosta (na primere g. Pinska) [New Sources on the History of the Holocaust (the Case of Pinsk)] (274-278); Rozenblat, Yevgenii S.: Kholokost v Zapadnoi Belarusi: Sotsialno-ekonomicheskiye aspekty [The Holocaust in Western Belarus: Social and Economic Aspects] (279-298); Shchukin, Vladimir: Sobytiya katastrofy na yuge Ukrainy (po materialam Nikolayevskoi oblasti, Ukraina) [Events of a Catastrophe in the Southern Ukraine (Based on the Nikolayev Region, Ukraine).] (299-304). *DR*

0017 Karády, Viktor: *Zsidóság Európában a modern korban: Társadalomtörténeti vázlat* [Jews in Europe in the Modern Age: A Brief Social History]. Trans.: László V. Tóth. Budapest: Új Mandátum, 2000. 503 pp. Based on unpublished French manuscripts.

Presents an analytical discourse on sociological processes and problems combined with a narrative of the main events in the history of European Jewry. Pp. 39-86 deal with the Middle Ages, including Christian anti-

Judaism and the expulsion from Spain, up to the 18th century. The rest of the book deals with the 18th-20th centuries. Describes the official antisemitism of the tsarist regime and of the united Romania (i.e. including territories which were once part of Hungary). Ch. 8 (pp. 333-373) deals with the theory and practice of modern antisemitism. Ch. 9 (pp. 377-414) relates the events of the Holocaust. The following chapter deals with the aftermath of the Holocaust, and describes the anti-Jewish policies of the Soviet Union and the former communist-bloc countries. *AV*

0018 Kotowski, Elke-Vera; Schoeps, Julius Hans; Wallenborn, Hiltrud, eds.: *Handbuch zur Geschichte der Juden in Europa. Vol. 1-2*. Darmstadt: Primus, 2001. 511; 507 pp.

Vol. 1 contains 21 essays dealing with Jewish history in specific countries or regions from the ancient or medieval period to the present, including references to antisemitism and/or the Holocaust. The 34 essays in vol. 2 deal with different aspects of Jewish religious, cultural, and everyday life. Partial contents of vol. 2: "Die Juden und die christliche Gesellschaft": Willoweit, Dietmar: Die rechtliche Stellung der Juden im Mittelalter (299-308); Wallenborn, Hiltrud: Zwischen Ausweisung und Aufklaerung: Juden in der christlichen Gesellschaft der fruehen Neuzeit (309-321); Hertz, Deborah: Konversion in Europa (322-335). "Judenfeindschaft": Frey, Winfried: Antijudaismus (367-378); Weinzierl, Erika: Moderner Antisemitismus von der Aufklaerung bis zum Nationalsozialismus (379-387); Friedlander, Henry: Schoa (388-408); Bergmann, Werner: Antisemitismus in Europa nach 1945 (409-424). *RW*

0019 Kreis, Rudolf: *Antisemitismus und Kirche: In den Gedaechtnisluecken deutscher Geschichte mit Heine, Freud, Kafka und Goldhagen*. Reinbek bei Hamburg: Rowohlt Taschenbuch Verlag, 1999. 333, [8] pp.

Asserts that it was the Church that purveyed the antisemitism that culminated in the Holocaust. It rejected the Law, which binds spirit to body, nature, life, and enjoins concern for the creation and for suffering fellow men. In its place it set the spirit alone, and concern only for the sufferings of Jesus and Mary. In the Middle Ages, the New Testament narrative was enhanced with gruesome details of imagined Jewish sadism. Antisemitism was transmitted by preaching, in legends and images, in the blood libel and host desecration accusation, in Shylock and Parsifal. Contends that Heine rightly feared the revival of the narrative by the Romantics and Idealists; he dreamed of it as the "passion flower," satirized it, and mourned the plight of the Jews under the domination of Christianity. Freud analyzed it in dreams and jokes. But, careful not to alienate Christians, he did not describe it directly but replaced the trinity Son – Mother – Jews by Son – Mother – Father: the Oedipus complex. Kafka's parables express the absence of a way out of the Jewish dilemma.

Goldhagen describes the results: German soldiers for whom it was entirely right and "normal" to kill Jews, and before that to torment and humiliate them as the Jews did Jesus. *RW*

0020 Littell, Franklin Hamlin: The Holocaust and the Christians. *Journal of Church and State* 41, 4 (Fall 1999) 725-738.

Based on a lecture presented at Mercer University, Atlanta, in February 1999. Discusses Christian responsibility for the Holocaust stemming from the Church's age-old antisemitism. Points to the 4th-century Constantinian unification of Church and state as a false start in Church history that institutionalized antisemitism. European Churches in the 19th and 20th centuries were in intellectual and spiritual decline, often lending support to the antisemitism of the state. Notes the failure of the German Church establishment to oppose the Nazis and the failure of the Free Churches to help save Jews. In regard to the Protestant Churches after the Holocaust, praises the 1980 Rhineland Declaration for its statement of repentance and its affirmation of the importance of Jewish survival. The Declaration recognizes the co-responsibility of German Christendom for the Holocaust. Stresses the need for personal spiritual repentance for "post-Auschwitz" Christians. *YC*

0021 Mason, David R.: A Christian Alternative to (Christian) Racism and Antisemitism. *Journal of Ecumenical Studies* 37, 2 (Spr 2000) 151-160.

Admits that Christianity bears responsibility for the spread of antisemitism and that the roots of much Christian anti-Jewish slander are located in the New Testament, especially in the Gospels. Nevertheless, argues that the fundamental principles of Christianity repudiate racism and antisemitism. Not one of the Gospels is an eyewitness account of the historical Jesus; they reflect the tensions between Jews and Christians in the Roman Empire after the Jewish revolt of 66-70. Disciples of Jesus (e.g. Peter, Paul) emphasized the universalism of the Christian message. Genuine Christianity rejects ethnocentrism and exclusivism; according to its vision, Jesus "re-presents" but does not exclusively constitute God, whose redemptive love relates to all human beings. *DR*

0022 Porter, Stanley E.; Pearson, Brook W.R., eds.: *Christian-Jewish Relations through the Centuries*. Sheffield: Sheffield Academic Press, 2000. 503 pp.

A collection of papers delivered at a conference held at the University of Surrey Roehampton, February 1999. Partial contents: Porter, Stanley E.; Pearson, Brook W.R.: Ancient Understandings of the Christian-Jewish Split [Relates to the development of Christian supersessionism, the Jewish view of Christians as apostates, and the separation of Christians from Jews in the minds of Romans.] (36-51); Horst, Pieter Willem van der:

Jews and Christians in Antioch at the End of the Fourth Century [Discusses John Chrysostom's anti-Jewish homilies and the "Apostolic Constitutions."] (228-238); Sherwood, Yvonne: 'Colonizing the Old Testament' or 'Representing Christian Interests Abroad': Jewish-Christian Relations across Old Testament Territory [Discusses Christian views of the Old Testament and the Jews as Other.] (255-281); Gearon, Liam: "The Statement": A Literary-Historical Reflection on Catholic-Jewish Relations [Shows how Brian Moore's novel, partly set in Vichy France, reflects pre- and post-Vatican II Catholic attitudes toward Jews.] (298-310); Pinsent, Pat: After Fagin: Jewishness and Children's Literature [Discusses opposition to antisemitism and treatment of the Holocaust in 20th-century children's literature.] (311-328); Wise, Irene: Images of Anti-Semitism in Nineteenth- and Twentieth-Century Popular Culture [Focuses on British and German popular literature, including for children.] (329-350); Gibson, Arthur: Philosophy of Psychotic Modernism: Wagner and Hitler (351-386); Blond, Louis P.: Metaphysics and the Holocaust (387-399); Wright, Melanie J.: The Nature and Significance of Relations between the Historic Peace Churches and Jews during and after the "Shoah" [Discusses the Quakers, German Mennonites, and Hutterites.] (400-425); Hayes, Michael A.: From "Nostra aetate" to "We Remember: A Reflection on the Shoah" (426-445); Jackson, Sue: Back to the Future: A Century of Jewish-Christian Relations in Poland [Reflections based on a trip to Poland in September 1998.] (446-464). *YC*

0023 Ruggeri, Alvaro: *La bimillenaria persecuzione cristiana degli ebrei.* Roma: Bardi, 1999. 225 pp.

Presents, in popular fashion aimed at a Catholic audience, a brief history of Christian persecution of the Jews in Europe during the past 2,000 years. Focuses on events in the Roman Empire, Spain, and Italy. Christian intolerance toward Jews was based on the myth, invented by the Church Fathers, that the Jews were deicides. The historical revision initiated by Vatican Council II and continued by Pope John Paul II indicates the correct path for a Catholic to take to overcome that error. *MZ*

0024 Siegert, Folker, ed.: *Israel als Gegenueber: Vom Alten Orient bis in die Gegenwart: Studien zur Geschichte eines wechselvollen Zusammenlebens.* Goettingen: Vandenhoeck & Ruprecht, 2000. 584 pp.

A collection of papers delivered at a conference in Muenster, May 1998. The following papers deal with antisemitism: Ben-Chanan, Yaacov: Rom und die Juden – die Juden und Rom (73-89); Rese, Martin: Wer war Israel als Gegenueber der ersten Heidenchristen? (147-157); Gueting, Eberhard: Kritik an den Jedaeern in Jerusalem: Literarkritische Beitraege zu einem unabgeschlossenen Gespraech ueber den Evangelisten Johannes (158-201); Khoury, Adel Theodor: Der fruehe Islam und die Juden (218-238);

Schreckenberg, Heinz: Zum Gegenueber von Israel und Kirche in einer spaetmittelalterlichen Adventspredigt: Roberto Caracciolo (1425-1495) (285-289); Kirn, Hans-Martin: Israel als Gegenueber der Reformatoren: Zur christlichen Sicht von Juden und Judentum im 16. Jahrhundert (290-321); Stolle, Volker: Israel als Gegenueber Martin Luthers – im Horizont seiner biblischen Hermeneutik (322-359); Wand, Klaus: Theodor Fritsch (1852-1933) – der vergessene Antisemit [Discusses, also, his journal "Hammer."] (458-488). *MZ*

0025 Sirges, Thomas; Schoendorf, Kurt Erich, eds.: *Hass, Verfolgung und Toleranz: Beitraege zum Schicksal der Juden von der Reformation bis in die Gegenwart*. Frankfurt a.M.: Peter Lang, 2000. 211 pp.

A collection based on papers delivered at a seminar of the Germanistische Institut at the University of Oslo, 1997, except for the paper by Jost Hermand which is based on a lecture delivered at the Institut in 1998. Contents: Schoendorf, Kurt Erich: Judenhass und Toleranz im Spiegel von Flugschriften und Einblattdrucken des 16. Jahrhunderts [Includes reproductions of illustrations.] (11-46); Stockinger, Ludwig: Toleranz und Ausgrenzung: Antijuedische Implikationen der Geschichtsphilosophie in der deutschen Literatur um 1800 (47-72); Sagmo, Ivar: Der Judenartikel in der norwegischen Verfassung von 1814 in der Sicht deutscher Kommentatoren [On barring immigration of Jews.] (73-87); Hamann, Brigitte: Antisemitismus im Wiener Fin de siècle und der junge Hitler (89-100); Schoendorf, Kurt Erich: Antisemitismus in der Sprache des Nationalsozialismus (101-117); Lorenz, Einhart: Nordeuropa und die juedischen Fluechtlinge aus Hitler-Deutschland nach 1933 (119-133); Hermand, Jost: Auschwitz und anderswo: Gedanken ueber politische Grossverbrechen (135-152); Vestli, Elin Nesje: Eine Annaeherung an Auschwitz als Ortschaft/"Zeitschaft" in Prosatexten von Jean Améry, Ruth Klueger, George Tabori und Peter Weiss (153-169); Sirges, Thomas: "Hitlers willige Vollstrecker": Die Rezeption des Goldhagen-Buchs in Deutschland (171-209). *RW*

0026 *Tirosh* 3 (1999).

This issue contains papers delivered at the third CIS Students' Conference on Jewish Studies, Moscow, July 1998. Partial contents: Zanemonets, Aleksandr: Greki i yevrei v sochineniyakh "poslednikh vizantiitsev": Odno antiiudeiskoye proizvedeniye Konstantinopolskovo patriarkha Gennadiya Skholariya [Greeks and Jews in the Writings of the "Last Byzantines": An Anti-Judaic Work by the Patriarch of Constantinople Gennadios Scholarios] (114-121); Zelenina, Galina: "Iskhod" 1492 goda: Analiz nekotorykh traktovok izgnaniya yevreyev iz Ispanii [The "Exodus" of 1492: An Analysis of Some Interpretations of the Expulsion of the Jews from Spain] (134-145); Markedonov, Sergei: S.G. Svatikov: Borba protiv antisemitskikh mifov [S.G. Svatikov: The Struggle against Antisemitic Myths] (155-166); Gruener, Frank: Protsess nad chlenami Yevreiskovo Antifashistskovo Komiteta i sovetskaya politika v Vostochnoi Yevrope (1948-1952): Nekotorye zamechaniya i voprosy [The Trial against the Members of the Jewish Antifascist Committee and Soviet Policy in Eastern Europe (1948-1952): Some Remarks and Questions] (188-199); Mitsel, Rita: Kiyevskaya konservatoriya v kontse 40-kh – nachale 50-kh godov: Metody i praktika borby s formalizmom i kosmopolitizmom [The Kiev Academy of Music in the Late 1940s-Early 1950s: Methods and Practice of the Struggle against Formalism and Cosmopolitanism] (200-210). *DR*

0027 *Yevreiskii mir: Sbornik 1944 goda* [Jewish World: A Collection of 1944]. Ierusalim: Gesharim; Moskva: Mosty kultury, 2001. 480 pp. First published in New York: Soyuz russkikh yevreyev, 1944.

Reprint of a collection of articles on Jewish history and culture. Partial contents: Polyakov-Litovtsev, S.: Agasferova legenda [Ahasuerus' Legend. Discusses the question whether the fate of the Jewish people in history, especially in World War II, is uniquely tragic in comparison with other peoples.] (77-98); Vishnyak, M.: Mezhdunarodnaya konventsiya protiv antisemitizma [An International Convention against Antisemitism] (99-116); Menes, A.: Katastrofa i vozrozhdeniye [The Catastrophe and Revival] (116-129); Shekhtman, I.: Sovetskoye yevreistvo v germano-sovetskoi voine [Soviet Jewry in the German-Soviet War] (227-246); Kulischer, Eugene Michel: Izgnaniye i deportatsiya yevreyev [Expulsion and Deportation of the Jews] (261-275); Goldshtein, A.: Obshchestvennye elementy v gitlerovskom antisemitizme [Elements of Public Interest in Hitler's Antisemitism] (327-338); Krol, Moisei A.: Iz vospominanii: Kishinyovskii pogrom 1903 g. i Kishinyovskii pogromnyi protsess [From the Memoirs: The Pogrom in Kishinev of 1903 and the Kishinev Pogrom Trial] (376-386); Lesin, A.: Epizody iz moyei zhizni (s poslesloviyem Ya. Kreplyaka) [Episodes from My Life (with an Afterword by Ya. Kreplyak). On the pogroms of 1881 and the reaction of Russian populists to them.] (390-394). *DR*

0028 Zippan, Claudia: Grabschaendungen im Kontext christlicher Kultur. *Jahrbuch fuer Antisemitismusforschung* 10 (2001) 13-35.

Discusses the centrality of death and funeral rites for the identity of every culture and religion, including the Christian and the Jewish. Identity is established by differentiation from the Other. For Christianity, particularly in New Testament times but continuing throughout history, it was especially difficult to differentiate itself from its parent religion, Judaism. One way Christians tried to do this was the symbolic destruction of Jewish identity through the desecration of cemeteries. *RW*

GENERAL: BY COUNTRY

Brazil

0029 Maio, Marcos Chor; Calaça, Carlos Eduardo: New Christians and Jews in Brazil: Migrations and Antisemitism. *Shofar* 19, 3 (Spr 2001) 73-85.

Describes Brazil as unique in regard to its treatment of Conversos after the expulsion from Portugal and of Jewish immigrants during the Nazi period. Notes that Brazil did not have a permanent Inquisition tribunal and that Conversos had the possibility of upward social mobility. While there was racism (as discussed variously by different historians), many Conversos succeeded in becoming plantation owners. Complexities are noted in regard to restrictions on Jewish immigration in the 1920s-30s. Although the Brazilian government initially opposed Jewish immigration, pragmatic foreign ministry officials often made exceptions for individual cases. U.S. pressure also resulted in Jews being allowed to enter Brazil. While thousands arrived in the country before 1930, many fewer were allowed to enter in the following decade. One factor that activated anti-Jewish prejudice was Jewish resistance to the ethos of Brazilian assimilationism. *YC*

Czech Lands

0030 Bránský, Jaroslav: *Židé v Boskovicích* [The Jews of Boskovice]. Boskovice: Klub přátel Boskovic; Nakl. Albert, 1999. 347, [20] pp. With a summary in German and English (pp. 319-344).

A history of the Jews of Boskovice (in German, Boskowitz), Moravia, from the 16th century. Describes changes in their social and legal status throughout the centuries. Pp. 74-83 discuss discrimination against Czech and Moravian Jews in the 18th century, especially the law issued by Charles VI in 1726 limiting the number of Jewish families granted residence permits, the internment of Jews in ghettos from 1727, and an attempt by Maria Theresia to expel the Jews in 1745. Shows how these events are reflected in municipal documents, and notes the antisemitic remarks of the Christian chroniclers. Pp. 133-137 relate to the Holocaust period. Describes the first arrests in 1940, the concentration of all the Jews of the district in the town's Jewish quarter in the fall of 1941, and their deportation to Terezín on 14-15 March 1942. Only 19 Jews survived. The local community ceased to exist in the 1950s. *JT*

0031 Vojtíšková, Marie: *Židé v České Lípě* [Jews in Česká Lípa]. Česká Lípa: Nakl. Vega, 1999. 123 pp. With a summary in German (pp. 117-120).

A history of the Jewish community in the town Česká Lípa (in German, Boehmisch Leipa) from the 16th century through the Holocaust. Relates the difficulties experienced by the community in periods of increased antisemitism: e.g. pp. 28-35 describe the pogrom committed by Austrian soldiers in December 1744 (during the Austro-Prussian war) as recorded in Jewish and municipal chronicles; pp. 59-62 report on the obstacles that the Jews encountered while attempting to rebuild and enlarge their synagogue in the 17th and in the 18th centuries. Pp. 86-92 deal with the Nazi period: the burning of the local synagogue in November 1938, Aryanization of Jewish property in 1940, and deportation of most of the Jewish community between 1943-45. Based on material from the municipal archive and the archive of the Kaunic ducal family. *JT*

Germany

0032 *Antisemitismus – die deutsche Normalitaet: Geschichte und Wirkungsweise des Vernichtungswahns*. Freiburg: ça ira, 2001. 288 pp. At head of title page: Arbeitskreis Kritik des deutschen Antisemitismus.

A collection of essays which point to the need to raise the issue and examine the roots of German antisemitism. Partial contents: Steidle, Hans: Vorindustrielle Judenfeindschaft am Beispiel Wuerzburgs [In the 12th-19th centuries.] (17-33); Bakonyi, Rainer; Woeldike, Andrea: Judenhass – 30 Silberpfennige christlich verzinst und deutsch eingetrieben [On anti-Jewish prejudice related to usury and economics in the medieval period.] (35-41); Hoffmann, Lutz: Der Antisemitismus als Baugeruest der deutschen Nation (43-58); Bakonyi, Rainer: Ueberlegungen zu Emanzipation, Assimilation und Antisemitismus (59-66); Woeldike, Andrea: Kapitalismus und deutscher Wahn (67-74); Scheit, Gerhard: Struktureller Antisemitismus: Martin Heidegger (75-86); Bakonyi, Rainer: NS-Deutschland und die Ermordung der Juden Europas (89-105); Rensmann, Lars: Dialektik des Antisemitismus: Zu Elementen und Urspruengen der nationalsozialistischen Judenverfolgung im Spiegel der Gesellschaftskritik Horkheimers und Adornos (107-132); Scheit, Gerhard: Immer das Selbe: Ueber Hannah Arendt und Martin Heidegger [A revised version of an article first published in "konkret" 10 (1999).] (133-148); Bakonyi, Rainer: Zwischen Schlussstrich und heimlichem Stolz auf den laengsten Schatten: Die deutsche Geschichtswissenschaft auf dem Weg zur normalen Nation [On the "Historikerstreit," Goldhagen, and historiography.] (155-165); Kuth, Ines; Woeldike, Andrea: Antisemitismus und Nationalsozialismus: Zu Moishe Postones "theoretischen Skizzen" (167-175); Schatz, Holger; Woeldike, Andrea: Nationalisierung der Arbeit, Antisemitismus und Vernichtung (177-188); Uebelacker, Karin; Weiss, Michael: Die fuenfziger und sechziger Jahre [On Holocaust reactions in West Germany.] (209-215); Haury, Thomas:

Der Antizionismus der Neuen Linken in der BRD: Sekundaerer Antisemitismus nach Auschwitz (217-229); Uebelacker, Karin; Semmel, Martin: Antisemitismus im neuen Deutschland (231-235). *ID*

0033 Berghahn, Klaus L.: *Grenzen der Toleranz: Juden und Christen im Zeitalter der Aufklaerung*. Koeln: Boehlau, 2000. viii, 304 pp.

Traces the progress of Jewish emancipation in Prussia from total lack of rights at the beginning of the 18th century to the Edict of Emancipation of 1812, analyzing the writings of advocates as well as of opponents. Perceives a change from antisemitism on religious grounds to the antisemitism of the philosophers of the Enlightenment, who accused the Jews of backwardness, superstition, the lack of a genuine spiritual religion, and separatism. Even the most enlightened advocates of Emancipation hoped that it would lead the Jews to assimilate fully in German Christian culture, giving up their own supposedly antiquated and worthless culture and religion. Around 1811, just as the Jews were about to receive the full rights of citizenship, a vicious religious-nationalistic antisemitism, always latent among the common people, reemerged also among intellectuals. Also describes changes in the way of life of Jews of wealth and culture. *RW*

0034 Nachama, Andreas; Schoeps, Julius Hans; Simon, Hermann, eds.: *Juden in Berlin*. Berlin: Henschel, 2001. 264 pp.

A richly illustrated history emphasizing Jewish cultural and religious life, socioeconomic conditions, and community organization, but also dealing with antisemitism and discrimination through the ages. Contents: Flumenbaum, Claudia-Ann: Von den Anfaengen bis 1789 (9-52); Schoeps, Julius Hans: Der Anpassungsprozess (1790-1870) (53-88); Schuetz, Chana C.: Die Kaiserzeit (1871-1918) (89-136); Brenner, Michael: Die Weimarer Jahre (1919-1932) (137-180); Simon, Hermann: Die Zeit des Nationalsozialismus (1933-1945) (181-220); Nachama, Andreas: Von 1945 bis zur Gegenwart (221-246). *RW*

0035 Raphaël, Freddy, ed.: *"...das Fluestern eines leisen Wehens...": Beitraege zu Kultur und Lebenswelt europaeischer Juden: Festschrift fuer Utz Jeggle*. Konstanz: UVK, 2001. 582 pp.

A collection of articles on Jewish history and culture in Europe, focusing on relations between Jews and non-Jews. The following essays deal with antisemitism: Schreiner, Klaus: Oeffentlicher Spott, gesellschaftliche Ausgrenzung, vorenthaltene Buergerrechte: Frankfurts Juden im Gesichtskreis Goethes, Boernes und Heines [On ritual murder accusations, anti-Jewish caricatures and stereotypes, social exclusion, and pogroms between 1241-1815.] (241-279); Klein, Peter K.: "Judenspottkarten" vom Kaiserreich zur NS-Zeit: Kontinuitaet und Umbruch judenfeindlicher Bildpolemik [On anti-Jewish caricatures on postcards; pp. 305-323 contain examples.] (281-323); Schmoll, Friedemann: Bewahrung und Vernichtung: Ueber Beziehungen zwischen Naturschutz und Antisemitismus in Deutschland (345-367); Lammers, Karl Christian: Die "Judenwissenschaft" im nationalsozialistischen Dritten Reich: Ueberlegungen zur "Forschungsabteilu ng Judenfrage" in Walter Franks "Reichsinstitut fuer Geschichte des neuen Deutschlands" und zu den Untersuchungen Tuebinger Professoren zur "Judenfrage" (369-391); Linke, Uli: The Politics of Blood: Antisemitic Imaginaries after 1945 [On the factors of blood, gender, and race in German politics of the 20th century.] (393-424); Hauser, Andrea: Die Praxis des Erinnerns in Israel: Einblicke in die Gedenkstaetten- und Museumsarbeit zur "Shoah" und ihre Konsequenzen fuer deutsche historische Museen (527-543). *ID*

0036 Serup-Bilfeldt, Kirsten: *Zwischen Dom und Davidstern: Juedisches Leben in Koeln von den Anfaengen bis heute*. Ed.: Ulrike Mast-Kirschning. Koeln: Kiepenheuer & Witsch, 2001. 240 pp.

A survey of Jewish life in Koeln from the Roman period to the present. Pp. 35-46 discuss the medieval period, including the persecution and expulsion of Jews during the Black Death epidemic beginning in 1349. No Jews lived in Koeln for the next 400 years. The 19th century was characterized by opposing trends: hope for emancipation as well as antisemitic outrages, among them the blood libel of Xanten in 1892. Pp. 121-206 deal with the Nazi period. Discusses the process of Aryanization, the activities of the Kulturbund, the "Kristallnacht" pogrom, the Yavne Jewish high school, headed by Erich Klibansky, who prepared the students for emigration (he was deported to Minsk in 1942, after having rescued 130 children). Between 1941-45, 11,000 Jews were deported from Koeln. Includes the story of Katharina Overath (née Meier), who saved the Bernauer family from deportation and was recognized by Yad Vashem as a Righteous Gentile. Includes numerous photographs and facsimiles, as well as inserts on specific sites connected with Jews in Koeln. *ID*

Italy

0037 Bonilauri, Franco; Maugeri, Vincenza, eds.: *Le Comunità ebraiche a Modena e a Carpi, dal medioevo all'età contemporanea*. Firenze: Giuntina, 1999. 283 pp.

A collection of papers presented at a conference in Modena – Carpi, Mai 1997. Partial contents: Calabi, Donatella: Dal quartiere ebraico alla costituzione del ghetto in Italia: Il caso di Modena (87-93); Zanardo, Andrea: Catecumeni e neofiti a Modena alla fine dell'antico regime (121-

139); Sarfatti, Michele: La persecuzione antiebraica negli anni 1936-1943 (233-239); Voigt, Klaus: I ragazzi di Villa Emma a Nonantola (241-269); Dogliani, Patrizia: Memoria e rappresentazione storica della deportazione ebraica in Italia e all'estero (271-283). *MZ*

0038 Colorni, Vittore: *Gazagà, ghetto, israeliti: Voci ebraiche dal Novissimo Digesto Italiano*. Mantova: Mantova Ebraica, 1998. 71 pp.

A reprint of items by Colorni which appeared in the journal "Novissimo Digesto Italiano." In 22 sections, describes the changing juridical status of the Jews in Italy as expressed in legislation from Roman times until 1947. Relates, also, to restrictions on Jewish life in the Roman Empire (4th-5th centuries), expulsions and enclosures between 1474-1541, reactions against emancipation between 1814-48, the fascist racial laws (1938-43), and the Nazi occupation (1943-45). *MZ*

Poland

0039 Duda, Eugeniusz: *The Jews of Cracow*. Trans.: Ewa Basiura. Kraków: Wydawnictwo "Hagada" and Argona-Jarden Jewish Bookshop, 1999. 127 pp. Originally published as "Krakowskie judaica" (Kraków: Wydawnictwo PTTK "Kraj," 1991).

0040 Galas, Michał, ed.: *Duchowość żydowska w Polsce: Materiały z międzynarodowej konferencji dedykowanej pamięci Profesora Chone Shmeruka, Kraków 26-28 kwietnia 1999* [Jewish Spirituality in Poland: Papers from the International Conference Dedicated to the Memory of Professor Chone Shmeruk, Kraków 26-28 April 1999]. Kraków: Księgarnia Akademicka, 2000. 431, [9] pp.

Partial contents: Kalik, Judith: Żydowska duchowość, religia i praktyki religijne w oczach Polaków w XVII-XVIII wieku [Jewish Spirituality, Religion and Religious Practices in the Eyes of Poles in the 17th-18th Centuries. Deals, also, with anti-Jewish polemical writings.] (81-90); Leiman, Sid Z.: Stosunek Jonatana Eibeschuetza do ruchu Frankistów w Polsce [Rabbi Jonathan Eibeschuetz's Attitude towards the Frankists] (229-236); Ben-Sasson, Havi: Historiografia ultraortodoksyjna i Holokaust – Mosze Prager jako studium przypadku [Ultraorthodox Historiography and the Holocaust – Moshe Prager as a Case Study] (391-403); Ratner, Tsila: Miejsca pamięci: Widoki nieobecności [Places of Memory: Landscapes of Non-Presence. About the image of Poland as the symbolic place of the Holocaust in the work of Israeli writers Yehudit Hendel, "Near Quiet Villages" and Eleonora Lev, "A Certain Kind of Orphanhood."] (405-420). *HV*

Russia, USSR, CIS

0041 Kaplanov, Rashid M.; Mochalova, Viktoriya V.; Chulkova, L.A., eds.: *Yevreiskaya kultura i kulturnye kontakty: Materialy shestoi yezhegodnoi mezhdunarodnoi mezhdistsiplinarnoi konferentsii po iudaike* [Jewish Culture and Cultural Contacts: Materials from the Sixth Annual International Interdisciplinary Conference on Jewish Studies]. Moskva: Sefer, 1999. 311 pp. Part 3 of the conference proceedings. See also part 2: "Istoriya yevreiskovo naroda" (1999).

Papers delivered at the conference held in Moscow. Partial contents: Belova, Olga V.: Obraz yevreya v narodnoi demonologii slavyan [The Image of the Jew in Slavic Popular Demonology] (86-99); Serov, Borys-N.: Obraz yevreyev v sochinenii I. Galyatovskovo "Messiya pravdivyi" [The Image of the Jews in the Work by J. Galatowski "Mesjasz Prawdziwy"] (100-114); Polyakova, Lyudmila: Oblik "ideinovo antisemita" v mirovozzrencheskoi sisteme klassika (I.S. Turgenev o B.M. Markeviche) [The Image of an "Antisemite on Principles" in the Worldview of a Classical Writer (I.S. Turgenev on B.M. Markevich)] (115-122); Katsis, Leonid F.: Otets Pavel (Florenskii), Lev Modenskii i "Shish" Alekseya Kruchyonykh (yeshchyo raz o dyavole i yevreyakh v rossiiskom kontekste) [Father Pavel (Florenskii), Leon da Modena, and "Shish" by Aleksey Kruchyonykh (Once More on the Devil and the Jews in the Russian Context)] (123-139). *DR*

0042 Postoutenko, Kirill: Imaginary Ethnicity: Jews and Russians in Russian Economical Mythology. *American Behavioral Scientist* 45, 2 (Oct 2001) 282-295.

In the 11th-18th centuries, Russian official mythology regarded the opposition between Russians and Jews as mainly religious. In the 19th century, with the consolidation of Russian national consciousness, this opposition came to be conceived as economic: the Russian was regarded as attached to the land, while the Jew was the embodiment of the money economy, thus a menace and a cosmopolitan. The persecution of Jews in the second part of the century mostly involved curtailing their economic flexibility rather than their religious freedom. At the same time, the new myth retained some religious undercurrents. This kind of mythology persisted into the Soviet period. *DR*

0043 Tabak, Yuri: Otnosheniye Russkoi Pravoslavnoi Tserkvi k yevreyam [The Attitude of the Russian Orthodox Church to the Jews]. *Dia-Logos* [1998-1999] (1999) 485-503.

Surveys the attitude of the Russian Orthodox Church toward the Jews and Judaism from the 10th century to the present. It was in many ways consonant with that of the Catholic Church, but also had some distinctive features. While Russian rulers have initiated and carried out anti-Jewish policies, the Church has never declared any definite position on the Jews. Thus, a harsh anti-Jewish position of some hierarchs has coexisted with a philosemitic attitude on the part of others. The veneration of Church Fathers, whose authority is incontestable in the Russian Church, has reinforced anti-Jewish traditions. Divides the Russian clergy and laity into four groups; only one is ready for Christian-Jewish dialogue, while two are harshly antisemitic. The leadership of the Church abstains from criticizing antisemitic declarations of some clerics, tolerates antisemitic publications printed by its publishing houses, and attacks pro-Judaic and liberal declarations of Church members. *DR*

Spain

0044 Bonnín, Pere: *Sangre judía: Españoles de ascendencia hebrea y antisemitismo cristiano*. Barcelona: Flor del Viento Ediciones, 1998. 382 pp.

An introductory chapter presents, in popular fashion, a portrait of Judaism. Pp. 99-244, "Una historia de desencuentros," recount the history of the Jews in Spain from Roman times to the present, and the stance of Christians and Spanish state institutions toward them. Pp. 353-374 contain a list of Spanish family names suspected of having Jewish origins, taken from lists of victims of the Inquisition, from censuses of Jewish quarters (juderías), and from other sources. *MZ*

0045 Rehrmann, Norbert; Koechert, Andreas, eds.: *Spanien und die Sepharden: Geschichte, Kultur, Literatur*. Tuebingen: Max Niemeyer, 1999. vi, 243 pp.

A collection based on papers delivered at a conference in Bremen, June 1997. Partial contents: Ben-Chanan, Yaacov: Juden im maurischen und christlichen Spanien (711-1492): Chancen und Scheitern einer Symbiose zwischen unterschiedlichen Kulturen (7-25); Bernecker, Walther L.: Die Vertreibung der Juden aus Spanien: Zur Diskussion ueber das "Dekadenz-Syndrom" (27-42); Franzbach, Martin: Zwischen Verdraengung und Verunglimpfung: Die Juden in der Literatur des "Siglo de Oro" (43-50); Zur Muehlen, Patrik von: Die Sepharden, der Holocaust und die iberische Fluchtroute (115-124); Rehrmann, Norbert: Kulturelles Gedaechtnis, nationale Identitaet und Literatur: Die Sephardenthematik in fiktionalen und essayistischen Texten des 19. und 20. Jahrhunderts [Deals mainly with antisemitic Spanish texts.] (195-222). *RW*

THE ANCIENT PERIOD

0046 Bar-Kochva, Bezalel: A Distorted Account of the Jewish "Origo" (C 1:305-311). *Internationales Josephus-Kolloquium, Amsterdam 2000,* ed. Juergen U. Kalms. Muenster: LIT, 2001. Pp. 11-28.

Discusses an antisemitic account of the life of the ancient Hebrews in Egypt and the Exodus, written by Lysimachus, an Alexandrian Greek, and found in Josephus' "Contra Apionem," Book 1, pp. 305-311. Lysimachus' work, written in the late 2nd century BCE, reflects tensions between Jews and Egyptians. He presents most of the elements of the anti-Jewish Exodus traditions found in Hellenistic and Roman literature. Josephus considers this work as the most hostile account of the origins of the Jews. The Jews are depicted as lepers and as people suffering from various diseases, who took refuge in Egyptian temples. They were held responsible for crop failures and, hence, needed to be purged; therefore, they were either drowned in the sea or sent into the wilderness. Some of them fled, under the leadership of Moses, who counseled them to be misanthropic and to loot the temples. The town they reached, Hierosyla (whence Jerusalem), supposedly meant "temple loot." *YC*

0047 Bieringer, Reimund; Pollefeyt, Didier; Vandecasteele-Vanneuville, Frederique, eds.: *Anti-Judaism and the Fourth Gospel: Papers of the Leuven Colloquium, 2000.* Assen: Royal Van Gorcum, 2001. xvi, 612 pp.

A collection of essays presented at a seminar held at the Katholieke Universiteit Leuven, January 2000. *YC*

0048 Eisenbaum, Pamela: Is Paul the Father of Misogyny and Antisemitism? *Cross Currents* 50, 4 (Win 2000-2001) 506-524.

Disputes the view (recently expressed, inter alia, by Daniel Boyarin) that Paul was the father of misogyny and antisemitism, and that he confused equality with sameness, although this may well have been true for many of his Christian followers. Views Paul as a Jew, not a founder of a new religion. Focuses on Galatians 3:28, "There is no longer Jew or Greek.. .," and concludes, in opposition to the common view in both the past and present, that Paul was not in favor of ending distinctions between people or of doing away with Jewish practices, e.g. circumcision. Stresses that Paul's mission was to the Gentiles; they had no need for circumcision or Jewish ritual. Paul was suggesting a union of male and female, as in Eden and as the nucleus of a family, and the combining of all peoples, Jews and Gentiles, into a family where the constituent members did not need to lose their identities. *YC*

0049 González Salinero, Raúl: *El antijudaísmo cristiano occidental (siglos IV y V)*. Madrid: Trotta, 2000. 318 pp. Based on the author's diss. – Universidad de Salamanca.

The Christianization of the Roman Empire (4th-5th centuries) resulted in progressive deterioration of the legal and social status of the Jews. Although they were still recognized as a "licita religio" (their legal existence guaranteed by the laws of the Empire), the Church gained influence on the legislation dealing with Jews. Access of Jews to public posts was prohibited, as well as possessing Christian slaves, testifying in trials or becoming lawyers. The Church isolated the Jews socially, prohibiting Christians from having any contact with them (marrying, sharing a meal, holding conversations, exchanging greetings, etc.). Through literature, sermons, and canonical law, the Church created and spread a degrading image of the Jew and of Judaism. For the Church, the existence of Judaism was justified only as "testes veritatis" – to prove the veracity of the Old Testament and the veracity of the promises contained in it which were fulfilled by Christianity, now the "Verus Israel." *MZ*

0050 Hennings, Ralph: Eusebius von Emesa und die Juden. *Zeitschrift fuer Antikes Christentum* 5, 2 (2001) 240-260.

Eusebius (ca. 300-ca. 359), in his Old Testament commentaries, makes liberal use of the Hebrew text and Jewish exegesis, but in his sermons he uses the same sources to show that the Church has superseded the Jews, who were cast off because of their rejection of Christ. This is proved by the destruction of the Temple and the cessation of prophecy, for which, he notes, the Jewish sages, too, blamed the Jews themselves. Although in Eusebius' time the Church triumphed through Constantine, the Jews clearly were still powerful enough in Syria to be perceived as a danger to Christianity. Eusebius does not advocate persecution or missionary work, but warns Christians to stay away from synagogues. *RW*

0051 Lerner, Berel Dov: No Happy Ending for Esther. *Jewish Bible Quarterly* 29, 1 (Jan-Mar 2001) 4-12.

The story in the Book of Esther may be read as a guide to the politics of Jewish life in the Diaspora. It tells of the dangers of antisemitism and how such dangers may be neutralized. It describes antisemitism's propaganda and real interests, but speaks also of the Jewish response. Discusses additions to the biblical story found in the Septuagint (Greek) translation, the Aramaic Targumim, and the Midrash. In traditional Jewish terms, God and His laws take precedence over the law of the land, and thus put Jews in a position of being accused of disloyalty and subversion in the Diaspora. Refers to Jewish "insularity" – e.g. prohibitions against sharing food, drink, or bed with non-Jews, and notes that the Jewish heroine

Esther was forced to commit such actions in order to save her people. Jewish sources emphasize "the tragic heart" of the story, the Jewess whose victimization is required; in the end, Esther remains trapped in the clutches of the pagan king. *YC*

0052 Rajak, Tessa: *The Jewish Dialogue with Greece and Rome: Studies in Cultural and Social Interaction*. Leiden: Brill, 2001. xix, 579 pp.

A collection of articles, most of them published previously. Ch. 26 (pp. 511-533), "Talking at Trypho: Christian Apologetic as Anti-Judaism in Justin's `Dialogue with Trypho the Jew'" appeared previously in "Apologetics in the Roman Empire: Pagans, Jews and Christians" (1999).

0053 *Studia Patristica* 35, 37 (2001).

Vols. 34-38 of this journal contain papers presented at the 13th International Conference on Patristic Studies, held in Oxford in 1999, edited by M.F. Wiles and E.J. Yarnold. The following papers deal with antisemitism: Vol. 35: Shepardson, Christine: Anti-Jewish Rhetoric and Intra-Christian Conflict in the Sermons of Ephrem Syrus (502-507). Vol. 37: Reynard, Jean: L'antijudaïsme de Grégoire de Nysse et du pseudo-Grégoire de Nysse (257-276); Hennings, Ralph: Eusebius von Emesa: Zwei Osterpredigten gegen die Juden (535-542). *SSC*

0054 Ulrich, Joerg: *Euseb von Caesarea und die Juden: Studien zur Rolle der Juden in der Theologie des Eusebius von Caesarea*. Berlin: Walter de Gruyter, 1999. 324 pp. Based on the author's Habilitationsschrift – Universitaet Erlangen, 1997.

Contends that Eusebius (ca. 263-339) was not hostile to the Jews, as some scholars hold, but only their opponent in theological dispute. He recognized Judaism as the source of Christianity and respected the Jews he knew in Caesarea. He blamed them, however, for deicide; for clinging to the old Law when Christ had now become the Law; and for claiming salvation exclusively for themselves, denying it to the Gentiles. For these sins the Jews were punished by the loss of Jerusalem and the Temple, and of the priesthood, the monarchy, and prophecy – three functions now lodged in Christ (not in the Church, as claimed in the doctrine of supersession). However, Eusebius does not mention their punishment in a tone of triumphalism but rather in pity and sorrow. Also discusses a sharply anti-Jewish letter by Constantine the Great quoted by Eusebius in his biography of the Emperor, arguing that it is genuine and not Eusebius's invention; and Constantine's anti-Jewish legislation, which was much more moderate than his utterances. Compares Eusebius with other Church Fathers whose attitudes to the Jews were similar. *RW*

0055 Weiss, Hans-Friedrich; Dobbeler, Axel von: Wo liegen die Wurzeln des christlichen Antijudaismus? *Zeitschrift fuer Neues Testament* 8 (2001) 35-47.

Presents two separate essays in the section of this issue called "Kontroverse," with an introduction by the editor on pp. 35-36. Weiss's essay, "Noch einmal: Zur Frage eines Antijudaismus bzw. Antipharisaeismus im Matthaeusevangelium" (pp. 37-41) is a response to a previous essay by von Dobbeler (published in ZNW 91, 2000) which is summarized in the editor's introduction: it interprets the mission for the conversion of the Gentiles in Mt. 28 and the mission for the reconstruction of Israel in Mt. 10 as complementary and not anti-Jewish, as asserted (in von Dobbeler's understanding) by Weiss and others. Weiss responds that he does not read anti-Judaism into Matthew but rather anti-Pharisaism. The conflict with Pharisaic Judaism caused the Early Church to repress the abstraction of Matthew's doctrine of the lasting mission of Israel; for the sake of post-Auschwitz Christian-Jewish dialogue it is important to revive this pro-Israel strand in the Gospel. In "Wo liegen die Wurzeln des christlichen Antijudaismus?" (pp. 42-47), von Dobbeler responds that New Testament texts that reflect intra-Jewish controversies and are not anti-Jewish in their original intention, are often misused and have given rise to anti-Jewish traditions that are as important as the original text. Ascribes the growing distance and hostility between Judaism and the Early Church to this internal strife; to Jewish rejection of Jesus; to practices unacceptable to Jews arising from the opening of the Church to pagans; and to the suppression of diversity in Judaism. *RW*

0056 Alcalá Galve, Ángel: Cristianos y judíos en Aragón: La disputa de Tortosa. *Inquisición y sociedad,* ed. Ángel de Prado Moura. Valladolid: Universidad de Valladolid, 1999. Pp. 27-63.

The disputation of Tortosa took place between February 1413 and November 1414. It was convened by Pope Benedict XIII (originally from Aragón), and was led by the Converso Jerónimo de Santa Fe (the former Joshua Lorki, rabbi and then personal doctor to the Pope). On the Jewish side, eight rabbis participated, led by Astruch Halevy. Jerónimo de Santa Fe appealed to talmudic and midrashic sources to demonstrate that Jesus was the Messiah, giving the texts an allegorical interpretation. He also used those talmudic texts containing negative statements about Christianity to prove the anti-Christian character of the Jewish religion. The rabbis offered a weak defense, giving the texts a literal interpretation. One after the other withdrew from the disputation and Jerónimo declared himself the winner. Some Jewish historians contend that the rabbis were not free to express their views. The disputation of Tortosa provoked a large wave of conversions in Aragón and in the rest of Spain. *MZ*

0057

עסיס, יום טוב; קפלן, יוסף, עורכים: דור גירוש ספרד: קובץ מאמרים. ירושלים: מרכז זלמן שזר לתולדות ישראל, 1999. טו, 322, 171 עמ'.

[Assis, Yom Tov; Kaplan, Yosef, eds.: *Jews and Conversos at the Time of the Expulsion: A Collection of Articles*. Jerusalem: Zalman Shazar Center for Jewish History, 1999. 15, 322, 171 pp.] Ten of the essays in the Hebrew section were translated from the Spanish by Shlomo Gitai.

תוכן חלקי: סוארז פרננדז, לואיס בילבאו: ערי קסטיליה והבעיה היהודית (3-11); הינוחוסה מונטאלבו, חוסה: הקהילות היהודיות במלכות ולנסיה בשלהי ימי-הביניים ערב הגירוש (13-34); אורפלי לוי, משה: לדמותה של היהדות על-פי "censura et confutatio libri Talmud" 1488 (63-83); ביינארט, חיים: גירוש ספרד: סוף מעשה במחשבה תחילה (99-122); מוטיס דולדר, מיגל אנחל: פירוקן הרשמי של הקהילות היהודיות בממלכת ארגוניה (123-154); בלסקו מרטינז, אסונסיון: גירוש היהודים מאדמות האצולה: אחוזות קבלריה ואלעזר של מסדר ההוספיטלרים (155-175); קרטה פרונדו, קרלוס: המלנכוליה בקרב אנוסי קסטיליה והגירוש של שנת 1492 (179-188); לדרו קסאדה, מיגל אנחל: יהודים ואנוסים באנדלוסיה, 1481-1508 (189-207); רבאדה אוברדו, מריה דל פילר: הזמת העדים של התביעה במשפטי האינקוויזיציה, 1483-1507 (209-220); יהלום, יוסף: אשמה והאשמה: תגובות לגירוש ספרד ולשמד בשירת הניצולים (273-286); בניטו רואנו, אלוי: השתלבותם מחדש של יהודים מגורשים בחברה הספרדית עד שנת 1500 (287-295).

A collection based on papers delivered at a conference held in Jerusalem in January 1992. They discuss the situation of the Jews and Conversos in Spain from 1391, the Inquisition, and the expulsion, as well as the fate of Portuguese Jews. Partial contents: Hebrew section: Suárez Fernández, Luis Bilbao: The Cities of Castile and the Jewish Problem (3-11); Hinojosa Montalvo, José: The Jewish Communities in the Kingdom of Valencia in the Late Middle Ages on the Eve of the Expulsion (13-34); Orfali Levi, Moisés: The Image of Judaism in "Censura et confutatio libri talmud," 1488 (63-83); Beinart, Haim: The Spanish Expulsion: From Design to Implementation (99-122); Motis Dolader, Miguel Ángel: The Official Dismantling of the Jewish Communities of the Crown of Aragón (123-154); Blasco Martínez, Asunción: The Expulsion of the Jews from the Lands of the Nobility: The Estates of Caballeria and Eleazar and the Order of the Hospitallers [On two families from Zaragoza.] (155-175); Carrete Parrondo, Carlos: Melancholy amongst the Conversos of Castile and the Expulsion of 1492 [Relates, especially, Conversos' reversion to Jewish practices, and expressions of stress and melancholy in regard to the actions of the Inquisition, based on Inquisition records from Cuenca.] (179-188); Ladero Quesada, Miguel Ángel: Jews and Conversos in Andalusia, 1481-1508 (189-207); Rábade Obradó, María del Pilar: Refutation of Testimonies of Witnesses for the Prosecution in Trials of the Inquisition, 1483-1507 (209-220); Yahalom, Joseph: Guilt and Blame: Responses to the Spanish Expulsion and the Forced Conversion in the Poetry of the Exiles (273-286); Benito Ruano, Eloy: The Re-integration of Expelled Jews into Spanish Society until 1500 (287-295).] English section: Carpenter, Dwayne Eugene: From Al-Burak to "Alboraycos": The Art of Transformation on the Eve of the Expulsion (25-37); Lunenfeld, Marvin: Instruments of Repression: The Marginalization of the Castilian Jewish Community by Isabella the Catholic's Officials, 1474-1492 (39-52); Hillgarth, Jocelyn N.: The Reactions of Catholic Intellectuals to the Jewish Presence in Spain during the Reign of the Catholic Monarchs [Especially Hernando de Talavera.] (53-64); Gampel, Benjamin Raphael: Ferdinand and Isabella and the Decline of Portuguese and Navarrese Jewries (65-92). *RE*

0058 Bell, Dean Phillip: *Sacred Communities: Jewish and Christian Identities in Fifteenth-Century Germany*. Boston: Brill Academic, 2001. xi, 301 pp.

Examines the world of Jewish-Christian relations by focusing on both Jewish and Christian interaction as well as Jewish and Christian forms of communal organization and identity, based on medieval sources. Ch. 4 (pp. 99-125), "Anti-Judaism: Between Religion and Community," gives an overview of anti-Judaism in late medieval Germany (see his discussion of this subject in his diss., University of California, 1995). *YC*

0059

בן-שלום, רם: קידוש השם ומרטירולוגיה יהודית בארגון ובקסטיליה בשנת קנ"א (1391): בין ספרד לאשכנז. תרביץ 70, 2 (ינ'-מרץ 2001) 227-282.

[Ben-Shalom, Ram: "Kiddush ha-Shem" and Jewish Martyrdom in Aragón and Castile in 1391: Between Spain and Ashkenaz. *Tarbiz* 70, 2 (Jan-Mar 2001) 227-282.]

Disputes research on the medieval period which stresses the dichotomy between Ashkenazic martyrdom and Sephardic conversion in the face of Christian persecution. Contends that there was also conversion in Ashkenaz, and that many Spanish Jews chose martyrdom in 1391. Discusses contemporary letters, halakhic literature, commentaries, and liturgical poetry which attest to the acceptance of the ideology of martyrdom by Spanish Jewry and to the actual martyrdom of many Jews. Refutes the idea that Jewish martyrs in Spain were influenced by Ashkenazi ideology, claiming that they were probably influenced by Christian martyrdom. Notes that there were cases of martyrdom among Spanish Jews in the 15th century as well, in spite of the many conversions which took place. *RE*

0060 Bennassar, Bartolomé: Portrait d'un fanatique – Torquemada. *L'Histoire* 259 (Nov 2001) 48-49, 52-55.

Inquires into the origins and veracity of the "black legend" of Torquemada, who is generally seen as representing all the excesses of the Inquisition in collective memory. Attempting to give a balanced picture of the first Grand Inquisitor, contends that he had no responsibility for the creation of the Inquisition but did participate in defining its methods and spirit, inspiring the expulsion of the Jews. Disputes Torquemada's role in the expulsion of Abraham Senior and Isaac Abravanel, in the trials of 1490-91, and in the La Guardia blood libel. Considers the number of victims attributed to Torquemada inflated. *EO*

0061 Brunschwig-Ségal, Annette: "Der jude habe si gemint": Verbotene Beziehungen zwischen Juden und Christen im Spaetmittelalter. *Judaica* 57, 3 (Sept 2001) 182-203.

Describes the trial in Zurich in 1415 of a Jew accused of sexual intercourse with a Christian maidservant employed in his home; of the maidservant, who had borne him two children; and of a Christian neighbor (possibly also in the Jew's service) who had assisted at both births and arranged for custody of the infants. Though all three insisted they did not know of the prohibition on sexual intercourse between Jews and Christians, this was obviously a lie. The Jew was sentenced to a large fine; the maidservant was driven through the city in a humiliating manner and then banished; and the neighbor was also humiliated. Mentions similar cases in other towns, and surveys canon law, local German secular law, and Jewish law on the matter. In many places the law prescribed the death penalty, but with one possible exception this penalty was never actually carried out. The accused could generally have it commuted to a fine. *RW*

0062 Claman, Henry N.: *Jewish Images in the Christian Church: Art as the Mirror of the Jewish-Christian Conflict, 200-1250 C.E.* Macon, GA: Mercer University Press, 2000. xi, 212, [16] pp.

Discusses images from the Old Testament and other Jewish themes in early Christian and medieval art in Western Europe. The doctrine of supersessionism appeared in Christianity very early on. In its developed form it stated that, properly interpreted, the Old Testament can prove the truth of Christianity; hence, it blamed the Jews for failure to understand their own Scriptures and for spiritual blindness. Despite the steady deterioration in the legal status of the Jews, and attacks on them by some Church hierarchs, Christian art before 800 contained supersessionist motifs only in mild forms. With the consolidation of the Church after 1000, the Jews came under increasing civil and social pressure. Christian art of the Romanesque-early Gothic epoch (church paintings and sculptures, stained glass, crosses, illustrations in literary works) was replete with supersessionist images, such as the "blind Synagoga." At the same time, the Church distinguished between "good Jews" like Abraham, Moses, and David, who were regarded as precursors of Christianity and were depicted with halos, and "bad Jews" of the contemporary epoch, who were to be ostracized. *DR*

0063

כהן, ג'רמי: ספר 'מחסום לפיהם של היהודים' של ריימונדוס מארטיני. מאה שערים: עיונים בעולמם הרוחני של ישראל בימי הביניים, לזכר יצחק טברסקי, עורכים עזרא פליישר ואחרים. ירושלים: הוצאת מאגנס, האוניברסיטה העברית, 2001. עמ' 279-296.

[Cohen, Jeremy: Raimundus Martini's "Capistrum Iudaeorum." *Me'ah She`arim: Iyunim be-olamam ha-ruhani shel Yisrael bi-yemei ha-beinayim, le-zekher Yitzhak Twersky* [One Hundred Gates: Studies in Medieval Jewish Spiritual Life, in Memory of Isadore Twersky], eds. Ezra Fleischer et al. Jerusalem: Magnes Press, Hebrew University, 2001. Pp. 279-296.]

Martini's "Capistrum Iudaeorum," written in 1267 but published only recently, instructs Christian preachers and missionaries in how to conduct a debate with Jews. The first half of the work contains quotes from the Bible which are believed to prove that the Messiah has already come; the second half contains refutations of arguments propounded by the Jews.

Martini's work was the first to make use of post-biblical Jewish literature, e.g. Talmud, Midrash, and medieval commentaries; he uses this literature, although he also criticizes it. Pablo Christiani had done the same in his debate with Nahmanides in 1263, and this work may actually be a result of that debate. Compares "Capistrum Iudaeorum," Martini's first anti-Jewish work, with his later "Pugio fidei adversus Mauros et Iudaeos" against Muslims and Jews, noting developments in his anti-Jewish views. *RE*

0064 Cohen, Mark R.; Somekh, Sasson: Interreligious Majalis in Early Fatimid Egypt. *The Majlis: Interreligious Encounters in Medieval Islam,* eds. Hava Lazarus-Yafeh et al. Wiesbaden: Harrassowitz, 1999. Pp. 128-136.

Provides evidence of the existence and nature of a "majlis" (discussion of competing faiths) in the court of the Fatimid vizier and former Jew Ibn Killis (late 10th century), who replicated the institution from the court of the Caliph al-Mu`izz. Christian-Arabic sources on the "majlis" have a decidedly anti-Jewish slant, but indicate that freedom of speech and non-recrimination against participants were the rule. A Muslim source depicts Ibn Killis' conversion and his unfavorable views of Jews as genuine. Cites the text of a Judeo-Arabic account by a Jewish scholar, found in the Cairo Geniza, which suggests that an Arabic translation of Hebrew prayers from Saadya Gaon's "Siddur" was ridiculed by Ibn Killis and other Muslims at the "majlis." This offensive treatment of Judaism was described as having shocked the Jews present, and perhaps it provided the impetus for the response found in the Geniza. This response to the polemical insults stressed the superiority of Judaism and its prayers. *YC*

0065

כהן, מרק ר. : בצל הסהר והצלב : היהודים בימי הביניים. תרגום : מיכל סלע. חיפה : הוצאת הספרים של אוניברסיטת חיפה ; לוד : זמורה-ביתן, 2001. 408 עמ׳.

[Cohen, Mark R.: *In the Shadow of the Crescent and the Cross: The Jews in the Middle Ages*. Trans.: Michal Sela. Haifa: University of Haifa Press; Lod: Zmora-Bitan, 2001. 408 pp.] Originally published as "Under Crescent and Cross: The Jews in the Middle Ages" (Princeton, NJ: Princeton University Press, 1994).

0066 Jordan, William Chester: *Ideology and Royal Power in Medieval France: Kingship, Crusades and the Jews*. Aldershot, Hants: Ashgate, 2001. x, [various pagings].

A collection of articles published previously. The following deal with antisemitism: Marian Devotion and the Talmud Trial of 1240 [Appeared in "Religionsgespraeche im Mittelalter" (1992).] (XI, 61-76); Problems of the Meat Market of Béziers 1240-1247: A Question of Anti-Semitism

[Appeared in "Revue des Etudes Juives" 135 (1976).] (XII, 31-49); Princely Identity and the Jews in Medieval France [Appeared in "From Witness to Witchcraft" (1996).] (XIII, 257-273); Home Again: The Jews in the Kingdom of France, 1315-1322 [Appeared in "The Stranger in Medieval Society" (1997).] (XIV, 27-45); Jews, Regalian Rights, and the Constitution in Medieval France [Appeared in the "AJS Review" 23 (1998).] (XV, 1-16); The Jews and the Transition to Papal Rule in the Comtat-Venaissin [Appeared in "Michael" 12 (1991).] (XVI, 213-232). *DR*

0067 Knapp, Fritz Peter: Heinrich von Langenstein, "Sermones Wiennenses ad Iudaeos convertendos": Die aeltesten aus dem deutschen Sprachraum erhaltenen Judenbekehrungspredigten: Praesentation und Interpretation eines Neufunds. *Mitteilungen des Instituts fuer Oesterreichische Geschichtsforschung* 109, 1-2 (2001) 105-117.

Discusses the Viennese theologian's rough drafts or notes for sermons to the Jews, dating from the end of the 14th or beginning of the 15th century. Although he used traditional arguments, such as God's punishment of the Jews for their rejection of the prophets and Jesus, and contemporary Jews' stubborn blindness to all teachings but those of their own tradition (purposely falsified by their ancestors who killed Jesus), Langenstein was open to dialogue. Not so his pupil Nikolaus von Dinkelsbuehl in a series of sermons (probably in 1420) to converted Jews – presumably Jews converted by force in the Viennese "gezerah." But Dinkelsbuehl's approach was mild when we remember that at this time Jews who would not convert were being burned at the stake. *RW*

0068 Krol, A.: Presledovaniya i izgnaniya yevreyev v Sredniye Veka v Severnoi i Tsentralnoi Yevrope: Ikh posledstviya [Persecutions and Expulsions of Jews in the Middle Ages in Northern and Central Europe: Its Aftermath]. *Drugoi narod* [Another People], Vol. 1, ed. Ye.S. Chekanovkina. Lugansk: Znaniye, 2000. Pp. 57-79.

Describes the pogroms in the Rhineland during the First Crusade and pogroms and other persecutions of Jews in Germany and France in the 13th-14th centuries. The persecutions ended with the expulsion of the Jews from England and France. In the aftermath of these events the legal and economic situation of the Jews in Europe deteriorated. *DR*

0069 Kuelzer, Andreas: *Disputationes Graecae contra Iudaeos: Untersuchungen zur byzantinischen antijuedischen Dialogliteratur und ihrem Judenbild*. Stuttgart: B.G. Teubner, 1999. 400 pp.

An imaginary dialogue between a Christian and a Jew was a popular form of Byzantine anti-Jewish polemics. The dialogues were not intended to convert Jews, and because of their authors' ignorance of Judaism and lack of any attempt to understand it, could hardly have done so; rather, they were handbooks for Christians to use in arguments with Jews, and were intended to strengthen Christians in their faith. Discusses 37 texts, sketching their contents and historical contexts. Arguments were based on allegorical interpretations of Old Testament passages seen as referring to Jesus, the Trinity, and supersession. Compares the Byzantine dialogues with Oriental Christian and Latin ones; the Latin ones used more rational arguments and were based on a better knowledge of Judaism, but were equally unlikely to convert anyone. *RW*

0070 Lohrmann, Klaus: *Die Wiener Juden im Mittelalter*. Berlin: Philo, 2000. 192 pp. (Geschichte der Juden in Wien, 1).

Describes the socioeconomic situation and communal organization of Jews in Vienna until the forced conversions, expulsion, and massacre of 1420-21. Chronicles the dukes' protection of the Jews and the restrictions imposed on the Jews by the Church. While up to the 13th century Jews and Christians were often business partners, in the 14th century relations soured: popular resentment at exploitation of the poor by the rich focused on the Jews and their "usury," and was reinforced by religious incitement. Attributes the "gezerah" of 1420 to the Jews' waning financial importance, to allegations of Jewish support for the Hussites, and to a doctrine developed by theologians at the University of Vienna who influenced Duke Albrecht V – i.e. faith must come before understanding, and so the Jews can be converted to Christianity, even by force, without the need of explaining it to them rationally first. *RW*

0071 Magin, Christine: *"Wie es umb der iuden recht stet": Der Status der Juden in spaetmittelalterlichen deutschen Rechtsbuechern*. Goettingen: Wallstein, 1999. 462 pp. Based on the author's diss. – Universitaet Goettingen, 1995.

Compares regulations concerning Jews in canon law, letters of privilege conferred by rulers, and 13th-14th century secular compilations reflecting regional or local customary law. These differed according to the differing interests of Church, rulers, and townspeople and the changing balance of power between them. The Church was determined to keep the Jews segregated and subservient; the temporal rulers, above all the Emperor, favored and protected them as a source of revenue; the towns initially wanted Jews as a stimulus to commerce. Compares regulations concerning Christian servants, dress codes, conversion, judicial proceedings, and trade, showing that they were more liberal toward Jews the less the compilers were influenced by Church law, although these liberal provisions were at times hedged in by restrictive conditions as distrust of Jews increased over the years. Concludes, however, that the declining status of the Jews is attributable not to legislation but to social change. *RW*

0072 Malkiel, David J.: Vestiges of Conflict in the Hebrew Crusade Chronicles. *Journal of Jewish Studies* 52, 2 (Fall 2001) 323-340.

Notes that despite the general ideology of martyrdom that permeates Hebrew chronicles of the First Crusade of 1096, traces of conflict are indicated in the actions and motivations of some of the Jews, including unheroic conduct, family conflict about how to react, and the "particular predicament of the young," who wanted to live. There are occasional expressions of dissent in regard to martyrdom and physical resistance. The options of flight or conversion are also mentioned, although these sometimes resulted in death. Stresses that such "inconvenient" material reinforces the view that the chronicles are trustworthy and that the chroniclers may be suggesting that martyrdom did not necessarily depend on bravery or religious idealism; it could have been applicable to all those who died voluntarily or even to all who were killed as Jews. *YC*

0073 Montes Romero-Camacho, Isabel: *Los judíos en la Edad Media española*. Madrid: Arco Libros, 2001. 79 pp.

Summarizes the history of the Jews in Spain from the 5th century until the expulsion in 1492. Points out, based on the views of Eloy Benito Ruano ("De la alteridad en la historia," Madrid, 1988), that the factor which determined the conflict between Jews and Christians was the perception by the Christians that the Jews were "the Other," a minority strange to the national social and cultural body, which would bring ruin to the entire nation. This perception became critical during the 14th and 15th centuries, leading to the expulsion of the Jews. *MZ*

0074 Muñoz Solla, Ricardo: Judíos y judaizantes de Caracena. *El Olivo* 53 (Jan-June 2001) 67-86.

An account of the life of Conversos in the village of Caracena (Soria) prior to 1492. As Caracena was a small village, it was difficult for the Conversos not to relate to the Jews (friends and relatives), sharing many aspects of life with them. This left them in a very weak position vis-à-vis the Old Christians, who often accused them of Judaizing. Based on testimonies of accusers at the tribunal of Sigüenza (which at that time had jurisdiction over Caracena) between 1496-98. *MZ*

0075 Petrukhin, Vladimir Y.: V.N. Tatishchev o yevreyakh v drevnei Rusi i sovremennaya istoriografiya [V.N. Tatishchev on the Jews in Ancient Russia and Modern Historiography]. *Vestnik Yevreiskovo Universiteta* 23 (2001) 7-20.

Critically examines information provided by Vasilii Tatishchev (1686-1750), in his "Russian History," on "Jewish dominance" in 10th-12th-century Kiev and on the pogrom there in 1113, which was followed by an expulsion of the Jews by Prince Vladimir Monomach. According to Tatishchev, in 1113, during an interregnum, the citizens of Kiev rose up against a pretender to the throne who favored the Jews, like the previous ruler Prince Svyatopolk. The uprising was accompanied by a pogrom. Vladimir Monomach, who succeeded to the throne, expelled the Jews from Russia "forever." This information cannot be found in the chronicles, and has been attributed to unknown sources used by Tatishchev or to his own inference. Nevertheless, some modern historians accept Tatishchev's version uncritically. Argues that Tatishchev based his account on earlier sources describing pre-1648 events in Ukraine. For Tatishchev, the attitudes toward Jews in these sources provided a model of how Jews should be treated. *DR*

0076 Resnick, Irven Michael: Medieval Roots of the Myth of Jewish Male Menses. *Harvard Theological Review* 93, 3 (July 2000) 241-263. An earlier version was published as a separate booklet: "On Roots of the Myth of Jewish Male Menses in Jacques de Vitry's History of Jerusalem" (Ramat-Gan: Bar-Ilan University, 1998).

0077 Rokeah, Zefira Entin, ed. & trans.: *Medieval English Jews and Royal Officials: Entries of Jewish Interest in the English Memoranda Rolls, 1266-1293*. Jerusalem: Hebrew University, Magnes Press, 2000. xxxvii, 507 pp.

Presents selections from the rolls relating to Jews, revealing government policy toward the Jews (including financial exploitation), financial affairs of both rich and poor Jews, and attitudes towards and actions against Jews (such as the frequent and false charges against Jews of debasing coinage, which led to the execution of almost 300 Jews). Discloses benefits from Jewish moneylending that accrued to the royal family. Reports on profits from the sale of Jews' property in Norfolk after they were robbed and murdered in the wake of the expulsion of 1290. *YC*

0078 Simonsohn, Shlomo: *The Jews in Sicily. Vol. Three: 1392-1414*. Leiden: Brill, 2001. xii, 1253-1812 pp.

A compilation of documents, found in various libraries and archives, pertaining to the history of the Jews in Sicily. Vol. 3 presents documents no. 1321-1879c. Each entry gives a brief description of the contents of the document in English, the date and the source of the document, and the original Latin text. The volume is accompanied by an index of persons, a geographical index, and a subject index. For antisemitism, see the subject index under badge, Jewish; jurisdiction, over Jews; persecution and harassment (of Jews); taxes. A significant number of documents issued by the king or members of the nobility instruct local authorities or clergymen to stop molesting or harassing the Jews. *SSC*

0079 Sklare, David: Responses to Islamic Polemics by Jewish Mutakallimun in the Tenth Century. *The Majlis: Interreligious Encounters in Medieval Islam,* eds. Hava Lazarus-Yafeh et al. Weisbaden: Harrassowitz, 1999. Pp. 137-161.

The milieu of 10th-century Baghdad, in contrast to that of Fatimid Egypt, was relatively tolerant and made possible the institution of free religious/theological debate between Muslims and Jews. Despite the accepted view that Jews did not write books to refute Islam, two are known wholly or partially – those of the Karaite Yusuf al-Basir and of the Rabbanite Samuel ben Hofni Gaon. These intellectuals were very familiar with Muslim culture, as were many of their fellow Jews. One danger of the "majlis" polemics was that Jews could be persuaded by the Muslim "mutakallimun" (theologians) and convert. Thus, there was an internal, as well as external (defensive) function to the Jewish manuals of debate written by al-Basir and Samuel. The latter was more cautious, apparently out of concern for the vulnerable position of Jews in Islamic society, even in Baghdad. He attempted to counter the claim that Islam had abrogated the Torah. Al-Basir boldly attempted to introduce doubt about the inimitability of the Qur'an. Jewish familiarity with Islamic culture both posed a threat to Jewish identity and provided tools to respond to this threat. *YC*

0080 Valdeón Baruque, Julio: Tolerancia e intolerancia en la Castilla bajomedieval. *Inquisición y sociedad,* ed. Ángel de Prado Moura. Valladolid: Universidad de Valladolid, 1999. Pp. 11-25.

A precarious balance between tolerance and hostility toward the Jews characterized medieval Spain. This balance came to an end in the middle of the 14th century, especially in Castile. The anti-Jewish propaganda of Enrique de Trastamara (engaged in a war against King Pedro I), the agitation against the Jews by the lower clergy, the anti-Jewish prejudices and stereotypes rooted in the populace, and the economic crisis (many Christians owed money to Jews) were factors leading to acts of violence against Jews in Castile. The violence reached a climax in the pogroms of 1391, resulting in the mass conversion of Jews. The 15th century saw the "Jewish problem" becoming the "Converso problem"; the old hatred against Jews was transferred into hatred of the New Christians, blending religious and socioeconomic arguments. *MZ*

THE EARLY MODERN PERIOD (1493-1788)

0081 Alpert, Michael: *Criptojudaísmo e Inquisición en los siglos XVII y XVIII: La Ley en la que quiere vivir y morir*. Barcelona: Ariel, 2001. 269 pp. Originally published as "Crypto-Judaism and the Spanish Inquisition" (Basingstoke, Hants: Palgrave, 2001).

0082 Alpert, Michael: *Crypto-Judaism and the Spanish Inquisition*. Basingstoke, Hants: Palgrave, 2001. x, 246 pp.

Examines the phenomenon of crypto-Judaism, how it began and how it developed. Relates the history of the Spanish Inquisition from the late 15th to the 18th centuries, in relation to its goal of rooting out and punishing Judaizers, i.e. Conversos who were charged with returning to Judaism. The accused were arrested, imprisoned, and tortured, after which most "repented" and were then punished according to the Church's rules. Approximately 40% of the Inquisition's victims were Judaizers; they were the most severely punished, often by flogging, sometimes forced to serve as galley rowers; 10% of them were killed. Inquisition archives also provide information about crypto-Jews who attempted to live double lives amidst great danger. *YC*

0083 Bell, Dean Phillip: Martin Luther and the Jews: The Reformation, Nazi Germany and Today. *The Solomon Goldman Lectures. Vol. VII,* ed. Dean Phillip Bell. Chicago: Spertus Institute of Jewish Studies Press, 1999. Pp. 155-187.

Re-examining the question of the existence of a "Sonderweg," or a straight line, between Luther's ideas and Hitler's, notes that Luther has been interpreted in varying ways from his own time through the Nazi period. Furthermore, Luther's view of the Jews was mainly theological rather than political, and he was less concerned with the Jews themselves than with using them as a foil for attacks on Christian society. With his stress on lineage, Luther was on the verge of making a racial distinction, but one that was still different from modern racism. Although there is no doubt that Luther was vehemently anti-Jewish in many of his theological discussions, in this respect he was a product of his age. He should not be held accountable for 20th-century misappropriation of his writings. *YC*

0084 Bernardini, Paolo; Fiering, Norman, eds.: *The Jews and the Expansion of Europe to the West, 1450 to 1800*. New York: Berghahn Books, 2001. xv, 567 pp.

A collection of papers delivered at a conference held at Brown University, Providence, RI, in June 1997. Partial contents: Katz, David S.: Israel in

America: The Wanderings of the Lost Ten Tribes from "Mikveigh Yisrael" to Timothy McVeigh [Appeared in Hebrew in "Zmanim" 74 (2001).] (107-122). The following six articles deal with the restrictions against and persecution of Conversos in early modern Portugal and colonial Latin America: Rowland, Robert: New Christian, Marrano, Jew [On Conversos in early modern Spain and especially in Portugal.] (125-148); Alberro, Solange: Crypto-Jews and the Mexican Holy Office in the Seventeenth Century (172-185); Uchmany, Eva Alexandra: The Participation of New Christians and Crypto-Jews in the Conquest, Colonization, and Trade of Spanish America, 1521-1660 [Deals, also, with restrictions against Conversos and Inquisitorial trials against Conversos in the whole of Spanish America.] (186-202); Boehm, Guenter: Crypto-Jews and New Christians in Colonial Peru and Chile (203-212); Novinsky, Anita Waingort: Marranos and the Inquisition: On the Gold Route in Minas Gerais, Brazil (215-241); Pieroni, Geraldo: Outcasts from the Kingdom: The Inquisition and the Banishment of New Christians to Brazil (242-251); Williams, James Homer: An Atlantic Perspective on the Jewish Struggle for Rights and Opportunities in Brazil, New Netherland, and New York [Deals, also, with Peter Stuyvesant's opposition to Jewish settlement in New Amsterdam.] (369-393). *DR*

0085 Braden, Jutta: *Hamburger Judenpolitik im Zeitalter lutherischer Orthodoxie, 1590-1710.* Hamburg: Christians, 2001. 606 pp.

The rights of Jews in Hamburg fluctuated in accordance with the power relationships between the city council, the citizenry, and the clergy. The council, which represented the upper class of merchants, wished to attract wealthy former Conversos because of their commercial connections and expertise. The clergy wanted a city united in Lutheran orthodoxy, and opposed the presence of the enemies of Christianity, the Jews. The citizenry's endemic antisemitism was fueled by the preachers; besides, the middle classes feared Jewish economic competition. The council negotiated with the clergy and with the Jews, who threatened to leave the city (whose prosperity depended largely on them). Successive contracts permitted Jews to meet for prayer privately and imposed other religious restrictions; the council enforced these flexibly, if at all. In 1697 the citizenry gained the upper hand and passed new restrictive regulations that caused many Sephardim to leave; but they also legalized the residency of the Ashkenazi Jews, who had lived in the city illegally through most of the century. In 1710, under the aegis of the Emperor, most of the new restrictions were canceled and the council's policy was restored. *RW*

0086 Braden, Jutta: Luthertum und Handelsinteressen: Die Judenpolitik des Hamburger Senats im 17. Jahrhundert. *Die Hamburger Kauffrau Glikl: Juedische Existenz in der Fruehen Neuzeit,* ed. Monika Richarz. Hamburg: Christians, 2001. Pp. 159-194.

Policies toward Jews in Hamburg during the 17th century were characterized by two opposing tendencies: economic liberalism and religious restriction. While the Senate was interested in the economic contribution of Jews (especially the Sephardim), the Lutheran Church treated them more harshly than any other minority. The governing body aspired to an "emancipation of the political aspect from the theological." The special status of Sephardic Jews, their economic skills, and their inclination toward secular culture, led to a break with the tradition of rejection of the Jews. For the first time, the town gates were opened to them at the end of the 17th century. The Sephardim's settlement paved the way for Ashkenazic Jews in Hamburg in that period. *ID*

0087 Câmara, Rubens R.: A Portuguese Inquisition Document. *Avotaynu* 17, 4 (Win 2001) 25-28. Originally appeared in Portuguese in "Gerações/ Brasil" 10, the journal of the Jewish Genealogical Society of Brazil.

Discusses the file of a case brought before the Lisbon Inquisition tribunal in 1618, from a genealogical point of view. Inês Henriques de Leão (born ca. 1590) was a member of an important Jewish and Converso family; she was related to the Aboabs and to the family of Baruch Spinoza. A number of members of her family, and their biographical information, are mentioned in the Inquisitorial record. Although she refused to plead guilty to being a Judaizer, she was declared a heretic and an apostate. After excommunication, she was delivered to secular justice, which could have executed her. Her denial of the testimony against her resulted in her being tortured. Final sentencing in 1620 included confiscation of her assets and incarceration. In 1621 she was released upon pleading that she was a good Christian and had been complying with the penitence imposed on her. *YC*

0088 Cerman, Ivo: Anti-Jewish Superstitions and the Expulsion of the Jews from Vienna in 1670. *Judaica Bohemiae* 36 [2000] (2001) 5-33.

Describes events in Vienna that preceded the expulsion of the Jews in 1670 and the reasons for the expulsion. Dismisses theories that the expulsion was a result of pressure exerted on Leopold I by the "Spanish faction" at his court, or that it evolved due to economic considerations. Argues that it was supernatural thinking and anti-Jewish superstitions (fear of Jewish magic, belief in their anti-Christian scheming, etc.) that effected the decision concerning the expulsion. These superstitions were shared by the two parties involved in the ousting of the Jews – the burghers (whose psychology was fully expressed by the city's mayor, Holzner) and the aristocracy; they were supported by the third involved party, the Church. The story of the Viennese expulsion helps to dismiss the historical models ascribing the leading cultural role (especially the civilizing and rationalizing role) to the absolutist court. Pp. 31-33 contain historical documents. *DR*

0089 Ferrand, Pierre: Lessing, Voltaire, and the Jews. *Midstream* 47, 6 (Sept-Oct 2001) 41-44.

Questions the usual depiction of Lessing as a philosemite and Voltaire as an antisemite (without, however, glossing over elements of obvious anti-semitism in the latter). Arrives at the conclusion that Lessing was a liberal when it came to Jews as individuals, if they were extremely assimilated, like his friend Moses Mendelssohn, but Lessing did not support any action to improve the civil status of the Jews in Germany. This is contrasted with Voltaire, whose revulsion for Jews did not prevent him from espousing civil rights for them. Provides information about Lessing's life and works to support this view, which indicates "the ineffectiveness of much of the respectable German liberalism in the century-and-a-half following his death." In contrast, France is hailed for having granted rights to the Jews, also as individuals, soon after Voltaire. *YC*

0090 Fijałkowski, Paweł: *Żydzi w województwach łęczyckim i rawskim w XV-XVIII w.* [Jews in the Łęczyckie and Rawskie Voivodships in the 15th-18th Centuries]. Warszawa: Żydowski Instytut Historyczny, 1999. 166 pp.

Presents a history of Jewish settlement in two regions of central Poland, analyzing the Jews' legal status and their economic and cultural activities. Pp. 48-60 deal with Christian-Jewish relations in the 16th-mid-17th centuries, and pp. 99-115 deal with the same topic up to the end of the 18th century. Discusses ritual murder and host desecration accusations in Rawa, Kłodawa, and Sochaczew in the mid-16th century, and later in other towns, among them Łęczyca. Argues that persecutions of the Jews were related to the developing Reformation movement in Poland. Calvinist and other Protestant communities sprung up, which provoked defensive reactions among the Catholics, who felt threatened and fought against "heretics," including Jews. The second reason for hostility was economic rivalry. Christian-Jewish antagonisms burgeoned during the wars of the 17th century, especially during the Swedish invasion. The nobility tried to accuse the minorities, among them Jews and Protestants, of responsibility for Poland's defeat. *HV*

0091 Gailus, Manfred: Die Erfindung des "Korn-Juden": Zur Geschichte eines antijuedischen Feindbildes des 18. und fruehen 19. Jahrhunderts. *Historische Zeitschrift* 272, 3 (June 2001) 597-622.

A revised version of Gailus' "Habilitationsvortrag" at the Technische Universitaet Berlin, February 1999. States that "Kornjude" was the pejorative term used for all speculators who bought up grain when it was cheap and sold it at exorbitant prices in times of scarcity, causing mass starvation. In reality, few of these speculators were Jews. However, the use of the term reinforced anti-Jewish sentiment, which erupted in riots in the wake of the hunger years 1816-17. *RW*

0092 Glassman, Bernard: *Protean Prejudice: Anti-Semitism in England's Age of Reason*. Atlanta: Scholars Press, 1998. xii, 250 pp.

In the 18th century, contemporary literature, the press, church sermons, etc., represented a broad spectrum of anti-Jewish prejudices which were shared by most of the enlightened populace. At the same time, there was philosemitism, which was largely based on the desire to convert the Jews. The debate over the Jew Bill of 1753 caused mass resentment in England; however, this resentment, as well as later upsurges of anti-Jewish sentiment, did not result in riots. Concludes that English society in the 18th century was prejudice-ridden, but compared to prejudices against various "others" (religious, ethnic, racial, etc.), antisemitism does not seem to have been an excessive or overwhelming passion; the Catholics, Quakers, Irish, Welsh, Blacks, etc., suffered no less than the Jews. In fact, the legal situation of English Jews before 1790 was much better than it was on the continent. *DR*

0093 Grodziski, Stanisław: The Jewish Question in Galicia: The Reforms of Maria Theresa and Joseph II, 1772-1790. *Polin* 12 (1999) 61-72. Appeared in Polish in "Lud żydowski w narodzie polskim" (1994).

Surveys consecutive laws introduced by Maria Theresa and Joseph II related to the legal status of the Jews in Galicia after the first partition of Poland (1772), comparing them to the legal situation of the Jews under Polish rule. The main goals of the new legislation were to reduce the number of the Jews in the Empire, to erase their ethnic and religious distinctiveness, and to Germanize the Polish Jews. German schools were founded. A whole range of laws aimed at liquidating external marks of Jewishness in costume, name, lifestyle, and language were introduced. A law intended to drive Jews from trade and crafts and settle them on farms did not succeed. Finally, the codification of previous laws in 1789 arranged matters in the spirit of enlightened absolutism. Jewish autonomy was weakened and compulsory military service was introduced. The special taxes for Jews were maintained. This legislation strongly affected Galician Jews. *HV*

0094 Herzig, Arno: Die christlichen Konfessionen und die Juden im Deutschland des 16. und 17. Jahrhunderts. *Die Hamburger Kauffrau Glikl: Juedische Existenz in der Fruehen Neuzeit*, ed. Monika Richarz. Hamburg: Christians, 2001. Pp. 147-158.

Discusses anti-Jewish attitudes of Christian confessions in early modern Germany. Luther argued that the Jews had squandered their chance by rejecting Jesus. He was ready to admit that before the Reformation they could not accept Christian belief because of falsification and misinterpretation of the biblical texts by the ancient Church, but after the Reformation nothing could justify their stubbornness and they must be expelled from Christian society. On the opposite extreme was Johannes Eck, who defended the ancient Church, quoting traditional anti-Jewish attitudes and contending that the Reformation was a Jewish enterprise. Jews were accused also of blasphemy, blood libels, usury, and sinful cohabitation. Each confession was proud of its conversion of Jews, forced in most cases. *ID*

0095 Horn, Maurycy: Stan i potrzeby badań nad stosunkami żydowsko-ukraińskimi w dawnej Rzeczypospolitej i podstawa źródlowa do tego zagadnienia [The State and Needs of Research on Jewish-Ukrainian Relations in Old Poland and the Sources for Such Research]. *Biuletyn Żydowskiego Instytutu Historycznego* 196 (Dec 2000) 461-484.

Analyzes the reasons for Ukrainian hostility toward Jews in the 17th-18th centuries. The main cause is seen in competition in trade and crafts and in the fact that Polish kings and proprietors of towns leased to the Jews the collection of income from their estates, customs and other goods. Jewish "arendars" were also used by the nobility to crush peasant revolts. The peasants' hatred of Jews led to pogroms and attacks against the Polish nobles. Mentions pogroms which occurred between 1591-1638, in which probably a few thousand Jews and Poles were killed. The Chmielnicki pogroms which began in 1648 are better documented. States that most of the Jews were killed between 1667-99, during the wars of Poland with Cossacks, Tatars, and Turkey. Surveys, as well, wars, revolts, and pogroms in the 18th century. Concludes that Polish, Jewish, and Ukrainian historiography differ in interpretation of the events and that there is not enough research on Ukrainian-Jewish relations in this period. *HV*

0096 Horowitz, Elliott: A "Dangerous Encounter": Thomas Coryate and the Swaggering Jews of Venice. *Journal of Jewish Studies* 52, 2 (Fall 2001) 341-353.

Discusses the predominant Christian image of the Jewish male as weak and effeminate that was fostered in Christian (and sometimes Jewish) literature and that continued to circulate up to modern times. Relates the account of Thomas Coryate, an Englishman who initiated a religious dispute with a rabbi in the ghetto of Venice in 1608, of how he fled from a Jewish crowd who threatened him because he "durst reprehend their religion." Contemporary authors and artists paradoxically depicted Coryate as effeminate. Contrasts this antisemitic image of the emasculating Jew with the medievally-based Christian idea that Jewish men were so like women that they too menstruated (as a punishment for killing

Christ). Coryate's contemporaries' association of Jews wanting to circumcise (read castrate) him is tied to Freud's noting this antisemitic fear. *YC*

0097

קליק, יהודית: הכנסייה הקתולית והיהודים בקרקוב ובקז'ימייז' עד חלוקות פולין. קראקא - קז'ימייז' - קרקוב: מחקרים בתולדות יהודי קרקוב, עורך אלחנן ריינר. תל-אביב: המרכז לחקר תולדות היהודים בפולין ומורשתם, המכון לחקר התפוצות, אוניברסיטת תל-אביב, 2001. עמ' 69-87.

[Kalik, Judith: The Catholic Church and the Jews in Kraków and Kazimierz in Pre-Partition Poland. *Kroke – Kazimierz – Kraków: Mehkarim be-toldot yehudei Kraków* [Kroke – Kazimierz – Kraków: Studies in the History of Kraków Jewry], ed. Elchanan Reiner. Tel-Aviv: Center for the History of Polish Jewry, Diaspora Research Institute, Tel-Aviv University, 2001. Pp. 69-87.]

Discusses Catholic-Jewish relations in Kraków and in nearby Kazimierz. In the 16th century Kraków became a publishing center. Publications included anti-Jewish polemics, much of which was written by faculty members of the Jagiellonian University. Mentions blood libels from the 15th century on, which were supported by the Church in the 18th century despite papal opposition. Reports on anti-Jewish laws passed by Catholic synods from the 16th century; these were largely based on papal bulls or decisions of earlier synods. The legislation was not drastic, but restricted Jews socially and economically while offering benefits to converts. Discusses, also, economic relations between Catholics and Jews, including Jewish leasing of lands and businesses from the Church, and loans made to individual Jews and Jewish communities by the Church. Although these relations sometimes caused conflict, they also restrained antisemitic tendencies to some extent. *RE*

0098 Kiessling, Rolf; Ullmann, Sabine, eds.: *Landjudentum im deutschen Suedwesten waehrend der Fruehen Neuzeit*. Berlin: Akademie Verlag, 1999. 401 pp.

A collection of articles on the history of rural Jewry in the area of Swabia-Bavaria in the early modern period. The following relate to antisemitism: Hoehnle, Susanne: Ortsherrschaft und juedische Gemeinde als Vertragspartner: Der Burgauer Rezess von 1717 fuer Ichenhausen (58-79); Koutná, Dana V.: Juden in fuggerischen Herrschaften (80-96); Holenstein, André: Bitten um den Schutz: Staatliche Judenpolitik und Lebensfuehrung von Juden im Lichte von Schutzsupplikationen aus der Markgrafschaft Baden(-Durlach) im 18. Jahrhundert (97-153); Schmoelz-Haeberlein, Michaela: Zwischen Integration und Ausgrenzung: Juden in der oberrheinischen Kleinstadt Emmendingen 1680-1800 (363-391). *ID*

0099 Klein, Birgit E.: Vom Hofjuden zum Metzger: Die Wandlungen des Levi von Bonn in der juedischen Erinnerung. *Neuer Anbruch: Zur deutsch-juedischen Geschichte und Kultur,* eds. Michael Brocke, Aubrey Pomerance, Andrea Schatz. Berlin: Metropol, 2001. Pp. 15-31.

Levi (ca. 1570-after 1621), also known as Judah Loeb son of Chaim, or Loeb Kraus, was appointed by the Elector of Koeln as supervisor over the Jews of the Electorate. He reported on decisions taken by a council of rabbis in 1603 in Frankfurt which obligated Jews to settle internal disputes only in Jewish courts. The Emperor regarded this as high treason and removed his protection from the Jews, who in consequence had no one to turn to during citizens' uprisings and pogroms. Shows how a misunderstanding turned Levi into a "shochet" in Jewish tradition. *RW*

0100 López Belinchón, Bernardo José: Conversos y nobleza o las desventuras de un corregidor. *Sefarad* 61, 1 (2001) 137-162.

During the 17th century, Portuguese Conversos in Spain held influential positions, protected by the Spanish nobility. Recounts the case of Jorge Enríquez, doctor of the Duce of Alba in the Castilian town of Alba de Tormes. His son-in-law, Andrés de Fonseca, was the corregidor of Alba, the highest authority in the town; other members of the family also held privileged positions. In 1622 Jorge Enríquez died. The influential Old Christians of Alba denounced Andrés de Fonseca to the Inquisition of Valladolid, accusing him and other members of the family of having carried out Enríquez' funeral in accordance with Jewish ritual. Fonseca and his family were imprisoned and condemned, but they were rehabilitated in 1624, helped by their wealth and influence. The Fonsecas moved to Madrid and there, again, fell victim to the Inquisition. In 1656 they escaped to Italy, where they lived openly as Jews. *MZ*

0101 *Maguén – Escudo* 119 (Apr-June 2001).

The section entitled "Antijudaísmo, Inquisición y Criptojudaísmo" includes the following essays: Garzón Serfaty, Moisés: La Inquisición en Venezuela (12-17); Novinsky, Anita Waingort: Consideraciones sobre los criptojudíos hispano-portugueses: El caso Brasil (18-26); Mizrahi, Rachel: Quinientos años de presencia judía en Brasil (37-44). *MZ*

0102 Marzalyuk, Ihar: Zhyd u belaruskai tradytsyi (nehatywnyya mentalnyya stereatypy i dosved supolnaha isnavannya) [Yid in Belorussian Tradition (Negative Mental Stereotypes and an Example of Coexistence]. *Yevrei Belarusi* 6 (2001) 25-37.

Shows that there are many anti-Jewish and anti-Judaic motifs in Belorussian writings of the 16th-18th centuries. Belief in Jewish witchcraft and ritual murder was widespread, as well as the motif that the Jew was causing damage to Christians economically. The population in some cities (Grodno, Mogilyov) tried to prevent Jewish settlement in them. Anti-Judaism also affected state policy. In 1495-1503 Jews were expelled from the Great Duchy of Lithuania; many restrictions were imposed on them after their return. However, anti-Jewish stereotypes did not prevent cooperation and normal relations between Belorussians and Jews. *DR*

0103 Meyers, Charles; Simms, Norman Toby, eds.: *Troubled Souls: Conversos, Crypto-Jews, and Other Confused Jewish Intellectuals from the Fourteenth through the Eighteenth Century*. Hamilton, N.Z.: Outrigger, 2001. vii, 194 pp.

A collection of essays on Conversos and their descendants, and their difficulties in coping with persecution and identity problems in Europe and elsewhere. Partial contents: Meyers, Charles: Dr. Hector Nuñes, a Converso in Elizabethan London: Dedicated Patriot and Despised Alien [Nuñes' commercial and other diverse activities in late-16th century England (ca. 1546-1591) failed to win him acceptance in English society because he was a Converso.] (111-128); da'Costa, Elizabeth Mendes: Montaigne and the Jewish Religion [Discusses the Converso ancestry of Michel de Montaigne (1533-1592) and his coded sympathy for the Jews.] (129-141); Jama, Sophie: The Date of Montaigne's Essays: A Hidden Identity [Presenting clues to Montaigne's alleged Converso antecedents, relates also to persecution of Jews and Conversos in his time.] (142-149); Simms, Norman Toby: Troubled Souls: The Inner and Outer Life of Jews in England after the Expulsion of 1290 (164-189). *YC*

0104 Mittlmeier, Christine: *Publizistik im Dienste antijuedischer Polemik: Spaetmittelalterliche und fruehneuzeitliche Flugschriften und Flugblaetter zu Hostienschaendungen*. Frankfurt a.M.: Peter Lang, 2000. 184 pp. Based on the author's diss. – Universitaet Muenchen, 1997.

Discusses in detail a broadsheet dealing with host desecration in 1477 in Passau (Nuernberg 1497/98), a chapbook on a host desecration in Sternberg (Magdeburg, 1493), and a chapbook on host desecration in Knobloch, 1510 (Frankfurt an der Oder, 1511), and refers to many other publications on the same and similar events. Analyzes their historical context and political purpose; often, they served as propaganda for the expulsion of Jews. The Church wished to propagate the doctrine that the consecrated host was the real body of Christ; thus the Jews, the original murderers of Christ, murdered him again and again in the host. They were seen as the embodiment of evil and in league with the Devil. These ideas were reinforced by the illustrations. The stories popularized by the chapbooks and broadsides retained their influence over the centuries up to the present day. *RW*

0105 Novinsky, Anita Waingort: Marranos and Marranism – a New Approach. *Jewish Studies* 40 [2000] (2001) 5-20.

Argues that a major gap in the study of Marranism has its source in the neglect of this phenomenon in Brazil, where 25-30% of the free white population in some areas were Marranos. Attributes their persecution less to religion than to racism and stresses, primarily, economic motivations. Considers that not enough attention has been paid to Marranos who did not return to Judaism. Shifts from historical considerations to a more typological approach, citing the French philosopher Edgar Morin, who discussed Marranism as a complex psychological experience of double identity, noting descendants of Marranos such as St. Theresa of Avila, Montaigne, and Spinoza, as well as Marrano types like Marx and Freud. Discusses, also, the Marrano dilemma of Heinrich Heine. *YC*

0106 Parello, Vincent: Los Mora de Quintanar de la Orden: Un criptojudaísmo familiar a finales del siglo XVI. *Sefarad* 61, 2 (2001) 395-418.

The large Converso Mora family in the Castilian town of Quintanar de la Orden enjoyed, between 1520-88, a period of quiet from the persecution of the Inquisition, which allowed them to continue observing Jewish practices and to flourish economically. During that time the Inquisition focused its efforts on fighting the Lutheran reform. But in 1588 the Inquisition of Cuenca (which at that time had jurisdiction over Quintanar de la Orden) discovered the existence of clandestine synagogues; between 1588-92, 63 members of the Mora clan were brought to trial and their community was dismantled. *MZ*

0107 Portnoy, Susana Likerman de: Cambios de conductas de las personas en los grupos familiares hispano-judíos ante la conversión y la instalación de la Inquisición. *Sefárdica* 12 (Apr 2001) 65-74.

Discusses patterns of behavior of the prominent Converso families of Juan Falcon (from Ciudad Real), Arias Dávila, and Nasi Mendes facing the persecution of the Inquisition during the 16th century. These families chose, respectively, to maintain the Jewish faith in secret, to assimilate into the Christian world, and to go into exile in order to preserve their Jewish identity. *MZ*

0108 Saraiva, António José: *The Marrano Factory: The Portuguese Inquisition and Its New Christians 1536-1765*. Translated, revised and augmented by H.P. Salomon and I.S.D. Sassoon. Leiden: Brill, 2001. xl, [16], 404 pp. Originally published as "Inquisição e cristãos-novos" (Lisboa: Editorial Estampa, 1969). A revised 5th edition appeared in 1985.

Presents the thesis that the Inquisition was established in Portugal in the 16th century by the upper classes who felt that they were being menaced, both materially and ideologically, by the new merchant bourgeoisie class, made up largely of New Christians. The English edition includes an "Introduction to the English Edition" (pp. ix-xiv); "A Word to the Reader" (pp. xv-xvii); "Prologue (Fifth Edition, 1985)" (pp. xix-xxi); and the "Introduction" (pp. xxiii-xl), partly revised and updated by the translators. Pp. 235-379 contain eight appendices, including the polemic between the author and I.S. Révah (mainly letters and interviews published in "Diario de Lisboa" in 1971), which appeared in the 1985 Portuguese edition. Révah accused Saraiva of libel and rejects the view that the Inquisition was a socio-economic phenomenon rather than an ethnic-religious one. *SSC*

0109 Schaub, Jean-Frédéric: *Les juifs du roi d'Espagne*. [Paris]: Hachette Littératures, 1999. 240 pp.

Ch. 3 (pp. 125-187) deals with the expulsion of the Jews of Oran in 1669. The existence of a Jewish community of about 500 persons in Oran between 1509-1669, established and tolerated by the Spanish king and legitimized by their status as royal interpreters and go-betweens, came to an abrupt end. Decreed by the Queen of Spain, Marianne of Austria, the expulsion was part of a policy aimed at boosting the authority of the Queen's favorite against claims to the throne by an illegitimate son of Philip IV. It also served to strengthen the head of the War Council, Marquis de los Vélez, who carried out the expulsion despite the Council's decision not to execute the expulsion order. Contends that the expulsion was not a direct manifestation of Spanish anti-Jewish ideology, although it appears to resonate with this component in Spanish history. Most of the expelled Jews settled in Nice. *EO*

0110 Singer, Alan Howard: *Aliens and Citizens: Jewish and Protestant Naturalization in the Making of the Modern British Nation, 1689-1753*. Diss. – University of Missouri, Columbia, 1999. 298 pp. Unseen.

Explores the controversies over naturalization policies between the Glorious Revolution and the failure of the Jewish Naturalization Act of 1753. After the Revolution, Whig governments frequently attempted to liberalize naturalization policies in Great Britain. Beginning in 1693, they attempted to pass legislation that would increase the immigration of Protestants escaping from Catholic dominions. It was thought that an influx of artisans would expand technical knowledge and cheapen the price of labor. After a series of failures to pass the legislation, the Pelham Administration in 1753 enacted a measure that was intended to make it easier for Jews to become naturalized citizens. It was supposed that a few wealthy Jews would increase Great Britain's stock of financial and commercial capital. This effort was blocked by a vicious campaign by merchants and City politicians to repeal the law by instilling negative public opinion. Politicians fearing for their political lives conceded. Along with examining the reasons why the Whigs were so persistent in their attempts at naturalization, explores the methods in which the opposition propagandized the supposed ill-effects of the legislation. Concentrating on the literature of the Jewish Naturalization controversy, demonstrates that the opposition, along with using traditional antisemitism, tapped into established notions of property and national identity to cast the Jews as an alien and destructive people.

0111 Tollet, Daniel, ed.: *Les textes judéophobes et judéophiles dans l'Europe chrétienne à l'époque moderne.* Paris: Presses Universitaires de France, 2000. xiii, 246 pp.

A collection of papers from a conference organized by the Centre d'études juives, held at the Université de Paris IV-Sorbonne, May 1995. Partial contents: Molette, Charles: Judéophilie et judéophobie au coeur de l'Eglise catholique dans deux textes à l'heure du Concile de Trente [On the catechism of the Council of Trent (1546-1566) and the bull of Pope Paul IV establishing the ghetto of Rome (1555).] (15-26); Escamilla-Colin, Michèle: Recherches sur les "traites judéophobes" espagnols des XVIe et XVIIe siècles [Analyzes, in particular, the "Discurso contra los judíos" (1623) by Vicente da Costa Mattos.] (27-49); Parente, Fausto: "Du tort qu'ont les Chrestiens de persécuter les Juifs": Quelques observations à propos du "philosémitisme" d'Isaac de La Peyrère (51-66); Tollet, Daniel: Le goupillon, le prétoire et la plume: Stéphane Żuchowski et l'accusation de crimes rituels en Pologne à la fin du XVIIe siècle et au début du XVIIIe siècle [Appeared previously in "Żydzi wśród chrześcijan w dobie szlacheckiej Rzeczypospolitej" (1996).] (67-81); Wilke, Carsten: Splendeurs et infortunes du talmudisme académique en Allemagne [On the views of Protestant scholars on the Talmud, both critical and apologetic.] (97-134); Ifrah, Lionel: Un sommet de la littérature antijudaïque: Le "Demurrer" de William Prynne (1656) [On his pamphlet against the return of the Jews to England.] (135-189); Fuks-Mansfeld, Renate Gertrud: Quelques réflexions à propos de la littérature pour et contre les Juifs et le judaïsme aux Pays-Bas septentrionaux (193-203); Link-Lenczowski, Andrzej Krzysztof: Les textes judéophiles et judéophobes en Pologne aux XVIIe et XVIIIe siècles (205-216); Filippini, Jean Pierre: Judéophilie et judéophobie dans la Toscane du XVIIIe siècle (217-227). *HV*

0112 Ullmann, Sabine: *Nachbarschaft und Konkurrenz: Juden und Christen in Doerfern der Markgrafschaft Burgau, 1650 bis 1750.* Goettingen: Vandenhoeck & Ruprecht, 1999. 563 pp. Based on the author's diss. – Universitaet Augsburg, 1997.

Discusses the history of rural Jewry in the margraviate Burgau between 1650-1750, including the structure of Jewish communal institutions, rights and privileges granted to the Jews, and economic and social interaction between Christians and Jews. However, there were also many foci of conflict between Christians and Jews, mainly in the religious and the legal domains. The common land was a major locus of disagreement because it served as an important economic base, especially for Jewish cattle-dealers. Christians were not ready to share land with Jews, whose social status was considered inferior. Anti-Jewish legends, and far-ranging cultural consequences ensuing from differences in religious rites, led to harsh anti-Jewish feelings, sometimes erupting in aggression, violence, and cemetery desecrations. These were expressions of an attitude which made the Jews the scapegoat for social and economic problems. *ID*

0113 Węgrzynek, Hanna: Deputacje Żydów polskich do Stolicy Apostolskiej w drugiej połowie XVIII w. [Missions of Polish Jews to the Holy See in the Second Half of the 18th Century]. *Kwartalnik Historii Żydów* 199 (2001) 319-326.

Polish Jews sent three delegations to Rome – in 1754, 1758, and 1761 – to complain about blood libel accusations and other persecutions of the Jews by Polish clergy. Surveys previous missions of the Jews to the Vatican and their effects. The immediate cause for sending the 1754 delegation was the trial and execution in Żytomierz of 13 Jews accused of killing a Christian child for ritual purposes. The delegations tried to obtain documents from the Vatican that would protect them from allegations relating to the use of Christian blood. One of the positive effects of the second delegation was an investigation conducted by Cardinal Ganganelli which confirmed that the accusations against the Jews were groundless. Concludes that the documents regarding this type of political activity on the part of Polish Jews indicate their extensive contacts with representatives of the Church on various levels. *HV*

0114 Zeldes, Nadia: The Last Will and Testament of a Sicilian Converso. *Revue des Etudes Juives* 159, 3-4 (July-Dec 2000) 447-459.

Analyzes the will, written in 1505, of the Sicilian Jew Manfredo La Muta, born as Muxa (Moses), who converted to Christianity after 1473 (the original Latin text of the will is appended). However, although he was old and ill, La Muta recovered. In 1512 he was accused by the Spanish Inquisition in Sicily of being a "Judaizer" and turned over to the secular authorities, who burned him at the stake. His wife, Elisabetta, who had converted also (not all spouses then did so), met a similar fate. Discusses a number of interesting features about this family. In contrast to the usual image of Sicilian Jews and Conversos, they were middle-class rather than poor; they also maintained contact with the wife's family, which remained

Jewish. Concludes that the evidence is insufficient to decide whether the convert was sincere or, in fact, a Judaizer. Proposes that his conversion may have been a reaction to severe anti-Jewish riots in Sicily in 1474. Since this conversion, the first one of its kind to come to light in this region, preceded the expulsion edict, it raises the question of the motives of other Sicilian converts. *YC*

0115 Zelenina, Galina: Izgnaniye 1492 goda v traktovke dvukh khronik XVI veka [The Expulsion of 1492 as Interpreted by Two 16th-Century Chronicles]. *Vestnik Yevreiskovo Universiteta* 24 (2001) 73-100.

Compares two chronicles: a Jewish one, "Shevet Yehuda" by Solomon Ibn Verga, and a Christian one, "Memorias del reinado de los Reyes Católicos" by Andrés Bernáldez. Despite a number of similarities between the texts, the chroniclers differ in their interpretations of the expulsion. For example, Ibn Verga believes that its cause was the king's wish to offer a sacrifice giving thanks to God for the capture of Granada, whereas Bernáldez ascribes the expulsion mainly to the bad influence of Jews on Conversos. According to Bernáldez, the Jews "senselessly" perceived their expulsion as a new exodus from Egypt, heralding the messianic era, while Ibn Verga regards it as banishment from a homeland. Ibn Verga does not regard the expulsion as a final catastrophe, nor does Bernáldez see it as the "final solution" of the Jewish and Converso problems. The Spanish chronicler does not believe that the "inherently corrupt" Jews would indeed adopt Christianity, and regards the Conversos as perennial heretics. *DR*

0116 Zimler, Richard: Identified as the Enemy: Being a Portuguese New Christian at the Time of "The Last Kabbalist of Lisbon." *European Judaism* 33, 1 (Spr 2000) 32-42.

Discusses, in popular style, the four traumatic stages in Portuguese Jewish history in the late 15th-16th centuries that led to the end of Jewish life in Portugal. First, the expulsion from Spain in 1492 brought over 100,000 Jewish refugees to Portugal. Many of them were impoverished and were sold into slavery. Second, in 1497 King Manuel forced 20,000 Jews to convert. The third major trauma was in 1506, when 2,000 Jews were attacked by the mob and burned. Finally, the Inquisition was inaugurated in Portugal in 1536 and it became almost impossible to practice Judaism in secret. The first auto-da-fé, in 1540, signified the end of Judaism in Portugal. Zimler is the author of the novel "The Last Kabbalist of Lisbon" (1998), which deals with this period in Portuguese Jewish history. *YC*

THE MODERN PERIOD (1789-2001)

General

0117 Belova, Olga V.: Otrazhenie etnokonfessionalnykh otnoshenii v slavyanskom folklore [The Reflection of Ethnoconfessional Relations in Slavic Folklore]. *Jews and Slavs, Vol. 6: Jerusalem in Slavic Culture,* eds. Wolf Moskovich, Oto Luthar, Samuel Schwarzband. Jerusalem: Center for Slavic Languages and Literatures, Hebrew University; Ljubljana: Scientific Research Center of the Slovenian Academy of Sciences and Arts, 1999. Pp. 11-21.

Describes popular views concerning Judaism – Jewish religious beliefs, cults, ritual objects, synagogues – widespread among Poles, Belorussians, Ukrainians, and other Slavic peoples, as reflected in their folklore. Discusses various beliefs, e.g. that the Jews re-enact the deicide in their cultic worship; that they worship idols, a golden calf, the sun, the moon, etc.; that Jesus was Russian or Polish, and only Judas was Jewish. Notes, also, belief in a Jewish connection with the Devil and with magic. *DR*

0118 Benbassa, Esther; Attias, Jean-Christophe, eds.: *La haine de soi: Difficiles identités*. Bruxelles: Editions Complexe, 2000. 306 pp.

A collection of articles on self-hatred aiming to widen the concept from the Jewish self-hatred developed by Theodor Lessing to self-hatred among other marginal groups and individuals. The following deal with antisemitism: Benoit, Martine-Sophie: Theodor Lessing et le concept de "haine de soi juive" (27-46); Le Rider, Jacques: Otto Weininger: Le cas par excellence de "haine de soi juive" (47-58); Lapierre, Nicole: Changement de nom: Le signe, la haine, le soi [Deals mainly with the changing of Jewish family names into gentile-sounding names in Germany in the 19th-early 20th centuries, and in France after 1945, showing how, in the first case, the aim was to create a distance from origins viewed as shameful, whereas the French Jews hoped to subvert an identity assigned to them without giving up their Jewishness.] (273-292). *EO*

0119 Benz, Wolfgang: *Bilder vom Juden: Studien zum alltaeglichen Antisemitismus*. Muenchen: C.H. Beck, 2001. 160 pp.

A collection of essays, most of them published previously. Contents: Antisemitische Bilder: Statt einer Einleitung (7-12); Das Bild vom maechtigen und reichen Juden [From the Middle Ages to the present. Appeared in "Reiz und Fremde juedischer Kultur" (2000).] (13-26); Das Konstrukt der juedischen Verschwoerung gegen die Welt: Die "Protokolle der Weisen von Zion" [Appeared in "Grosse Verschwoerungen" (1998).] (27-43); Die juedische Erfahrung: Die Legende von der deutsch-juedischen Symbiose vor 1933 (44-56); Judenfeindschaft als Zeitgeist: Theodor Fontane und die wilhelminische Gesellschaft [Appeared in "Theodor Fontane, am Ende des Jahrhunderts" (Wuerzburg: Koenigshausen und Neumann, 2000).] (57-69); "Man muss sich schaemen, Jude zu sein": Tucholsky, die Juden in Deutschland und der Antisemitismus ab 1933 [Appeared in "Jahrbuch fuer Antisemitismus-forschung" 5 (1996).] (70-85); Mythos Anne Frank [Appeared in "Die Zeit" (September 1998).] (86-95); Judenfeindschaft aus Abwehr: Aktualitaet und Tradition des Antisemitismus in der Schweiz [Appeared in "Judaica" 56 (2000).] (96-109); Zwischen Antisemitismus und Philosemitismus: Juden in Deutschland nach 1945 (110-128); Antisemitismusforschung als gesellschaftliche Notwendigkeit und akademische Anstrengung [Appeared in "Wissenschaft vom Judentum" (2000).] (129-142). *ID*

0120 Bilewicz, Michał; Pawlisz, Bogna, eds.: *Żydzi i komunizm* [Jews and Communism]. Warszawa: Stowarzyszenie "Jidełe," 2000. 175 pp.

This special issue of "Jidełe" discusses problems related to the participation of the Jews in the communist movement in the USSR and Poland; antisemitism was one of the causes of their engagement. Partial contents: Schatz, Jaff: Zagadka pokolenia żydowskich komunistów [The Enigma of the Generation of Jewish Communists. Translation of an extract from his book "The Generation: The Rise and Fall of the Jewish Communists of Poland" (1991).] (10-28); Smolar, Aleksander: Tabu i niewinność [Tabu and Innocence. An extract from his article first published in "Aneks" 41-42 (1986).] (52-62); Deutscher, Isaac: Rewolucja rosyjska a kwestia żydowska [The Russian Revolution and the Jewish Question. A lecture delivered at the London School of Economics in 1964.] (88-102); Kersten, Krystyna: Żydzi – władza komunistów [Jews – Power of Communists. Appeared previously in her book "Polacy, Żydzi, komunizm – anatomia półprawd 1939-68" (1992).] (130-141). *HV*

0121 Borchmeyer, Dieter; Maayani, Ami; Vill, Susanne, eds.: *Richard Wagner und die Juden*. Trans.: Dorothee Goebel. Stuttgart: J.B. Metzler, 2000. 354 pp. Some of the articles have been translated from the English.

A collection based on papers delivered at a conference in Bayreuth, August 1998. Each article is followed by a summary by Wolf-Daniel Hartwich of the discussion which followed each paper. Partial contents: Friedlaender, Saul: Bayreuth und der Erloesungsantisemitismus (8-19); Borchmeyer, Dieter: Heinrich Heine – Richard Wagner: Analyse einer Affinitaet (20-33); Fischer, Jens Malte: Richard Wagners "Das Judentum in der Musik": Entstehung – Kontext – Wirkung (35-52); Bermbach, Udo:

Das aesthetische Motiv in Wagners Antisemitismus: "Das Judentum in der Musik" im Kontext der "Zuercher Kunstschriften" (55-76); Danuser, Hermann: Universalitaet oder Partikularitaet? Zur Frage antisemitischer Charakterzeichnung in Wagners Werk (79-100); Yovel, Yirmiyahu: "Nietzsche contra Wagner" und die Juden (123-142); Large, David Clay: Ein Spiegelbild des Meisters? Die Rassenlehre von Houston Stewart Chamberlain (144-158); Fulcher, Jane F.: Wagner, der Antisemitismus und die Auseinandersetzung zwischen der "traditionellen" und der faschistischen Rechten in Frankreich (160-176); Vaget, Hans Rudolf: Wieviel "Hitler" ist in Wagner? Anmerkungen zu Hitler, Wagner und Thomas Mann (178-204); Rose, Paul Lawrence: Wagner und Hitler – nach dem Holocaust (223-236); Gay, Peter: Wagner aus psychoanalytischer Sicht (251-260); Sheffi, Na`ama: Wagner in Israel: Vom Verbot bis zur Schaffung eines politischen Symbols, 1938-1997 (328-345). *RW*

0122 Burgio, Alberto, ed.: *Nel nome della razza: Il razzismo nella storia d'Italia, 1870-1945*. Bologna: Il Mulino, 1999. 565 pp.

A collection of papers delivered at a conference in Bologna, November 1997. Partial contents: Germinario, Francesco: Latinità, antimeridionalismo e antisemitismo negli scritti giovanili di Paolo Orano (1895-1911) (105-114); Miccoli, Giovanni: Santa Sede, "questione ebraica" e antisemitismo alla fine dell'Ottocento (215-246); Frigessi, Delia: Cattaneo, Lombroso e la questione ebraica (247-264); Catalan, Tullia: Antisemitismo politico e lotta nazionale a Trieste (1897-1907) (265-277); Urettini, Luigi: Stereotipi antisemiti ne "Il Mulo" (1907-1925) (293-308); Minerbi, Alessandra: Tra solidarietà e timori: Gli ebrei italiani di fronte all'arrivo dei profughi ebrei dalla Germania nazista (309-319); Sarfatti, Michele: Il razzismo fascista nella sua concretezza: La definizione di "ebreo" e la collocazione di questi nella costruenda gerarchia razziale (321-332); Bon, Silva: Scuole ebraiche a Trieste negli anni del razzismo di Stato (333-345); Guerrini, Irene; Pluviano, Marco: La propaganda antisemita fascista nell'America del Sud (347-357); Losurdo, Domenico: La crisi del processo di emancipazione degli ebrei: Per un'analisi comparata [In late 19th-early 20th century Europe.] (359-377); Rossi-Doria, Anna: Antisemitismo e antifemminismo nella cultura positivistica [In the late 19th-early 20th century, especially in Italy.] (455-473). *MZ*

0123 Filler, Susan M.: Mahler as a Jew in the Literature. *Shofar* 18, 4 (Sum 2000) 62-78.

Discusses musicological writings about the effect of Mahler's Jewish background on his work – antisemitic ones (both in Mahler's time and in the Nazi period) and neutral or philosemitic ones (including in the postwar period). The writings often reflect the positions of their authors:

assimilated or Nazi-threatened Jews, refugees, opportunists, or even Nazi bureaucrats. Some authors were consistent in their views, while others changed them; e.g. after World War II some authors expunged racist ideas from their works. Antisemitism was expressed by some of the composer's contemporaries and even by an American Jew in 1920. The most vicious anti-Mahler work was by a Nazi Party member who wrote a work on "Music and Race." Today there is not yet a consensus on the effect of Mahler's Jewish background on his work, either in terms of style or "Jewish being." *YC*

0124 Gessner, Erich Josef, ed.: *Altenstadt und Osterberg: Sammelband zur Geschichte juedischer Gemeinden im Landkreis Neu-Ulm*. 2nd ed. Neu-Ulm: Landkreis, 2001. [9], 108, 79, 52 pp. First published in 1995.

Comprises three sections, by different authors, written at different times, relating to the history of two Jewish communities in the area of Neu-Ulm, Germany. Stresses the unstable life of the Jews in the early modern period, depending on the capriciousness of the sovereigns, whose protection they required. Contents: Rose, Hermann: Geschichtliches der Israelitischen Kultusgemeinde Altenstadt [First published by Rose in 1931. Traces the problematic position of the "Schutzjude" in the 18th century; he was forced to pay high taxes for a residence permit, and lived constantly under the threat of expulsion.] (108 pp.); Sommer, Karin: Die Juden von Altenstadt: Zum Alltagsleben in einem Judendorf von ca. 1900 bis 1942 [M.A. thesis, Muenchen, 1982. Pp. 40-79 discuss the Nazi period.] (79 pp.); Ecker-Offenhaeusser, Ute: Die Judengemeinde von Osterberg: Juedisches Leben in Osterberg – die Entwicklung der israelitischen Kultusgemeinde [Published in Brackenheim, 1995. Discusses, also, restrictions against Jews in the area in the 19th century.] (38 pp.). *ID*

0125 Gilbert, Martin: *The Jews in the Twentieth Century*. New York: Schocken Books, 2001. 376 pp.

A history of Jews in the 20th century in the style of an album, accompanied by numerous photographs. Mentions, inter alia, antisemitism in pre-World War I Europe; pogroms in revolutionary Russia and in interwar Eastern Europe; the Nazi rise to power and anti-Jewish measures in Germany, Austria, and the annexed Czech Lands. Ch. 5 (pp. 196-237), "The Second World War," deals mainly with the Holocaust, presenting the Jews not only as victims of the great massacre, but also as resisters and participants in warfare. Mentions, also, the Kielce pogrom of 1946, the Eichmann trial, and attempts of the Soviet authorities to suppress the Jewish national revival in the USSR. *DR*

0126 Hagemeister, Michael: Der Mythos der "Protokolle der Weisen von Zion". *Verschwoerungstheorien: Anthropologische Konstanten – historische Varianten,* eds. Ute Caumanns, Mathias Niendorf. Osnabrueck: fibre, 2001. Pp. 89-101.

Wonders at the fascination of the poorly written and confused "Protocols," which seems to influence people who have never read the work by its mere existence and supposed authenticity. Points out that it describes not only a conspiracy but also a negative Utopia, a benevolent Jewish dictatorship that prefigures the modern totalitarian state. Part of its fascination comes from the mysterious accounts of its origin. On the other hand, historiographic accounts of its origins by serious scholars are also insufficiently based on fact. Disputes, particularly, the allegation that the original was composed in French by Russian agents in Paris. *RW*

0127

כ"ץ, דוד ס.: ישראל באמריקה: ממקווה ישראל לטימותי מקווה. זמנים 74 (אביב 2001) 84-93.

[Katz, David S., 1953-: Israel in America: From Mikveh Israel to Timothy McVeigh. *Zmanim* 74 (Spr 2001) 84-93.] Appeared in English in "The Jews and the Expansion of Europe to the West, 1450 to 1800" (2001).

Contends that Menasseh ben Israel's "Mikveh Yisrael" ("The Hope of Israel," 1650), which claimed that there were members of the ten lost tribes living in South America, influenced messianic and racist movements. Inter alia, John Wilson's 19th-century theory that the Anglo-Saxons were descendants of the tribe of Ephraim led to the founding of the British Israelites. The movement was short-lived in England, but became popular in the U.S., where the Americans were viewed as descendants of the tribe of Menasseh. The movement saw the Jews (the tribe of Yehudah) as attempting to usurp the position of the British Israelites as the chosen people, and it merged with the antisemitic Christian Identity movement in the 1950s. Discusses Christian Identity's association with the Aryan Nations, whose ideology is based on that of the British Israelites. *RE*

0128 Kertzer, David Israel: *The Popes against the Jews: The Vatican's Role in the Rise of Modern Anti-Semitism.* New York: Alfred A. Knopf, 2001. 356 pp. Appeared in Great Britain as "Unholy War: The Vatican's Role in the Rise of Modern Anti-Semitism" (Basingstoke, Hants: Macmillan, 2001).

With the emergence of modern antisemitism in the 19th century, the Catholic Church made a strong distinction between ecclesiastical anti-Judaism and "pagan" antisemitism. The Church admitted its share in the former, but negated its support of the latter – the Jew-hatred that ultimately led to the Holocaust. Argues that historical scrutiny does not support this claim. After the defeat of Napoleon, the Papal States returned their Jews to the ghettos and staunchly opposed their emancipation, which was a symbol for the dangers of modernity. As modern antisemitic movements took shape in the 19th century, the Church was a major player in them, warning people of the "Jewish peril." Its stance toward ritual murder accusations in the 19th-20th centuries was ambivalent at best; nor was the Church free of racism. Highlights the role of Catholic periodicals such as "La Civiltà Cattolica" and "L'Osservatore Romano" in supporting antisemitism. Dwells on the role of the Vatican's representatives in France, Austria, and interwar Poland in backing antisemitism. *DR*

0129 Levenson, Alan Todd: The German Peace Movement and the Jews: An Unexplored Nexus. *Leo Baeck Institute Year Book* 46 (2001) 277-301.

The German-Austrian peace movement of the late 19th-early 20th centuries attracted many Jews to its ranks; from its inception, it opposed antisemitism. The movement was home to a number of outspoken anti-antisemites; many of its activists favored a progressive resolution of the "Jewish question." Discusses the views of Bertha von Suttner (1843-1914), Ludwig Quidde (1858-1941), Heinrich von Coudenhove-Kalergi (1859-1906), and Friedrich Wilhelm Foerster (1869-1966). These people were intellectuals representing both the Austrian aristocracy and the German bourgeoisie. The anti-antisemitism of von Suttner and Quidde stemmed from their secularism and liberalism, while the other two were devout Christians. Organizations like the Deutsche Friedensgesellschaft became a framework in which Jews and non-Jews could collaborate as equals. *DR*

0130 MacMaster, Neil: *Racism in Europe, 1870-2000.* Basingstoke, Hants: Palgrave, 2001. viii, 248 pp.

A historical study focusing on antisemitism and anti-Black racism, noting parallels between them, such as the use of negative stereotypes. The main bearers of racism are "Whites," whose views have been affected by Social Darwinism and eugenics. Ch. 3 (pp. 86-114), "The Rise of Political Anti-Semitism," contains brief sections on various aspects of anti-Jewish racism (ideological, economic, religious, etc.), as well as antisemitism in Germany and Austria, Eastern Europe, and Britain and France until 1914. Ch. 5 (pp. 140-165), "Anti-Semitism in the Nazi Era," reports on antisemitism in Germany, Poland, Italy, France, Hungary, and the USSR between 1914-45. The conclusion (pp. 209-223) discusses the phenomenon of antisemitism in countries without Jews, and how antisemitism is often covert, with new racist movements focusing on "democratic" defense against immigrants and other "aliens." *YC*

0131

מרגולין, רון , עורך. הקונגרס העולמי למדעי היהדות, 12 (1997): דברי הקונגרס. חטיבה ה: החברה היהודית בת-זמננו. ירושלים: האיגוד העולמי למדעי היהדות, 2001. 268, 176 עמ'.

[Margolin, Ron, ed.: *World Congress of Jewish Studies, 12th (1997): Proceedings. Division E: Contemporary Jewish Society*. Jerusalem: World Union of Jewish Studies, 2001. 268, 176 pp.

תוכן חלקי: אורביטו נידם, יעל: פרשת 'ילדי וילה אמה' בראי הצלת ילדים בתקופת השואה (154-145); לוין, דב: המערך הרפואי של יהודי הארצות הבלטיות בשואה, 1941-1945 (166-155).

Papers presented at the quadrennial congress held in Jerusalem, July-August 1997. Partial contents: Hebrew section: [Orvieto Nidam, Iael: The Case of the 'Children of the Villa Emma' in the Light of Rescue of Children during the Holocaust Period (145-154); Levin, Dov: The Medical System of the Jews of the Baltic States in the Holocaust, 1941-1945 [In the towns, the ghettos, and the labor camps.] (155-166).] Non-Hebrew section: Levenson, Alan Todd: Herzl's Gentile Supporters [On Carl Friedrich Heman, Camilla Theimer, Hermann Bahr, and Bertha and Arthur von Suttner; discusses, also, antisemitism.] (1-11); Halbronn, Jacques E.; Bliman-Halbronn, Larissa: The Term "Protocols," from the Zionist Congresses to "The Protocols of the Elders of Zion" and the Reception of the Russian "Protocols" in Central Europe before 1917 (23-31); Kadosh, Sandra (Sara) Berliant: Youth Aliyah Policies and the Rescue of Jewish Children from Europe 1939-1942 (95-103); Rother, Bernd: Did the Nazis Think That Sephardim Were Jews? [On attempts to exempt the Sephardi Jews from Nazi persecution in the Netherlands and in France.] (105-113); Hoffman, Stefani: The Impact of the Collapse of the Soviet Union on Attitudes toward Jews in Russian Politics (145-152). *SSC*

0132 Penslar, Derek Jonathan: *Shylock's Children: Economics and Jewish Identity in Modern Europe*. Berkeley: University of California Press, 2001. xi, 374 pp.

Examines how Jews in modern Europe perceived and accounted for their economic difference, viewed from the perspective that the "Jewish question" was socioeconomic and not only political. Discusses the development of a Jewish political economy and ideas of Jewish identity, especially in relation to the German Jewish bourgeoisie. Ch. 1 (pp. 11-49), "Jews, Paupers, and Other Savages: The Economic Image of the Jew in Western Europe, 1648-1848," discusses the stereotype of the Jew as parasitical, dangerous, and outside the established economic order. Examines Jewish responses to these negative depictions. One response was Jewish philanthropy as an attempt to correct "problematic" aspects of Jewish economic life. Jewish social policies that emerged to deal with issues raised by antisemitism included social engineering in the form of mass colonization – in Palestine, Argentina, and the USSR. The last period discussed ends in 1933. In this context, connects the desire for a new type of Jew with internalization of economic antisemitism. Concludes that in the post-World War II period, the "Jewish question" as understood in earlier European history no longer exists. *YC*

0133 Pflegerl, Siegfried: *Ist Antisemitismus heilbar? Zur Bearbeitung einer fatalen Tradition*. Frankfurt a.M.: Peter Lang, 2001. 260 pp.

After presenting a theory of social structure and of the relations between majority and minorities, traces the history of antisemitism in the light of this theory. Maps the political camps in interwar Austria and the type and degree of their antisemitism. Chronicles the history of the Final Solution and cites historians' attempts to explain its genesis. Analyzes antisemitism in present-day Austria and Germany on both Right and Left. Discusses the demand of the majority to suppress memory of the Holocaust, which creates a chasm between this majority and the Jewish minority. Describes Jewish reactions to majority pressure through attempts at assimilation, leading to identity conflict and self-hatred, and conversely through Zionism. Concludes that the solution to discrimination is based on recognizing the unity of mankind, rooted in the "absolute, infinite essentiality" while preserving complexity. *RW*

0134 Reder, Melvyn W.: The Anti-Semitism of Some Eminent Economists. *History of Political Economy* 32, 4 (2000) 833-856.

Finds a degree of antisemitism in the attitudes of three economists – John Maynard Keynes, Joseph Schumpeter, and Friedrich Hayek – which is related to their belonging to the educated elite. Their "ambivalent antisemitism" was manifested in some contexts and toward some individuals but not in others. The prejudice of Keynes involved negative stereotypes of Jews. These economists' contemporaries, and later biographers, tended to dismiss the possibility that these intellectuals were antisemitic, largely because they supported some Jewish colleagues. While their prejudice did not affect their economic theories, it may have impacted on their views about the functioning of academic institutions. Hayek's antisemitism was revealed in his views on the immigration to Austria of Ostjuden and his justification of restrictions on them. Discusses a change in public attitudes that became less tolerant of antisemitism after 1945. Attributes this to changes in American universities. Among the factors was increased demand for both quantity and quality in academic staff, which increased the number of places for Jews, and made antisemitism less acceptable since Jews had more choice of employment.

See the response by Ronald Hamowy, "A Note on Hayek and Anti-Semitism" [Ibid. 34, 1 (2002) 255-260] which claims that Hayek was "pro-Semitic." See also Reder's reply (ibid., 261-272), which rejects Hamowy's arguments, stressing Hayek's hostility toward Jews. *YC*

0135 Rensmann, Lars: *Kritische Theorie ueber den Antisemitismus: Studien zu Struktur, Erklaerungspotential und Aktualitaet.* Hamburg: Argument Verlag, 1998. 386 pp.

Expounds the Frankfurt School's theory of antisemitism as a paranoia rooted in the authoritarian personality, produced by socialization in the authoritarian society. Comments that, while the theory is still largely valid, it requires some modification in the light of newer sociological and psychological insights as well as the actual forms taken by antisemitism in Germany today. After a long postwar period of latency, antisemitism has reappeared, usually in code but also overtly and violently, and increasingly in the younger generation; this in spite of a more liberal socialization in West Germany since the 1960s. In former East Germany, authoritarianism does not seem to have led to antisemitism. An important element in post-Holocaust antisemitism, even more potent today, is "secondary antisemitism," already diagnosed by the Frankfurt School: antisemitism arising from the need to repress the memory of the Holocaust and German guilt. This emerged in the almost unanimous condemnation of Goldhagen for exposing the antisemitic motivation of "ordinary Germans" who participated in the extermination of the Jews. *RW*

0136 Stone, Dan: An 'Entirely Tactless Nietzschean Jew': Oscar Levy's Critique of Western Civilization. *Journal of Contemporary History* 36, 2 (Apr 2001) 271-292.

Discusses the views of Oscar Levy (1867-1946), a German-born Jew who became the leading Nietzschean in England. Levy followed his mentor in blaming the Jews for the degeneration of Western civilization. In addition to suffering from self-hatred, Levy was also partial to the conspiracy theory, blaming the Jews for the economic and political miseries of the world. Like Nietzsche, he believed in an aristocratic ideal. Initially, he sympathized with Italian fascism as a bulwark against Bolshevism, but he came to reject both fascism and Nazism. Although he thought that criticism of the Jews might save them, he considered the Jews responsible even for their own enemies, the Nazis. Finally, after attacking Jews, Christians, communists, fascists, and democrats, Levy remained a solitary Nietzschean, remote from all major contemporary political philosophies. *YC*

0137 Stratton, Jon: *Coming Out Jewish: Constructing Ambivalent Identities.* London: Routledge, 2000. ix, 341 pp. Appeared in the U.S. as "Coming

Out Jewish: On the Impossibility of Jewish Assimilation" (New York: Routledge, 2000).

Explores Jewish identity in its modern and postmodern phases, mainly in Europe, Australia, and the USA, focusing on assimilation and the ambivalent status of the Jew. Ch. 3 (pp. 84-114), "Ghetto Thinking and Everyday Life," notes that Ashkenazi Jews fear an antagonistic world and pass on this fear to their children even when there is no longer anything to fear. Ch. 4 (pp. 117-136), "Jews, Representation and the Modern State," links modern antisemitism to the rise of the nation-state and its attempt to create a monolithic national identity. Ch. 7 (pp. 195-219), "Jews, Race and the White Australia Policy," appeared in "Jewries at the Frontier" (1999) and, in a shorter version, in "Journal of Australian Studies" 50-51 (1996); deals with Australian Jews being considered non-White or White in regard to local racism. Ch. 8 (pp. 220-250), "Jews and Multiculturalism in Australia," appeared in "Diaspora" 5 (1996); notes that from the 1930s Ashkenazi Jews were unwelcome but finally, in the postwar period, were accepted as an ethnic group whose nationality was associated with the Holocaust that had affected their people's identity. *YC*

0138 Tewes, Henning; Wright, Jonathan, eds.: *Liberalism, Anti-Semitism, and Democracy: Essays in Honour of Peter Pulzer*. Oxford: Oxford University Press, 2001. viii, 300 pp.

A "Festschrift" for Pulzer, a scholar who has specialized in German and Austrian history, especially the interrelations between antisemitism and failures of liberalism. Partial contents: Pulzer, Peter George Julius: From Danube to Isis: A Career in Two Cultures [Pulzer, born in 1929 to a Jewish family, immigrated from Vienna to Britain in February 1939 and grew up in London.] (13-25); Tewes, Henning: The Committed Observer of a Janus-Faced Century: Peter Pulzer on Liberalism, Anti-Semitism, and Democracy [Pp. 28-36 discuss Pulzer's research on pre-World War I Jewish-German history.] (26-46); Markovits, Andrei S.: Peter Pulzer's Writing on Political Anti-Semitism and the Jewish Question in Germany and Austria: An Assesssment (47-59); Pelinka, Anton: Anti-Semitism and Ethno-Nationalism as Determining Factors for Austria's Political Culture at the "Fin de Siècle" (63-75); Palmowski, Jan: Between Dependence and Influence: Jews and Liberalism in Frankfurt am Main, 1864-1933 [Explores the relationship between liberalism and antisemitism in an urban setting.] (76-101); Wright, Jonathan: Liberalism and Anti-Semitism in Germany: The Case of Gustav Stresemann [On the career of the liberal politician (1878-1929) who was opposed to antisemitism.] (102-126). *YC*

0139 Thing, Morten: *Den historiske jøde: Essays & ordbog* [The Historical Jew: Essays & Dictionary]. København: Forum, 2001. 269 pp.

The essay on pp. 70-94, "Kan tekster draebe?" ("Can Texts Kill?"), discusses the origins of the "Protocols of the Elders of Zion" and the alleged world Jewish conspiracy. In Denmark the "Protocols" was published in September 1920. A lively debate followed; efforts were made, in certain circles, to connect the "Protocols" with the Bolsheviks and their idea of world domination. Relates to the current renaissance of the "Protocols" all over the world. Pp. 95-186, "Den evige jøde" ("The Eternal Jew"), deals with the mythical figure in literature. Quite a number of European thinkers, writers, and artists have used this figure in their works, among others well-known Danes such as Kierkegaard and Andersen. Central questions in the debate are: Are the Jews eternal? Do they stand out as different, exotic, oriental, or are they dangerous? Are the Jews, as opposed to other peoples or nations, in fact outside of History? The book ends (pp. 187-259) with a dictionary of terms related to Judaism. *RSL*

0140 Traverso, Enzo: *Nach Auschwitz: Die Linke und die Aufarbeitung des NS-Voelkermords*. Trans.: Paul B. Kleiser, Ulla Varchmin. Koeln: ISP, 2000. 220 pp.

A collection of ten essays, all of them published previously. Six of them appeared in English in his collection "Understanding the Nazi Genocide: Marxism after Auschwitz" (1999). Partial contents: Auschwitz, Marx und das 20. Jahrhundert [Appeared in French in "Critique Communiste" 143 (1995); in German in "Ausblicke auf das vergangene Jahrhundert: Die Politik der internationalen Arbeiterbewegung von 1900 bis 2000" (Hamburg: VSA, 1996.] (13-33); Der Antisemitismus als kultureller Code: Frankreich und Deutschland im Vergleich [Appeared in French in his book "Pour une critique de la barbarie moderne: Ecrits sur l'histoire des Juifs et de l'antisémitisme" (1996).] (34-48); Lehren einer Niederlage: Die deutsche Linke angesichts des Antisemitismus der Nazis [Appeared in French in "Mauvais temps" 6-7 (March 2000).] (49-59); Die Blindheit der Intellektuellen: Die "Ueberlegungen zur Judenfrage" von Jean-Paul Sartre und ihre Kritiker [Appeared in English in "October" 87 (1999).] (60-78); Von der Frankfurter Schule zu Ernest Mandel: Fragestellungen und Sackgassen des Marxismus hinsichtlich Auschwitz [Appeared in French as an afterword to the second edition of his book "Les marxistes et la question juive" (1997).] (79-104); Die Einzigartikeit von Auschwitz: Hypothesen, Probleme und Irrwege der historischen Forschung [Appeared in French in his book "Pour une critique de la barbarie moderne: Ecrits sur l'histoire des Juifs et de l'antisémitisme" (Lausanne: Editions Page Deux, 1997) and in "Parler des camps, penser les génocides" (Paris, 1999).] (63-78); Das Vermaechtnis: Der Aufstand im Warschauer Ghetto [Appeared in French in "La Revue" (Apr-June 1993).] (124-135); Die Shoah, der Historiker und der oeffentliche Gebrauch der Geschichte: Zur Kontroverse um Goldhagens Buch "Hitlers willige Vollstrecker"

[Appeared in French in "L'Homme et la Société" 125 (July-Sept 1997).] (136-152); Sozialismus und Ethik: Norman Geras und die politische Philosophie nach Auschwitz [Appeared in French in "Critique Communiste" 154 (Winter 1999).] (178-194); Zum Schluss (195-198). *RW*

0141 Traverso, Enzo: *Understanding the Nazi Genocide: Marxism after Auschwitz*. Trans.: Peter Drucker. London: Pluto Press, with the International Institute for Research and Education, 1999. viii, 152 pp.

A collection of six essays, all of them published previously, five of them translated for this collection.

0142 Tych, Feliks: *Długi cień Zagłady: Szkice historyczne* [The Long Shadow of the Holocaust: Historical Studies]. Warszawa: Żydowski Instytut Historyczny, 1999. 167 pp.

A collection of essays, most of them published previously. Partial contents: Świadkowie Shoah: Zagłada Żydów w polskich pamiętnikach i wspomnieniach [Witnesses of Shoah: The Extermination of Jews in Polish Diaries, Memoirs and Reminiscences. An abridged English version appeared in the "Biuletyn Żydowskiego Instytutu Historycznego" 192 (1999).] (9-54); Ocaleni z Zagłady i ich ocena postawy społeczeństwa polskiego w okresie okupacji: Zapomniane świadectwo z roku 1944 [Survivors of the Holocaust and Their Views on the Poles' Attitudes during the Occupation: A Forgotten Testimony from 1944. Presents and comments on a letter published in "Gazeta Lubelska" in November 1944, written by a Holocaust survivor who relates to the lack of moral values of many Poles who denounced the Jews. First published in "Almanach Żydowski" (1997-98).] (55-67); Obraz Zagłady Żydów w potocznej świadomości historycznej w Polsce [The Perception of the Holocaust of the Jews in Common Historical Consciousness in Poland. Appeared in "BŻIH" 189 (1999).] (69-93); Problem Zagłady w polskich podręcznikach szkolnych do nauczania historii [The Problem of the Holocaust in Polish Textbooks for Teaching History. Appeared in "BŻIH" 188 (1998).] (95-117); Kilka uwag o Marcu 1968 [Some Remarks about March 1968. A revised version of a paper delivered at a conference in Warsaw, March 1998.] (119-136); Zakres odpowiedzialności społeczeństwa niemieckiego za Zagładę Żydów: Reakcje w Niemczech na książkę Daniela J. Goldhagena o źródłach Holocaustu [The Scope of Responsibility of German Society for the Extermination of the Jews: Reactions in Germany to the Book by Daniel J. Goldhagen about the Causes of the Holocaust. First appeared in "BŻIH" 180 (1996).] (149-167). *HV*

1789-2001: By Country

Argentina

0143 *Indice* 20 (2000).

This issue is entitled "Discriminación En torno de los unos y los otros." Partial contents: Galante, Miguel; Jmelnizky, Adrián: El primer peronismo y los migrantes de posguerra vinculados a la Shoá (Holocausto), 1946-1950 (51-92); Galante, Miguel; Jmelnizky, Adrián: Selección de documentos sobre las respuestas del estado argentino ante las solicitudes de ingreso al país durante la segunda posguerra (93-137); Braylan, Marisa et al.: Informe sobre antisemitismo en la Argentina 1998 [Appeared in English as a separate booklet: "Report on Anti-Semitism in Argentina – 1998" (2000).] (155-195); Ladwig-Winters, Simone: El rol de los abogados judíos en el desarrollo del derecho en Alemania, y su destino ulterior hasta mil novecientos cuarenta y cinco (277-296); Braylan, Marisa et al.: Informe sobre la situación de los detenidos-desaparecidos judiós durante el genocidio perpetrado en Argentina (297-361). *MZ*

0144 Silber, Daniel: Sobre el antisemitismo en la Argentina. *La cuestión judía: Seminario y... – un seminario, debates, análisis y opiniones sobre el tema.* Buenos Aires: Ediciones Cuadernos Marxistas, 2001. Pp. 52-77.

Discusses antisemitism in Argentina from 1889 until the present, giving examples from literature, education and politics. From the beginning of Jewish immigration (which was an important part of the European mass emigration) the ruling land oligarchy had a hostile attitude toward the Jews. They considered them as a paradigm of both the foreigner and the revolutionary, a danger to Argentinean national identity and institutions. This stereotype spread to other sectors of the population, and, fueled by political and economic crisis, led to discrimination and violence. *MZ*

China

0145 Chou, Hsun: *Chinese Perceptions of the 'Jews' and Judaism: A History of the Youtai.* Richmond, Surrey: Curzon, 2001. x, 202 pp. Based on the author's diss. – University of London. On the title page, the author's name is given as Zhou Xun.

Shows how various social groups in China, in different periods (from 1605 to 1997), viewed the "Jews" as a homogeneous group in order to differentiate themselves from them. In China a negative attitude toward Jews was first connected to the images of Jews introduced by Catholic and Protestant missionaries in the 19th century. Chinese modernizers cited Jewish models for a national renaissance. Eugenics in China, as opposed to Germany, often viewed the Jews as a superior race, although Jews were also considered disease-prone. Other stereotypes were of Jews as intellectuals and as successful businessmen. Zionists were sometimes perceived positively as nationalists, but at other times as imperialists. In Japanese-occupied China, the Japanese (not the White Russian population or the Nazis) are blamed for anti-Jewish ideology and actions. Before the founding of Israel, there was sympathy for the Jews as victims of racial discrimination, but afterwards they were seen as tools of imperialism. Zionists were sometimes seen as capitalists and sometimes as victims of imperialism and fascism (especially under Mao). Recently, the stereotype of Jews as rich and successful has been touted as an inspiration for aspiring young Chinese capitalists. *YC*

Denmark

0146 Wagner, Thorsten: Jøder og andre danskere: Den nyere antisemitismeforskning og dens implikationer for dansk historieskrivning: En forskningsoversigt [Jews and Other Danes: The New Research on Antisemitism and Its Implications for Danish Historiography: An Overview]. *Nordisk Judaistik* 22, 2 (2001) 157-192.

Discusses recent research on antisemitism and the applicability of this research to a critical investigation of the relations between Jews and non-Jews in 19th-20th-century Denmark, a topic which has been neglected until recently. The reevaluation of the continuing impact of religious anti-Judaism on "modern" forms of antisemitism, and a new appreciation of the cultural dimensions of anti-Jewish stereotyping, have been instrumental in a new understanding of antisemitism in Denmark. The new generation of research calls for a rewriting of Danish history: the liberal democratic political culture that informed the formation of the Danish state did not do without exclusionist practices; on the contrary, it rejected cultural or ethnic heterogeneity. Instead of earlier historiography which exonerated Danish antisemitism through self-serving comparisons (e.g. the heroic rescue of the Jews in World War II), the new historiography promises new insights into the development of Danish national identity, which vascillated between inclusion and exclusion. *SSC*

France

0147 Fulcher, Jane F.: A Political Barometer of Twentieth-Century France: Wagner as Jew or Anti-Semite. *Musical Quarterly* 84, 1 (Spr 2000) 41-57.

Shows how the French Right often used Wagner's antisemitic ideas to stress the need to purge French music of foreign, i.e."Jewish," tendencies. This approach began with the composer Vincent d'Indy and the monarchist Action Française at the time of the Dreyfus Affair. The nationalist Ligue de la Patrie Française found Wagner's antisemitism compatible with their values. D'Indy composed a Wagner-inspired antisemitic opera. World War I led to a partial rejection of Wagner as a "Boche" and a "Jew" whose alleged Jewish origin was blamed for his "Oriental style" by the right-wing Action Française. However, d'Indy continued to defend Wagner, finding allies among the fascist sympathizers in France. D'Indy's book "Richard Wagner et son influence sur l'art musical aujourd'hui" (1930) exemplified the influence of the German antisemitic composer on French political culture of the Right. This trend continued with Lucien Rebatet, who stressed Wagner's idea of Jewish composers "polluting" French music. Such ideas still have influence today. *YC*

Germany

0148 Baumann, Ulrich: *Zerstoerte Nachbarschaften: Christen und Juden in badischen Landgemeinden, 1862-1940*. Hamburg: Doelling und Galitz, 2000. 341, [16] pp. Based on the author's diss. – Universitaet Freiburg im Breisgau, 1998.

Analyzes relations between Christians and Jews in the 18 villages of southern Baden with large Jewish populations ("Judendoerfer") in the context of economic and political changes. Despite opposition in the 1860s to Jewish emancipation, both populations realized the importance of working together; Christian peasants and Jewish traders depended on each other, a dependency sometimes creating intimacy and sometimes resentment. Peasants saw Jews as unable and unwilling to do "real" (i.e. physical) work, and after emancipation as insolent. Antisemitic agitators and propaganda coming from northern Germany found fertile ground in the south; in the 1890s there were incidents of violence. Many joined the Nazi Party even before 1933, when it displaced the old leadership that had worked for coexistence. But despite Nazi efforts to prevent contact with Jews, peasants remained loyal to Jewish cattle-dealers until 1938. Describes the persecution of Jews and the "Kristallnacht" pogrom. In October 1940 all the Jews of Baden were deported to Gurs; the survivors were deported in 1942 to extermination camps. *RW*

0149 Beckmann, Volker: *Die juedische Bevoelkerung der Landkreise Luebbecke und Halle i.W.: Vom Vormaerz bis zur Befreiung vom Faschismus (1815-1945)*. Lage: Verlag Hans Jacobs, 2001. 594 pp. Based on the author's diss. – Universitaet Bielefeld, 2001.

Surveys the history of Jewish communities in the rural district of Luebbecke and Halle (Westphalia) between 1815-1945. Pp. 240-318 discuss local discrimination and exclusion policies toward the Jews between 1849-58. Afterwards, the growing antisemitic movement emanating from Berlin led to an increase in the power of antisemitic political parties. In elections held between 1924-30, these parties gradually obtained a majority of the votes. Pp. 420-557 discuss the Nazi period: boycott, Aryanization, discrimination, exclusion, and persecution between 1933-38, culminating in the "Kristallnacht" pogrom. The Jews' reactions involved spiritual and cultural resistance, solidarity, and emigration, but between 1939-41 emigration became more and more difficult. Jews were taken for forced labor and were crowded into "Jewish houses." In March 1942 deportations began. At least 165 Jews from the area were deported to concentration camps; 22 survived. *ID*

0150 Bereswill, Mechthild; Wagner, Leonie, eds.: *Buergerliche Frauenbewegung und Antisemitismus.* Tuebingen: Diskord, 1998. 125 pp.

A collection based on papers delivered at a workshop at the Archiv der deutschen Frauenbewegung in Kassel, December 1996. Partial contents: Krohn, Helga: Aufbruch in eine neue Gesellschaft? Erwartungen juedischer Frauen an die deutsche Frauenbewegung und die Grenzen der Zusammenarbeit (13-43); Bereswill, Mechthild; Wagner, Leonie: "Eine rein persoenliche Angelegenheit": Antisemitismus und politische Oeffentlichkeit als Konfliktfeld im "Bund Deutscher Frauenvereine" [First appeared in "Die Philosophin" 15 (1997); in English in the "Journal of Genocide Research" 1, 2 (1999).] (45-63); Omran, Susanne: "Woran erkenen wir die Prostituierte?" Sittlichkeit, Grossstadtdiskurs und Antisemitismus im Kontext der Frauenbewegung (65-87); Henze, Dagmar: Antijudaismus – Antisemitismus: Zum Zusammenhang von christlichen Traditionen und antijuedischen Stereotypen (89-99); Gehmacher, Johanna: Die Eine und der Andere: Moderner Antisemitismus als Gesschlechtergeschichte (101-120). *RW*

0151 Golomb, Jacob; Wistrich, Robert Solomon: Nietzsche's Politics, Fascism and the Jews. *Nietzsche-Studien* 30 (2001) 305-321.

Argues that it is a mistake to hold Nietzsche even indirectly responsible for Auschwitz. He opposed both racism and antisemitism, which were crucial to Nazism. He was also a fierce individualist and, hence, any propagandistic use of his ideas involved drastic distortion of them. Although he was very critical of Judaism and even more so of Christianity, he saw a possible positive role for the former in Europe. He also viewed nationalism as very dangerous, particularly in Germany. This makes him an unlikely forerunner of fascism. Rather, he should be seen as a "tragic prophet of the spiritual vacuum that gave birth to the totalitarian abysses of the twentieth century." *YC*

0152 Gotzmann, Andreas; Liedtke, Rainer; Rahden, Till van, eds.: *Juden, Buerger, Deutsche: Zur Geschichte von Vielfalt und Differenz, 1800-1933*. Tuebingen: Mohr Siebeck, 2001. ix, 444 pp. (Schriftenreihe wissenschaftlicher Abhandlungen des Leo Baeck Instituts, 63).

Partial contents: Sieg, Ulrich: Der Preis des Bildungsstrebens: Juedische Geisteswissenschaftler im Kaiserreich (67-95); Hoffmann, Stefan-Ludwig: Buerger zweier Welten? Juden und Freimaurer im 19. Jahrhundert [An English translation appeared in "German History" 18 (2000).] (97-119); Reitmayer, Morten: Zwischen Abgrenzung und Ausgrenzung: Juedische Grossbankiers und der Antisemitismus im deutschen Kaiserreich (147-170); Lindner, Erik: Deutsche Juden und die buergerlich-nationale Festkultur: Die Schiller- und Fichtefeiern von 1859 und 1862 [The Jews celebrated Fichte as the champion of German nationalism, and ignored his antisemitism. While some observers justified this stance, declaring Fichte's antisemitism outmoded, antisemites jeered.] (171-191); Haibl, Michaela: Im Widerschein der Wirklichkeit: Die Verbuergerlichung und Akkulturation deutscher Juden in illustrierten Zeitschriften zwischen 1850 und 1900 [Including both serious portraits and genre pictures and antisemitic caricatures.] (217-237 + 8 pp. of plates); Liepach, Martin: Das Krisenbewusstsein des juedischen Buergertums in den "Goldenen Zwanzigern" (395-417). *RW*

0153 Loader, Colin: Werner Sombart's "The Jews and Modern Capitalism." *Society* 39, 1 (Nov-Dec 2001) 71-77.

Before 1933, Sombart's "The Jews and Modern Capitalism" (1911) was viewed solely as a work on the origins of modern capitalism. Since World War II, many scholars have focused on Sombart's antisemitism and its place in the history of German antisemitism. Sombart ascribed to the Jews some group characteristics (rationalism, abstractness of spirit, etc.) which had been viewed negatively in the 17th-18th centuries but became necessary for the capitalist of the later epoch. He regarded these traits as "racial," but he used the word "race" in the sense of ethnicity. He made many positive statements about the Jews and insisted on the preservation of the German-Jewish symbiosis, so useful for Germany. In the 1930s, Sombart came to support Nazism because he believed that the Jewish spirit had ceased to be complementary and had instead become dominant. Sombart's views were supported by Zionists and some Jewish intellectuals, but were branded as antisemitic by assimilationists. The antisemites in Germany were also divided on how to perceive his ideas. *DR*

0154 Moebius, Sascha: Arbeiterbewegung, Zionismus und Antisemitismus vor 1933. *Hundert Jahre Zionismus – Befreiung oder Unterdrueckung? Beitraege der Gegentagung zum Herzl-Jubilaeum.* Koeln: ISP, 1998. Pp. 23-26.

Traces the changing attitudes of the socialist movement to antisemitism (mainly in Germany), stating that while the early socialists tended to be antisemitic, they opposed antisemitism when it became identified with movements hostile to their own. But neither the Social Democrats nor the communists developed a strong and consistent policy on antisemitism, and they did not stand up against the radical right antisemitic parties of the 1920s. *RW*

0155 Omran, Susanne: *Frauenbewegung und "Judenfrage": Diskurse um Rasse und Geschlecht nach 1900.* Frankfurt a.M.: Campus, 2000. 522 pp. Based on the author's diss. – Universitaet Dortmund, 1999.

Analyzes attitudes of the German feminist movement to social and cultural developments considered threatening to women in their role as nurturers of life; all of these, in one way or another, were perceived as involving Jews. It was claimed that urbanization fosters neuroticism, to which Jews, an urban population, are especially prone; that the prostitution that flourishes in cities is in large part the work of the Jewish white slave trade; that capitalism, pioneered by Jews, stands in opposition to woman consumers; that the solidarity and selfless service of women in wartime is undermined by profiteers, many of them allegedly Jews; and that assimilated German Jews, estranged from their own tradition, threaten German demographic vitality by corrupting values and by setting a trend of low birthrates and small families. Expounds the thought of the Jewish feminist Henriette Fuerth (1861-1938), who stressed the positive value of these supposedly negative phenomena. She called on Jews to be proud of their Jewishness, while learning from antisemites what faults they need to correct, and on Jewish mothers to raise their children to cope with the greater demands society would make of them than of the non-Jews. *RW*

0156 Reich-Ranicki, Marcel: *Mein Leben.* Stuttgart: Deutsche Verlags-Anstalt, 1999. 566 pp.

The prominent literary critic (b. 1920) writes of his boyhood in Nazi Berlin, and describes how in October 1938 he was expelled with his family to Poland. They lived in Warsaw, and later moved into the ghetto. Writes about musical performances (he reported on them in the ghetto newspaper) and about the Jewish Council, where he worked as head of the translation and correspondence department. He took down the protocol of the meeting of 22 July 1942, at which SS commander Hermann Hoefle demanded the deportation of thousands of Jews daily, and which resulted in the suicide of Adam Czerniaków. "Useful Jews," including employees of the Jewish Council and their spouses, were exempt. Reich-Ranicki married in the ghetto; the couple escaped to the "Aryan side" and found shelter with Poles until the liberation. The rest of his family perished. In Poland until 1958 (where he adopted the pen-name Ranicki), and subsequently in West Germany, the author worked as a literary critic, advancing to literary editor of the "Frankfurter Allgemeine Zeitung" and moderator of a popular literary TV program. Speculates whether hostility toward him is due to resentment of his criticism or to antisemitism. Describes his reactions to the "historians' debate" and to Martin Walser's call for an end to Germans' obsession with guilt for the Holocaust. *RW*

0157 *Stories of an Exhibition: Two Millennia of German Jewish History*. Berlin: Jewish Museum Berlin, 2001. 224 pp. "Text – Ingke Brodersen and Ruediger Dammann, in collaboration with the researchers of the Permanent Exhibition."

A richly illustrated volume accompanying the exhibitions at the newly-opened Jewish Museum in Berlin. On antisemitism, see "'The Dream of Political Unity': The Struggle for Emancipation" (88-95), "'Awakened from the Dream': The Completion and Destruction of Emancipation" (138-145), "'Pick up the Shield of David': The Reaction to National Socialist Persecution" (148-159), "Exile and Holocaust: The Exhibition in the Axes" [the Axes are in the basement of the Museum] (182-185), "The Gallery of the Missing" [on items which would have been part of the Museum's displays, but were lost due to the Holocaust] (186-189). *SSC*

0158 Urban-Fahr, Susanne: Eine Geschichte ohne Ende? Juedische Reaktionen auf den Antisemitismus. *Tribuene* 158 (2001) 120-131.

Describes the spread of antisemitism in Germany since the 1880s and its recent resurgence. Mentions Jewish solutions such as Zionism, Jewish cultural revival, or emigration. The dominant reaction, however, was the founding in 1893 of the Centralverein and its activities – at first the attempt to educate the public through pamphlets and lectures, in addition to legal actions against antisemites; from 1929, also anti-Nazi propaganda and support of anti-Nazi political parties. Today, Jewish leaders in Germany are expected to function as spokesmen against xenophobia in general, and to reassure the public that German democracy is still intact. But it is important to fight antisemitism specifically, and non-Jews must participate in this task. *RW*

0159 Wistrich, Robert Solomon: Was Nietzsche a Fascist Thinker? *Partisan Review* 68, 2 (Spr 2001) 201-217.

Contends that the myth of Nietzsche as a precursor of Italian fascism and German Nazism – invented and supported by some Nazi, communist, and conservative intellectuals – is a gross distortion of the truth. In particular, the accusations that Nietzsche was an antisemite and that his works were an inspiration for the Nazi Holocaust are unsupported. Nietzsche attacked Judeo-Christian ethics and other Judaic elements in Christianity, but his views were free of racism. He despised antisemitism and the German nationalism of his time, he scorned the Aryan myth, and was in fact a cosmopolitan. It was his fierce anti-Christian and anti-decadent rhetoric, as well as his provocative language, that gave the Nazis an opportunity to harness citations from Nietzsche for their cause. *DR*

Hungary

0160 Braham, Randolph L., ed.: *Tanulmányok a Holokausztról. [Vol.] I.* [Studies on the Holocaust. Vol. I]. Budapest: Balassi Kiadó, 2001. 227 pp.

Contents: Braham, Randolph L.: Magyarország keresztény egyházai és a holokauszt [The Hungarian Christian Churches and the Holocaust] (9-36); Róbert, Péter: A holokauszt a magyar sajtóban [The Holocaust in the Hungarian Press.] (37-71); Földes, Anna: A holokauszt a magyar (próza)irodalom tükrében [The Holocaust in Light of Hungarian (Prose) Literature.] (73-122); Karsai, László: Új magyar egyetemi és középiskolai tankönyvek a holokausztról [New University and Secondary School Textbooks on the Holocaust.] (123-157); Varga, László: A holokauszt és a rendszerváltás Magyarországon [The Holocaust and the Change in Government in Hungary. Deals with the period after the collapse of communism.] (159-199). Includes an appendix: Fontosabb zsidó tárgyú törvények és rendeletek Magyarországon (1849-2000) [The Main Laws and Regulations Relating to the Jews (1849-2000)] (203-213). *MA*

0161 Gyurgyák, János: *A zsidókérdés Magyarországon: Politikai eszmetörténet* [The Jewish Question in Hungary: Theory of Ideas from a Political Point of View]. Budapest: Osiris, 2001. 788 pp.

Examines the "Jewish question" in Hungary from the aspect of political history of ideas, from the 18th century to the present. Describes polemics by politicians, literary personalities, historians, and thinkers, quoting extensively from their writings. Pp. 25-208 present a historical background. Deals with conservative and liberal attitudes concerning the immigration and the assimilation of Jews. Describes the extreme-right movements from Istóczy to Szálasi. Devotes special attention to the

Horthy era. Examines the "urbanist-folkist" debate from the beginning of the 20th century through the new generation of urbanists in the 1920s-30s, pointing out that the "folkists" provided the ideology for the antisemitic policies of the Horthy era. Discusses the postwar years, up to the revival of the "urbanist-folkist" debate in the early 1990s. *AV*

0162 Ungvári, Tamás: *Ahasvérus és Shylock: A "zsidókérdés" Magyarországon* [Ahasuerus and Shylock: The "Jewish Question" in Hungary]. Budapest: Akadémiai Kiadó, 1999. 407 pp.

Investigates the "Jewish question" in Hungary from the Reform Age in the first half of the 19th century to the present, from the point of view of cultural history, history of ideas, sociology, and literary history. Changes in ideas are traced from the Enlightenment to Hungarian neo-Nazism. Describes disputes about the "Jewish question" conducted by politicians and intellectuals, conservative and liberal approaches to the assimilation of Jews (e.g. Széchenyi and Kossuth), the pogroms in 1848 and their background, Istóczy's Antisemitic Party in the 1880s and the Tiszaeszlár blood libel in 1882, the pogroms that followed, and how antisemitism permeated the agrarian movement and nationalistic ideologies in the 1890s. Focuses on the rise of antisemitism in the interwar period and its roots. Assimilation was a failure, as demonstrated by a blend of popular and political antisemitism, pogroms, and anti-Jewish organizations and laws. The views of historian Gyula Szekfű and the debate between the "urbanists" and the "folkists" are treated at great length. Relates, also, to the reappearance of antisemitism in the 1990s. *AV*

0163 Ungvári, Tamás: *The "Jewish Question" in Europe: The Case of Hungary.* Trans.: Tamás Ungvári, Nóra Arató, Judit Zinner. Boulder, CO: Social Science Monographs; Highland Lakes, NJ: Atlantic Research and Publications, distributed by Columbia University Press, New York, 2000. 358 pp. Originally published as "Ahasvérus és Shylock: A 'zsidókérdés' Magyarországon" (Budapest: Akadémiai Kiadó, 1999).

Islamic World

0164 Shivtiel, Avihai: A Twentieth Century Blood Libel and Its Reflection in the Arabic Version of August Rohling's "Der Talmudjude." *Hebrew Language and Jewish Studies,* eds. Aharon Ben-David, Isaac Gluska. Netanya: Association for Society and Culture, 2001. Pp. lxxxiii-xc.

Discusses the Arabic translation (Cairo, 1899 and 1938; 2nd edition, Beirut, 1968 and 1970) of the antisemitic "Der Talmudjude" (Vienna, 1871) by August Rohling, which presented "evidence" that the Talmud encouraged the killing of non-Jews. The translation includes Charles

Lorraine's account of the court proceedings of the Damascus Affair. Shivtiel states that such blood libels were "imported" to the Middle East from the time of Napoleon. Although the translator, Yusuf Hanna Nasr Allah, claims to be objective, he repeats antisemitic canards, such as that the Jews bribe officials in order to free themselves from punishment for ritual murder. Gives examples of a "wave" of antisemitic works published in Cairo, Beirut, and Damascus in the second half of the 19th century. Such works often compare ritual murders with the "murder" of Jesus. *YC*

Italy

0165 Fontana, Laura; Giovagnoli, Giorgio, eds.: *I nemici sono gli "altri": Convegno sull'Olocausto*. Firenze: Giuntina, 1999. 209 pp.

A collection of papers delivered at a conference in Rimini, February 1999. Partial contents: Luzzatto Voghera, Gadi: Lo stereotipo dell'ebreo nell'Italia contemporanea (31-41); Giovagnoli, Giorgio: I giovani di oggi tra indifferenza e disinformazione: Risultati di un sondaggio sullo sterminio svolto nelle scuole superiori riminesi [Pp. 65-122 present the questionnaire and results.] (43-122); Goldstaub, Adriana Ventriglia: Emarginazione e pregiudizi contemporanei (123-132); Pezzetti, Marcello: Dei campi di concentramento e dei lager della morte (133-144); Fontana, Laura: Il destino dei bambini sotto il Terzo Reich [Deals, also, with the fate of Jewish children in the concentration camps.] (151-165); Colombo, Furio: Memoria dell'Olocausto: Perchè preservarla (173-181). *MZ*

Latin America

0166

בז'רנו, מרגלית ואחרים, עורכים: יהדות אמריקה הלטינית: קובץ מאמרים לכבוד חיים אבני. ירושלים: אמיל"ט - אגודת מחקר יהדות לטינו-אמריקה; המכון ליהדות זמננו, האוניברסיטה העברית, 2001. 343 עמ'.

[Bejarano, Margalit et al., eds.: *Latin American Jewry: Essays in Honor of Haim Avni*. Jerusalem: Amilat – Research Association for Latin American Jewry; Institute of Contemporary Jewry, Hebrew University, 2001. 343 pp.]

תוכן חלקי: מילגרם, אברהם: האם התיישבות יהודית-חקלאית בברזיל בתקופת השואה הייתה אפשרית? (113-97); בן-דרור, גרסיאלה: ועידת אוויאן: העיתונות הקתולית הארגנטינית ועיצוב דעת הקהל (140-127); בקל, אליסיה גוחמן דה: מיעוטים, מדינה ותנועות לאומניות במעמד הבינוני של מקסיקו: ליגה אנטי-סינית ואנטי-יהודית (המאה העשרים) (156-141); בוקסר דה ליוורנט, יהודית: מקורות ללגיטימציה של הנוכחות

היהודית במקסיקו: הצבעתה של מקסיקו בעד המשוואה ציונות=גזענות והשפעתה על הקהילה היהודית (184-160).

Partial contents: [Milgram, Avraham: Was Jewish Agricultural Settlement in Brazil Possible during the Holocaust? [A Portuguese version was presented at the 10th World Congress of Jewish Studies (1989).] (97-113); Ben-Dror, Graciela: The Evian Conference: The Argentinian Catholic Press and the Shaping of Public Opinion [Appeared in Spanish in "Judaica Latinoamericana" 2 (1993).] (127-140); Backal, Alicia Gojman de: Minorities, State and Nationalist Movements among the Middle Class of Mexico: The Anti-Chinese and Anti-Jewish League (20th Century) [A Spanish version was presented at the 9th World Congress of Jewish Studies (1985) and appeared in "Judaica Latinoamericana" 1 (1988).] (141-156); Bokser de Liwerant, Judit: Sources for Legitimization of the Jewish Presence in Mexico: Mexico's Vote for the Equation of Zionism=Racism and Its Influence on the Jewish Community [Appeared in Spanish in "Judaica Latinoamericana" 3 (1997).] (160-184).] *RE*

Poland

0167 Cooper, Leo: *In the Shadow of the Polish Eagle: The Poles, the Holocaust and Beyond*. Basingstoke, Hants: Palgrave, 2000. xiv, 255 pp.

A study of antisemitism in Poland in the 20th century. Describes the increase in antisemitism between 1918-39. Reflects on the question of the role of Polish antisemitism in the Holocaust. During the war, most Poles either collaborated with the Nazis or at least approved of the economic measures against the Jews. The rightist anti-German resistance (e.g. Narodowe Siły Zbrojne) regarded Jews as enemies of Poland and sometimes cooperated with the Nazis in the murders. The Armia Krajowa was also antisemitic, and sometimes killed Jews. Many Poles profited from the Holocaust. The Catholic Church was silent. Discusses, also, the Poles who rescued Jews and their motivations. After the war there was a wave of murders of Jews who had survived the Holocaust or were repatriated from the USSR. Antisemitism remained a political factor in Poland both in the communist and post-communist periods. *DR*

0168 Tonini, Carla: *Operazione Madagascar: La questione ebraica in Polonia, 1918-1968*. Bologna: CLUEB, 1999. 285 pp.

Examines relations between Poles and Jews from the end of World War I until 1968. States that the combination of traditional Catholic antisemitic stereotypes, rooted in the mentality of the Polish people, the recurrent economic crisis in an overpopulated country, and the fragile and changing international position of Poland were the fertile ground for the rise of hostile attitudes toward the Jews. Asserts that antisemitism was used by the Right and Left as a weapon for their political struggles. Focuses on the "Madagascar" project, a plan of the nationalist right to achieve the emigration of a million Jews from Poland in 1936. Nothing ever came of this plan. Pp. 135-192 deal with the fate of Polish Jews under Nazi and Soviet occupation, and pp. 193-269 deal with antisemitism under the postwar communist regime [the latter is a revised and enlarged version of an article which appeared in "Storia Contemporanea" 24 (1993)]. *MZ*

0169 Wapiński, Roman, ed.: *Polacy i sąsiedzi – dystanse i przenikanie kultur. Vol. II* [Poles and Their Neighbors – Cultural Distance and Mutual Influences. Vol. II]. Gdańsk: Wydawnictwo "Stepan design," 2001. 350 pp. At head of title page: Prace Zakładu Historii Myśli i Kultury Politycznej Instytutu Historii Uniwersytetu Gdańskiego.

A collection of essays on mutual relations between Poles and their neighbors in the 19th-20th centuries. Partial contents: Lewalski, Krzysztof: Antysemityzm czy antyjudaizm: Refleksje na temat stosunku chrześcijan wobec Żydów w XIX wieku na ziemiach polskich [Antisemitism or Anti-Judaism: Reflections on Christians' Attitudes toward Jews on Polish Territories in the 19th Century] (116-132); Skóra, Wojciech: Dramat Żydów polskich w międzywojennym Szczecinie na tle polityki polskiej i niemieckiej (1933-1938) [The Tragedy of the Polish Jews in Interwar Szczecin against the Background of Polish and German Policies (1933-1938)] (229-253); Berendt, Grzegorz: Żydzi jako podmiot i przedmiot wydarzeń polskiego Października 1956 roku [The Jews as Subject and Object of the Events of October 1956 in Poland] (268-315); Tebinka, Jacek: Stosunek dyplomacji Wielkiej Brytanii do wydarzeń 1968 roku w Polsce [The Attitude of British Diplomacy toward the Events of 1968 in Poland] (316-350). *HV*

Russia, USSR, CIS

(see also Ukraine)

0170 Gereben, Ágnes: *Antiszemitizmus a Szovjetunióban* [Antisemitism in the Soviet Union]. Budapest: PolgArt, 2000. 623 pp.

Pp. 15-70 deal with the roots of antisemitism in Russia before 1917 (the Black Hundred, the "Protocols of the Elders of Zion," the pogroms). The rest of the book deals with antisemitism in the USSR, basically until 1953. Relates to the pogroms during the Civil War, the closure of Jewish institutions and seizure of their assets by the state, the establishment of agricultural colonies for Jews in Crimea, and the Birobidjan plan. Discusses causes for the rise of antisemitism and describes the official attitude. Pp. 125-221 deal with the Holocaust in Soviet territories under

German occupation. The rest of the book deals with the Jewish Anti-Fascist Committee, the trials against its members in 1952, the Slánský trial, the Doctors' Plot, and the whole postwar campaign by the regime against Jews, including loss of jobs, arrests, and executions with or without trial. A numerus clausus policy also made both study and work difficult for Jews. *AV*

0171 Madievskii, Samson A.: Schutz von den Bolschewiki? Juden und die russische Revolution. *Tribuene* 158 (2001) 157-166.

Describes the oppression of Jews in tsarist Russia, which led many Jews to emigrate and some young Jews to join revolutionary movements. During World War I, the Russian army killed or expelled Jews near the front lines. Nevertheless, during the Revolution many Jews initially supported the Whites; but these rejected the Jews and removed them from their army. White Russian troops carried out pogroms (as did other troops, including Cossacks on the side of the Reds until the Bolsheviks established their authority and stopped them). Faced with the alternative of Bolshevism or pogroms, it was natural that Jews, even Jewish bourgeois, chose Bolshevism. *RW*

Sweden

0172 Andersson, Lars Mikael: *En jude är en jude är en jude...: Representationer av "juden" i svensk skämtpress omkring 1900-1930* [A Jew Is a Jew Is a Jew...: Representations of "Jews" in the Swedish Comic Press 1900-1930]. Diss. – Lunds Universitet, 2000. 621 pp. Unseen.

Argues that antisemitism was clearly evident in Sweden at the turn of the century. Through analysis of the comic press, which was a crucial element in mass culture and political life, demonstrates how the Swedish national identity was constructed in opposition to "the Jew." In the decades before and after 1900, antisemitism had very little to do with actual Jews in Sweden. It was rather a question of a need created by the process of modernization and of how the established antisemitic discourse identified the "Jew" as the villain. The comics provide insights into the complex processes by which national identities were formed; they also largely contributed to their creation. In comics, film and literature, the penetration of the antisemitic discourse is seen most clearly. "The Jew" is represented as the antithesis of "the Swede."

Ukraine

0173 Chernova, Lyubov E., ed.: *Yevreiskaya mysl skvoz veka: Sbornik nauchnykh trudov po iudaike, yevreiskoi istorii i kulture. Vol. 3* [Jewish Thought through the Ages: A Collection of Scientific Papers on Judaica, Jewish History and Culture. Vol. 3]. Dnepropetrovsk: Dnepropetrovskii Institut Sotsialnykh i Obshchinnykh Rabotnikov, 1999. 203 pp.

A collection of papers delivered at an international conference on Jewish studies held in Dnepropetrovsk. Partial contents: Shaikin, Iosif M.: Mezhnatsionalnye otnosheniya v zemledelcheskikh koloniyakh na yuge Ukrainy [Interethnic Relations in Agricultural Colonies in the Southern Ukraine. Mentions pogroms in Jewish colonies in the late 19th-early 20th centuries.] (131-137); Gentselman, Zh.: Religioznaya zhizn yevreyev Ukrainy nachala XX stoletiya [The Religious Life of Ukrainian Jews in the Early 20th Century. Mainly on the Beilis affair.] (144-149); Kozerod, Oleg V.: Viiskovi zlochyni armiyi Denikina proty yevreyiv na terytoriyi UNR u 1919 r. [The Military Crimes against Jews of Denikin's Army in the Area of the UNR in 1919] (170-175); Yelisavetskii, Ster Ya.: Vosstaniya v getto i lageryakh Zapadnoi Ukrainy [Uprisings in Ghettos and Camps of Western Ukraine] (182-188); Litvak, S.: Tragicheskaya uchast yevreyev Novozlatopolskovo raiona Zaporozhskoi oblasti v 1941-43 godakh (po materialam arkhivnovo fonda) [The Tragic Fate of the Jews of the Novozlatopol District of the Zaporozhye Region in 1941-43. On material in archival collections.] (188-191). *DR*

0174 Khiterer, Viktoriya Ya.: *Dokumenty po yevreiskoi istorii XVI-XX vekov v Kiyevskikh arkhivakh* [Documents on Jewish History in the 16th-20th Centuries in Kiev's Archives]. Kiev: Institut Iudaiki; Moskva: Mosty kultury, 2001. 223 pp.

Mentions, inter alia, collections of documents pertaining to tsarist anti-Jewish policies, the activities of the Black Hundred, the Beilis affair and other blood libels, pogroms in Ukraine during the Civil War period, the Holocaust, and the postwar anti-Jewish policies of the Ukrainian Soviet authorities. *DR*

Uruguay

0175 *Antisemitismo en Uruguay: Raíces, discursos, imágenes (1870-1940).* Montevideo: Trilce, 2000. 224 pp.

Contents: Feldman, Miguel: El antisemitismo en Uruguay durante los años treinta (9-30); Camou, María Magdalena: Nazismo en Uruguay (1930-1940) (31-59); Abend, Gabriel: Zola, los judíos y el Novecientos:

Repercusiones en Uruguay del affaire Dreyfus a través de la prensa (60-128); Aldrighi, Clara: La ideología antisemita en Uruguay: Su contexto católico y conservador (1870-1940) (129-224). *MZ*

USA

0176 Benson, Evelyn Rose: *As We See Ourselves: Jewish Women in Nursing.* Indianapolis: Sigma Theta Tau International Honor Society of Nursing, Center Nursing Pub., 2001. xii, 196 pp.

Ch. 8 (pp. 119-139), "As Others See Us: 'I Never Experienced Anti-Semitism' versus 'Thank God There Are No Jews among Us'," deals with antisemitism experienced by 20th-century American Jewish nurses, based on an informal survey. While many nurses did not encounter antisemitism in their professional lives, some did experience antisemitism in the form of prejudice, insensitivity, and discrimination. The latter reported on cases of hostility, ignorance, insensitivity (including jokes and lack of respect for religious needs), and stereotyping. Notes, also, instances of Christian prejudice in what some have referred to as "a Christian profession." Jewish nurses occasionally encountered quotas or rejection for jobs due to antisemitism. Many repulsed antisemitism by asserting their Jewishness. *YC*

0177 Korman, Gerd: When Heredity Met the Bacterium: Quarantines in New York and Danzig, 1898-1921. *Leo Baeck Institute Year Book* 46 (2001) 243-276.

The germ theory of disease accepted in medicine in the late 19th century coexisted in the public health arena with old-time convictions and practices. Many immigrants to the U.S., among them Jews from East Central Europe, were regarded as bearers of diseases inherent to them. The image of the Jewish immigrant as a biological threat to the American nation influenced the behavior of public health officials and was reinforced by their practices. Examines the quarantine policy applied to immigrants entering the U.S., which in the absence of effective medical tests and drugs was the main tool for preventing epidemics. Compares quarantines imposed on American soldiers returning from Cuba in 1898 with those imposed on Jewish immigrants. Unlike the soldiers, Jews were held responsible for any contagious illness discovered; all sorts of anti-Jewish expressions made their way into official reports. Examines, also, quarantines imposed by the U.S. in Danzig for prospective immigrants after World War I. In the reports of the American officials, Jews were the only group identified with contagious diseases and described as bearing threatening social, political, and economic characteristics. *DR*

1789-1918: General

0178 Hart, Mitchell Bryan: Racial Science, Social Science, and the Politics of Jewish Assimilation. *Isis* 90, 2 (June 1999) 268-297.

In the late 19th-early 20th centuries, when the majority of scholars spoke about Jewry in racial terms or as a sick and physically degenerate community, many Jewish authors resorted to the same discourse. However, while non-Jewish scholars used racism and a medical approach in order to buttress their antisemitism, Jewish social scientists used them for their own, pro-Jewish agendas. Zionists resorted to racial arguments in order to demonstrate that Jews cannot live with non-Jews. On the other hand, assimilationists used the concept of Jewry as a diseased body as a tool in their struggle against antisemitism, claiming that if the Jews had favorable conditions (like those in the USA) they would lose all signs of physical "degeneration." Anthropologists such as Franz Boas and Maurice Fishberg used this idea in order to refute arguments of antisemites and nativists in the USA who advocated immigration restrictions on Jews. *DR*

0179 Mack, Michael: The Metaphysics of Eating: Jewish Dietary Law and Hegel's Social Theory. *Philosophy & Social Criticism* 27, 5 (2001) 59-88.

Analyzes how "Jewishness" functions as a scapegoat for the apparently unbridgeable gap between spirit and matter in Hegel's social and aesthetic theory, which had roots in the idealistic philosophy of Kant. Hegel criticized the Jews as "materialistic" and connected this with their dietary laws, especially their avoidance of consuming blood. For the philosopher, this reflected their unwillingness to sacrifice themselves to something higher than themselves, in particular the state. Although he did not wish to exclude the Jews from the modern state, Hegel viewed the Jews as "immutable" and having no place in his idealistic conception of the body politic. Concludes that Hegel's anti-Judaism has the potential to lead to political and pseudo-scientific antisemitism. *YC*

0180 Zieger, Karl, ed.: *Emile Zola, "J'accuse...!" Réactions nationales et internationales*. [Valenciennes]: Presses Universitaires de Valenciennes, 1999. 161 pp.

Papers presented at an international colloquium held at the University of Valenciennes in March 1998. Contents: Zola, Emile: J'accuse...! Lettre à M. Félix Faure, Président de la République (13-22); Pagès, Alain: "J'accuse...!" Radiographie d'un article de presse (25-34); Oriol, Philippe: Les réactions politiques au "J'accuse!" (35-47); Delmaire, Danielle: Réceptions de "J'accuse!" au pays de "Germinal" (49-61); Cahm, Eric: "J'accuse...!" Vue du monde anglo-saxon, d'après la presse quotidienne anglaise et américaine (63-77); Zieger, Karl: "Marianne brise le miroir qui lui renvoie les rides de son visage": "J'accuse!" et la presse de langue allemande (79-95); Séverine: Impressions d'audiences [Reprint of a series of articles published in the Brussels journal "Petit Bleu" in February 1898 by Séverine, pen-name of Caroline Rémy Guebhard (1855-1929).] (97-148); Oriol, Philippe; Pagès, Alain: Bibliographie du centenaire (151-161). *HV*

1789-1918: Australia

0181 Hyams, Bernard: What Did They Think of the Jews? The Early Years of the Sydney "Bulletin." *Australian Jewish Historical Society Journal* 15, 4 (June 2001) 547-563.

The "Bulletin" commenced publication in January 1880. Examination of the issues published between 1880-85 shows that the early editions of the magazine displayed a lively interest in Jewish people and issues; its comments were even sympathetic. Later, however, attacks on local Jews as well as on world Jewry appeared which went well beyond innocent banter on ethnic and religious matters. The attacks focused on Jewish "racial" traits and aspects of socioeconomic prejudice (e.g. Jewish monetary greed and unscrupulousness). The "Bulletin"'s most strident antisemitism, at the turn of the century, has been linked to the Boer War, stemming from allegations that British actions were motivated by Jewish financial interests. States that this theme occurred even earlier, when a contingent of troops from New South Wales was sent to assist the British Expeditionary Force in Sudan in March 1885. From 1884 on, the motif of Jewish undue financial influence on the state appeared in the "Bulletin." *DR*

1789-1918: Europe

General

0182 Hrytsak, Yaroslav: A Ukrainian Answer to the Galician Ethnic Triangle: The Case of Ivan Franko. *Polin* 12 (1999) 137-146.

The Ukrainian poet and writer Ivan Franko (1856-1916) was often accused of being an antisemite. However, his views were in fact ambivalent – although his political statements and actions were at times filled with nationalist emotion and resentment, he never clearly articulated any anti-Jewish program, and he consistently struggled against xenophobia. Nevertheless, Franko, following Drahomanov, a Ukrainian political thinker, advocated limited integration of Jews, i.e. the granting of civil rights to those Jews who expressed solidarity with Ukrainian "popular ideals"; for the rest of the Jews, Franko supported a gradual and planned emigration. He thought that Ukrainians should support Jewish efforts to create their own independent state, and in return he expected the Jews not to impede the struggle for Ukrainian national liberation. *DR*

0183 Katsis, Leonid F.: Yevropeiskaya kulturnaya yudofobiya v zerkale "Shuma vremeni" Osipa Mandelshtama [European Cultural Judaeophobia as Reflected in Osip Mandelshtam's "The Noise of Time"]. *Vestnik Yevreiskovo Universiteta* 23 (2001) 77-118.

Examining Mandelshtam's autobiographical work "Shum vremeni" ("The Noise of Time," 1909), finds that some of its chapters are influenced by antisemitic writings popular in his time and by attempts to respond to them. Relates, in particular, to allusions to Wagner's "Judentum in der Musik" and to the article "Estrada" ("Musical Stage"), published in 1909 by the Russian German music critic Emil Medtner (under the pseudonym Volfing). Both Wagner and Medtner believed that German-Jewish musical synthesis, e.g. through Jews performing German musical works, was impossible. Mandelshtam disputed this thesis, referring, especially, to the same musicians (Jozef Hofman, etc.) against whom Medtner wrote. Influences of Otto Weininger and anti-antisemitic writings by Nordau and Buber may also be discerned in "The Noise of Time." Shows that the chapter "Judaean Chaos," in which Mandelshtam wrote degradingly about the synagogue and rabbi, described not an ordinary visit to a synagogue, but the ceremony of dedication of the St. Petersburg synagogue in 1893, at which the city's Jewish elite was present. *DR*

Austria

0184 Fellner, Guenter: Vom Judenhut zum Trachtenhut? Diskurse ueber Kleidung und Politik in Salzburg (1800-1900). *Archiv fuer Kulturgeschichte* 83, 2 (2001) 331-376.

Traces attitudes to folk costume as expressed in Salzburg journals at the turn of the 19th century, during the 1848 revolution, and at the turn of the 20th century, and their relationship to social attitudes and ethnic identity. In parallel, traces attitudes to Jews; they were banned from Salzburg until their emancipation in 1867. Thereafter, there was tension between their acceptance, with pressure for assimilation (expressed, inter alia, by their adoption of folk costume) on the one hand, and their growing exclusion and isolation on the other. Soon after the Anschluss in 1938, Jews were forbidden to wear folk costume. Pp. 353-360 trace the development of the priest Josef Anton Schoepf, editor of a Salzburg Catholic periodical and professor of ecclesiastical history and law at the University of Salzburg, from conventional anti-Jewish prejudice to militant opposition to antisemitism; he was one of the few prominent churchmen active in the Austrian Verein zur Abwehr des Antisemitismus. *RW*

0185 Furst, Lilian R.: Freud and Vienna. *Virginia Quarterly Review* 77, 1 (Win 2001) 49-62.

Discusses the conflicted relationship between Freud and Vienna, the city where he lived and worked for 77 years before leaving under Nazi threat in 1938. He remained an outsider, especially as a Jew. There was a higher percentage of Jews (11%) in Vienna than in any other Central European city. Antisemitism increased under Mayor Karl Lueger and due to the influence of Georg von Schoenerer. However, Jews were scapegoated even before they became disproportionately prominent in the professions and in cultural life. Some of the Jews converted. Freud adopted a secular ("konfessionlos") stance, but continued to identify as a Jew. While he was ambivalent about his adopted city, the latter was largely negative toward him. His pioneering works were either ignored or considered pornographic and dangerous to society. He was feared as a breaker of taboos. Vienna resented this Jewish outsider who dared to challenge its self-image as a city of culture by focusing on sexuality and other "base" aspects of life. *YC*

0186 Staudacher, Anna Lea: *Wegen juedischer Religion – Findelhaus: Zwangstaufen in Wien, 1816-1868. Vol. 1-2*. Frankfurt a.M.: Peter Lang, 2001. 496; 586 pp.

Discusses and analyzes forced conversion in Vienna between 1818-68. Argues that during this period more than 2,500 Jewish children born in the Viennese hospital Wiener Gebaerhaus, a maternity clinic, were separated from their mothers and converted to Christianity. Until 1843 the names of the children were changed; some were transferred to the Findelhaus (foundling hospital) without a name. Ca. 80% of these children died due to illness. When a Jewish mother wanted to see her child, she had to first show her own certificate of baptism. States that most of the mothers of these converted children were lower class; often they were impecunious and unmarried girls who came to Vienna to work as servants. Christians viewed the conversion of Jewish children against the wishes of their parents, practiced for centuries, as justified, for the salvation of "innocent souls." Forced conversions became illegal in 1868. Pp. 29-445 in vol. 2 contain a concordance of the mothers' names with biograpical information, and pp. 447-586 present a list of the names of the children. *ID*

Belarus

0187 Bukhovets, Oleg G.: The Image of Jews in Byelorussia: Petitions as a Source for Popular Consciousness in the Early Twentieth Century. *International Review of Social History: Supplement* 9 (2001) 171-184.

A study based on examination of 662 petitions sent to the tsar and the State Duma from the Belorussian regions during the First Russian Revolution in 1905-07, found in the archives of St. Petersburg, Moscow, and Belarus. Only 69 petitions, or 10.4%, touch upon or mention Jewishness directly, and another 24 address Jewishness implicitly. Five petitions express a positive attitude toward Jews, eight express a neutral attitude, and the rest may be characterized as antisemitic in various degrees. The core image of the Jew, as seen by the authors of the 69 petitions, is of a rich and successful person. Notes that the petitions that express the most negative attitude toward Jews express an even more negative attitude toward non-Jewish landlords, etc. The antisemitism expressed in the petitions is a form of hatred of everything alien and heterodox. Contends that the "silence" on Jews and Jewishness of the other 569 petitions shows that the general attitude toward Jews was neutral or indifferent, i.e. the level of antisemitism in Belorussia was low in this period. *DR*

Czech Lands

0188 Frankl, Michal: The Background of the Hilsner Case: Political Antisemitism and Allegations of Ritual Murder 1896-1900. *Judaica Bohemiae* 36 [2000] (2001) 34-118.

The Hilsner case in 1898 was the only court trial to take place in Central and Western Europe after 1867 which de facto confirmed the allegation of Jewish ritual murder and sentenced the defendant for such a crime. This was possible due to the conspicuous rise of antisemitism in the Czech Lands in 1896-1900. The antisemitic movement in the Czech Lands developed in the 1870s-80s and was represented by nationalist groups, the strongest being the Young Czech Party, and by the conservative Catholic movement and its main vehicle, the Christian Social Party. The Czech press also spread antisemitic ideas. Antisemitism was regarded by the nationalists and the conservatives as a slogan which could be used to unite the politically fragmented Czech society and to denigrate the rival Social Democratic Party. Antisemitism reached a peak in 1897, prior to and during the elections to the Reichsrat. The elections were followed by the emergence of new antisemitic organizations (among them the short-lived National Self-Defence League) and by anti-Jewish riots in Prague and elsewhere. This was the atmosphere in which the Hilsner case arose. *DR*

0189 Kieval, Hillel Joseph: *Languages of Community: The Jewish Experience in the Czech Lands*. Berkeley: University of California Press, 2000. xi, 311 pp.

Despite the growing assimilation of the Jews with Czech language and culture in the late 19th-early 20th centuries, many sectors of the Czech national movement regarded the Jews as a Germanizing element. Ch. 8 (pp. 181-197), "Death and the Nation: Ritual Murder as Political Discourse in the Czech Lands" [an earlier version appeared in "Jewish History" 10 (1996)], describes cases of blood libels in 1892-1902 and reflects on the role played by ritual murder discourse in the rhetoric of Czech nationalism. Concludes that the ritual murder accusation against the Jews, as well as belief in their sexual licentiousness, added psychological dimensions to the rhetoric of economic nationalism – e.g. accusations of economic exploitation and "linguistic treason." Ch. 9 (pp. 198-216), "Masaryk and Czech Jewry: The Ambiguities of Friendship," disputes the myth of Masaryk as a staunch opponent of antisemitism. Throughout his career Masaryk could not overcome antisemitic prejudices adopted in his childhood; he paradoxically combined support for the Jews of Bohemia with expressions of resentment and distrust. *DR*

0190 Lapáček, Jiří: Protižidovské bouře v Kojetíně v roce 1887 [Anti-Jewish Riots in Kojetín in 1887]. *Židé a Morava: Sborník příspěvků přednesených na konferenci konané 11. listopadu 1998 v Kroměříži* [The Jews and Moravia: A Collection of Contributions to the Conference Held on 11 November 1998 in Kroměříž], ed. Petr Pálka. Kroměříž: Muzeum Kroměřížska, 1999. Pp. 3-16.

Describes the events that followed the disappearance of a young Christian girl named Marie Krybusová from Kojetín, Moravia, on 24 April 1887. The Jewish community was accused of ritual murder; a small crowd of people caused some damage in the Jewish quarter. After the body of the girl was found in the river, the possibility of a criminal act was rejected; it was thought that she committed suicide. The incident continued to incite unrest amongst the populace. When more violence erupted, a small army unit was temporarily stationed in the town. As a result, 50 people involved in the riots were convicted by the district court in Olomouc. Some minor incidents followed three years later, accompanied by social unrest. Another ritual murder accusation arose in the town in March 1918. *JT*

0191 Spyra, Janusz: Jewish Rights of Residence in Cieszyn Silesia, 1742-1848. *Polin* 14 (2001) 31-48.

Discusses the history of restrictions on Jews, especially in terms of residence and marriage, in the southern part of Silesia between 1742-1848. While the Habsburg rulers were often hostile to any Jewish presence, they agreed to a restricted presence for financial purposes (e.g. taxes and economic franchises). Local authorities tended to be more flexible, also for economic reasons. Traces the history of restrictive legislation, and its repeal or disregard, and notes its effect on the Jewish community in terms of religion (scattered Jews were unable to create communal institutions) and on family life. A basic goal of the legislation was to restrict the size of the Jewish population. With the revolutions of 1848, special Jewish taxes and limitations on residence and marriage were abolished. However, Franz Joseph I revoked these changes; only after 1859 did the Jews of Cieszyn Silesia have their equal rights confirmed. *YC*

France

0192 Cosnier, Colette; Hélard, André: *Rennes et Dreyfus en 1899: Une ville, un procès*. Paris: Horay, 1999. 399 pp.

A study of the repercussions on the city of Rennes of the second court martial of Dreyfus which took place there in August 1899. A conservative, bourgeois city of 70,000 inhabitants, Rennes was considered an unfortunate choice by the populace; only a handful of intellectuals, students, and workers supported Dreyfus. Recounts how the demonstrations, presence of celebrities, secret agents, and local and foreign journalists affected life in the city. Describes various expressions of antisemitism, hatred between Dreyfusards and anti-Dreyfusards, and exceptional actions taken by the Dreyfusards grouped around Victor Basch and the Ligue des Droits de l'Homme. The relationships between politics, justice, spectacle, business, news, public opinion, and new press practices were brought into play in the summer of 1899, turning Rennes into a microcosm foreshadowing attitudes and behavior typical of the 20th century. *EO*

0193 *Dreyfus & Zola: A Moment in the Conscience of the World: An Exhibition from the Lorraine Beitler Collection*. [Toronto]: University of Toronto Art Centre, 2000. 47 pp.

The catalogue of an exhibition held at the University of Toronto Art Centre, 15 September-9 December 2000. Contains facsimiles from newspapers, posters, books, postcards, etc. about the Dreyfus Affair and the role of Emile Zola. Many of the depictions are antisemitic. An essay by Lorraine Beitler (pp. 11-21), "The Media, Art and the Imperatives of Social Justice," emphasizes the role of the media – uncensored newspapers (some of which were strongly antisemitic), caricatures, and even photography – in either espousing justice or berating Dreyfus and the subversive values he was taken to represent. *YC*

0194 Dreyfus, Alfred: *Carnets (1899-1907): Après le procès de Rennes*. Ed.: Philippe Oriol. Paris: Calmann-Lévy, 1998. ix, 468 pp.

Presents the original version of Dreyfus' memoirs which chronicle the period from the second indictment until the official pardon. Justifies the publication of the final two chapters, relating to dissent within the Dreyfusard camp, which were deleted in all previous versions. Believes that the five notebooks which comprise the unexpurgated version of Dreyfus' "Cinq années de ma vie" could enlighten the public about the reasons for the inevitable split between those who viewed Dreyfus as a person vs. those for whom he was a symbol of the struggle for justice in France. Pp. 27-285 contain the unabridged text by Dreyfus. Pp. 287-438 include the editor's explanatory notes. *TS*

0195 Guiffan, Jean: *La Bretagne et l'affaire Dreyfus*. Rennes: Terre de Brume, 1999. 271 pp.

Examines the impact of the Dreyfus Affair in Brittany between 1894-1906, based on press reports. A reputedly monarchist and reactionary rural province with a strong landed aristocracy and influential Catholic clergy, Brittany was divided between an overwhelming anti-Dreyfusard majority and a small cautious pro-Dreyfusard group that began to develop only after the appearance of "J'accuse" in January 1898. While many of the antisemitic anti-Dreyfusard demonstrations in Nantes, Rennes, and smaller Breton towns from 1898 onward degenerated into violence against Jewish establishments and even individuals, the first local branch of the Ligue des Droits de l'Homme was established in Rennes in January 1899. States that although antisemitism was common in Brittany well before the Affair, the local population often sided against Dreyfus for patriotic rather than antisemitic reasons. Concludes that it is time to incorporate the Affair into local monographs where its memory has been repressed. *EO*

0196 Oriol, Philippe, ed.: *Bernard Lazare, anarchiste et nationaliste juif*. Paris: Honoré Champion, 1999. 373 pp.

A collection of articles about Bernard Lazare and texts and correspondence by him. Partial contents: Oriol, Philippe: Bernard Lazare anarchiste [Discusses, also, the Dreyfus Affair as an important step in Lazare's evolution, and Lazare's anarchistic "antisemitism."] (17-101); Sandrel, Carole: Bernard Lazare journaliste et anarchiste (127-149); Iancu, Carol: Bernard Lazare et les Juifs de Roumanie [On antisemitism in Romania and Lazare's writings in favor of granting rights to Romania's Jews.] (181-191); Correspondance avec Edouard Drumont (227-240); Mémoire de Bernard Lazare sur ses activités pendant l'affaire Dreyfus (241-268). *HV*

Germany

0197 Armbrecht, Sabine: "Die Loesung der Judenfrage": Der Weg zur Sozialdemokratie – Julius Moses' "Juedische Epoche" (1902-1910). *Sachor* 11 (2001) 63-77.

Moses (1868-1942) was a physician, well-known for his championship of social medicine, and as a speaker, writer, and member of the Reichstag for the Social Democrats. He was also concerned all his life with the "Jewish question." Between 1902-10 he founded, published, and edited the "General-Anzeiger fuer die gesamten Interessen des Judentums," which dealt with every aspect of Jewish life and culture, but especially with the demand that Jews be accepted as equal citizens without having to give up their Jewishness. In 1907 he also published "Die Loesung der Judenfrage: Eine Rundfrage," a compilation of the replies of prominent Germans whose views he had solicited. Politically, he was first a Liberal, but became disillusioned with the Liberals' antisemitism and concluded that only the Social Democrats offered a solution to the "Jewish question." In 1910 he joined the party and gave up the editing of the "General-Anzeiger" in favor of political activity. The Nazis deported him in 1942 to Theresienstadt, where he died soon after. *RW*

0198 Bauer, Alfredo: Die Stellung der Juden in der ersten Haelfte des XIX. Jahrhunderts und Heines Uebertritt zum Christentum. *Heine-Jahrbuch* 39 (2000) 184-191.

Contends that in Germany (unlike Eastern Europe), Jewish self-emancipation from outdated and sterile Orthodoxy took place not within Jewish culture but through assimilation to German culture; Heine exemplifies this process. Suggests that assimilation should have resulted in acceptance; that this did not happen was due to economic competition in a bourgeois society that had begun to decay before it even reached full fruition, and thus could not meet the needs of all its members. Notes that antisemites opposed assimilation of Jews as part of their opposition to emancipated bourgeois society. *RW*

0199 Ferrari Zumbini, Massimo: *Le radici del male: L'antisemitismo in Germania da Bismarck a Hitler*. Bologna: Il Mulino, 2001. 1128 pp.

Describes the rise of antisemitism in Imperial Germany. States that the emancipation of the Jews as well as their economic, social, and political improvement in the Bismarck era, was counterbalanced by the growth of antisemitic movements which tried to win a place in the parliament in order to turn German policy and legislation against the Jews. They did not succeed, but they left their imprint on German society. Theodor Fritsch (1852-1933) understood that for antisemitism to succeed a long process was necessary, and that process had to evolve outside of parliament. He published the antisemitic newspaper "Der Hammer," synthesized antisemitic ideology in the "Handbuch der Judenfrage" (1907), and dedicated his energies to unifying the scattered antisemitic groups. He updated the traditional hostile image of the Jews from the eternal, wandering Jew, strange to society and culture, to the "mutant Jew," one who is able to assimilate. He equated these Jews to viruses, spreading an irrational fear of Jews. This new view of the Jew was augmented by the influx of millons of Eastern European Jews inundating Germany on their way to America. *MZ*

0200 Fischer, Jens Malte: *Richard Wagners "Das Judentum in der Musik": Eine kritische Dokumentation als Beitrag zur Geschichte des Antisemitismus.* Frankfurt a.M.: Insel Verlag, 2000. 380 pp.

Wagner's essay of late 1850 took up a debate that had been waged in German music periodicals all that year; he also used many of the concepts and terms of his predecessors. What was new in his essay was a "proto-racism," an implication that the Jews' alleged negative qualities were fixed in their nature. Wagner was influenced by the general atmosphere among German intellectuals and by his failures in Paris, which he blamed on the domination of musical life by Jews and especially by Meyerbeer. In 1850 the essay attracted little attention. In 1869 Wagner republished it, this time with a preface and a long postscript. By then, he had developed a paranoid belief in a Jewish conspiracy against him. The republication was greeted by a storm of protest. But ten years later the antisemitic movement honored Wagner as its precursor, though he distanced himself from it because of its vulgarity. Assesses the influence of the essay up to and through the Nazi period and even afterwards. Pp. 139-196 contain the 1850 version of the essay, with footnotes indicating the changes in the 1869 version, together with the preface and postscript of the 1869 version. Pp. 197-356 contain reactions of Wagner's contemporaries. *RW*

0201 Hartston, Barnet Peretz: *Judaism on Trial: Antisemitism in the German Courtroom (1870-1895)*. Diss. – University of California, San Diego, 1999. 393 pp. Unseen.

The decades after German unification were characterized by a few sensational antisemitic trials as well as smaller cases involving public discussion of the "Jewish question." Based primarily on newspaper accounts, government files, and court documents, examines the nature of these trials, assesses their role in German politics, and evaluates their effect on the overall development of German antisemitism. A majority of the cases were slander trials involving prominent Jews and antisemitic agitators. These trials provided individual agitators with a valuable public forum for spreading their own political propaganda. Also discusses the use of ritual-murder trials and many other incidental criminal proceedings as anti-Jewish propaganda. Frequent discussion of the trials in the national press kept public attention focused firmly on the "Jewish question." Such cases allowed the otherwise small antisemitic parties and political press to exert a disproportionate influence on the course of German politics.

0202 Hein, Lorenz: Gabriel Riesser und sein Kampf um die Gleichstellung der Juden. *Freiburger Rundbrief* 8, 1 (2001) 5-12.

Sketches the life and career of Riesser (1806-1863) and discusses his tracts for the civic equality of the Jews as human beings and as Germans, tracing the origins of his ideas to Luther (freedom of conscience), Lessing, and Herder. Against irrational hatred, he called for love of one's fellow man; against the demand for assimilation he called for respect for Jews who remained loyal to their religion. *RW*

0203 Hirsch, Erika: Assimilation und Antisemitismus im Kaiserreich. *Schriften der Erich-Muehsam-Gesellschaft* 21 (2001) 136-148.

Describes the normality of assimilation in Wilhelmine Germany, noting that it could coexist with Jewish religion and the consciousness of being Jewish. However, this consciousness was heightened by growing antisemitism, causing especially younger Jews to leave German organizations in which they were no longer welcome and join Jewish-national ones. Traces the history of antisemitism in Germany before the First World War, commenting that although no Jew could escape it, its dangerous consequences were not recognized and it never diminished the patriotism of the prewar generation and its identification with fatherland and Kaiser. *RW*

0204 Hoffmann, Stefan-Ludwig: Brothers or Strangers? Jews and Freemasons in Nineteenth-Century Germany. *German History* 18, 2 (2000) 143-161. Translated from German: "Juden, Buerger, Deutsche: Zur Geschichte von Vielfalt und Differenz, 1800-1933" (2001).

Between the end of the 18th century and 1914, the universalist and inclusive discourse of the Enlightenment turned into the exclusive discourse of nationalism and antisemitism. This process can be demonstrated in relation to the 19th-century Masonic lodges. The early lodges were conceived as fraternities, disregarding social cleavages and thus unifying humankind. Even then, many lodges opposed Jewish membership because of alleged Jewish "lack of sociability." The 1840s-70s saw a gradual increase in the participation of Jews in the lodges. From the 1880s to the 1910s an antisemitic, albeit not racist, mood permeated the lodges – religious Jews could not join, and potential Jewish members were expected to undergo a cultural conversion to Christianity. The number of Jewish Freemasons then decreased. *DR*

0205 Kampmann, Christoph: Protest gegen die Obrigkeit? Zur Deutung der judenfeindlichen Unruhen waehrend des Vormaerz. *Historisches Jahrbuch* 121 (2001) 471-500.

Argues that the "Hep!-Hep!" riots of 1819 were protests against Jewish emancipation; i.e. Jews were not, as other historians have claimed, scapegoats in a general revolt against state authority; rather, there was opposition to state authority when it favored emancipation of the Jews. The Congress of Vienna left emancipation to the decision of the individual Laender; the riots broke out in those places that had begun to lift the restrictions on Jews. Demonstrates this process in the case of Wuerzburg, Frankfurt, and Hamburg. Quotes contemporaries who almost unanimously attributed the riots to opposition to emancipation. Denies the relevance of comparison with England because of the negligible Jewish population there; a more relevant comparison would be with the riots against Papists in England. *RW*

0206 Pfahl-Traughber, Armin: Aufklaerung und Antisemitismus: Kants, Lessings und Fichtes Auffassungen zu den Juden. *Tribuene* 158 (2001) 168-181.

Examines the attitudes of three representatives of the German Enlightenment toward Jews. Kant's remarks are marginal: he rejects Judaism as a religion of revelation, not reason, and really not a religion at all; and he notes as a strange fact that the Jews, especially in Poland, are a nation mostly of merchants and known for cheating. Lessing, in his dramas, pleads for mutual respect between Christians and Jews and respect for their religions. Fichte, in his Enlightenment phase, sees the Jews as a state within the state, and the mightiest enemy of the peoples amongst whom they live and whom they exploit, because of their dual morality. They must by no means be granted citizenship rights, unless it were possible "to cut off their heads and replace them with others in which there is not one Jewish idea." In conclusion, points out that all of the hostile references to Jews by thinkers of the Enlightenment are incidental; none wrote a work expressly dedicated to this theme. They upheld traditional prejudices, even though these conflicted with the values of the Enlightenment. *RW*

0207 Puetz, Peter: Nietzsche und der Antisemitismus. *Nietzsche-Studien* 30 (2001) 295-304.

Interprets a passage from "Die Geburt der Tragoedie," in which Nietzsche relates the Aryan myth of Prometheus to the Jewish myth of the Fall of

Man, as implying the inferiority of Judaism. Notes that Hitler also refers to the Prometheus myth in a similar vein. Yet Nietzsche wrote scathingly against antisemites. Explains that the early Nietzsche, under the influence of the mentality of his time and the antisemitism of his mentors Schopenhauer and Wagner, was indeed antisemitic; but he soon saw the vulgar stupidity of antisemitism, and it was one of the causes of his break with Wagner. Also discusses the opposition between Athens and Jerusalem, and Nietzsche's physiological characterization of antisemitism as a "disease." *RW*

0208 Ruerup, Miriam: Juedische Studentenverbindungen im Kaiserreich: Organisationen zur Abwehr des Antisemitismus auf "studentische Art." *Jahrbuch fuer Antisemitismusforschung* 10 (2001) 113-137.

The Jewish student fraternities were founded in the 1890s, after the German fraternities increasingly excluded Jews; their original purpose was to combat antisemitism and maintain Jewish honor. Jewish students seeking integration into German society joined the Kartell-Convent (KC); those aspiring to greater national-Jewish identity joined one of two organizations with Zionist tendencies. All adopted the customs of the German fraternities, particularly dueling; challenge to a duel was considered essential to the maintenance of Jewish honor. The Jewish-national fraternities also represented Jewish interests in campus politics. Remarks that day-to-day internal fraternity affairs increasingly took precedence over the fight against antisemitism. *RW*

0209 Ruerup, Miriam: *Juedische Studentenverbindungen im Kaiserreich von ihrer Gruendung bis zum Ersten Weltkrieg – zwischen Anpassung und Eigenstaendigkeit.* M.A. thesis – Technische Universitaet Berlin, 1999. 137 pp.

Jewish fraternities were founded in reaction to the exclusion of Jewish students from German fraternities in the 1890s. Both Zionist and non-Zionist fraternities adopted the traditional behavior of German students' societies, including their customs of drinking and dueling. The Jewish fraternities were founded due to increasing antisemitism; however, the Jewish students adopted, of all things, the rituals of that part of German society which was responsible for the diffusion of antisemitic ideology as a cultural code. *ID*

0210 Schramm-Haeder, Ulrike: *Jeder erfreuet sich der Gleichheit vor dem Gesetze, nur nicht der Jude: Die Emanzipation der Juden in Sachsen-Weimar-Eisenach (1823-1850).* Muenchen: Urban & Fischer, 2001. 298 pp. Based on the author's diss. – Friedrich-Schiller-Universitaet Jena, 2000.

Discusses the process of acquisition of equal rights for the Jews in the Grand Duchy Sachsen-Weimar-Eisenach between 1823-50. The way to emancipation was complex, influenced by decisions on various levels, e.g. by the government, parliament, administration, and local authorities. In 1823 a "Juden-Ordnung" was adopted, which accorded some rights to Jews; however, it was restricted to the domains of trade and authorization of marriage. Jews had to fight for amelioration of their rights, mainly by recourse to juridical debate. The revolution in 1848 had a positive influence on the situation of the Jews and led in 1850 to civic equality. *ID*

0211 Sudhoff, Arne: Agitation und Mobilisierung laendlicher Bevoelkerung im ausgehenden 19. Jahrhundert: Die kurhessische Zeitung "Reichsherold" im Schnittpunkt von Antisemitismus und Agrargesellschaft. *Aschkenas* 11, 1 (2001) 87-120.

The "Reichsherold" was founded by Otto Boeckel in 1887 to support his campaign for election to the Reichstag on an antisemitic, pro-peasant platform, and continued as his organ until the end of his political career in 1894. Boeckel soon passed from general articles on the baneful influence of the Jews, especially on the peasantry, to a "practical antisemitism": guidance on the part of the organization of cooperatives to free peasants from the grasp of Jewish traders and moneylenders. The newspaper also published serial novels that conveyed the same message. *RW*

0212 Volkov, Shulamit: *Das juedische Projekt der Moderne: Zehn Essays*. Muenchen: C.H. Beck, 2001. 246 pp.

A collection of essays, all published previously and revised for this volume, dealing mainly with the acculturation of German Jews during the Wilhelmine period. The following deal with antisemitism: Nationalismus, Antisemitismus und die deutsche Geschichtsschreibung [Appeared in "Nation und Gesellschaft in Deutschland" (1996).] (49-61); Antisemitismus und Antifeminismus: Soziale Norm oder kultureller Code [Appeared in Hebrew in "Zmanim" 46-47 (1993).] (62-81); Sprache als Ort der Auseinandersetzung mit Juden und Judentum in Deutschland, 1780-1933 [Appeared in "Juedische Intellektuelle und die Philologien in Deutschland, 1871-1933" (2001).] (82-96). *RW*

0213 Wawrzyn, Heidemarie: *Vaterland statt Menschenrecht: Formen der Judenfeindschaft in den Frauenbewegungen des Deutschen Kaiserreiches*. Marburg: Diagonal-Verlag, 1999. 275 pp. Based on the author's diss. – Universitaet Bremen, 1998.

Describes the aims and the organizational history of each of the main women's organizations in Wilhelmine Germany. Most of these characterized themselves as "German and Christian." While Jewish women held leadership positions in most of them, their Christian "sisters," often intimate friends in private life, ignored their identity as Jews and demanded the same of them. The Juedische Frauenbund was likewise never mentioned with the other confessional associations. Women proud of their Jewishness protested against this attitude, though they did not see it as antisemitic. Subtle antisemitism also appears in the evasiveness of Christian leaders when asked to defend Jewish colleagues against antisemitic attacks. Suggests that the organizations' early abandonment of a feminism based on equal rights, in favor of cultivation of the special qualities of women, paved the way for other types of discrimination such as antisemitism. *RW*

Italy

0214 Fuellenbach, Elias H.: Nachgedanken zur Seligsprechung von Pius IX. *Freiburger Rundbrief* 8, 1 (2001) 28-32.

Explains the indignation aroused in both Jews and non-Jews by the recent beatification of Pope Pius IX (1792-1878). Not only was he anti-Jewish, in the Catholic theological tradition, but he also had words of praise for supporters of the new, more virulent type of antisemitism; in sermons and speeches he accused the Jews of cupidity and of responsibility for revolution and liberalism. The climax of his persecution of the Jews was his condonement of the abduction of the six-year-old Jewish child Edgardo Mortara who was then raised in the Church, in the face of protests from around the world and even of Catholic theologians. *RW*

0215 Gentile, Emilio: The Struggle for Modernity: Echoes of the Dreyfus Affair in Italian Political Culture, 1898-1912. *Journal of Contemporary History* 33, 4 (Oct 1998) 497-511.

An earlier version was presented at the international conference "L'Affaire Dreyfus cent ans après," Hebrew University, November 1994. During 1898-99 Italy underwent a serious economic, social, and political crisis which seemed to threaten the progress of a liberal democracy in the making. At the beginning of the 20th century, Italy renewed its march towards democracy. Events in Italy influenced the way in which the Dreyfus case was perceived; the myth of modernity was the main theme that, either directly or indirectly, was at the center of the considerations of Italian political culture regarding the Dreyfus case. Examines analyses published in the Italian press of the time. Most commentators considered antisemitism one of the main reasons for the Dreyfus Affair. *SSC*

0216 Mattioli, Aram: Nichts als Hunde: Ein besonders duesteres Kapitel der neueren Kirchengeschichte: Bis 1870 zwang der Papst Roms Juden, im

Ghetto zu leben. *Evangelischer Arbeitskreis Kirche und Israel in Hessen und Nassau: Materialdienst* 2 (Apr 2000) 15-17.

From 1555 the Jews of Rome had been forced by the popes who ruled in the Papal States to live in a ghetto under conditions among the worst in Europe. After 18 years of emancipation by the French (1796-1814) they were forced back into the ghetto, subject to laws that barred them from all but the lowest-paying occupations, segregated them from Christians, and required them to attend sermons for their conversion. Pius IX, in his first two years as pope, eased many restrictions and pulled down the ghetto walls. But after 1848 he became reactionary: he reimposed the restrictions, reerected the ghetto walls, and intensified the pressure for conversion. The Mortara affair aroused protests throughout the Western world and helped bring about the end of the pope's temporal power. In 1870 the Italian revolutionary forces conquered Rome and emancipated the Jews. *RW*

0217 Salvadori, Roberto G.: *1799: Gli ebrei italiani nella bufera antigiacobina.* Firenze: Giuntina, 1999. 124 pp.

The invasion of northern and central Italy by Napoleon's army in 1796 resulted in an exacerbation of the chronic economic crisis of northern Italy and in a deterioration of the already harsh living conditions of the Italian peasants. Between 1796-99 several uprisings took place throughout the country against the occupiers. The Jews, who had been favored by the French with dissolution of the ghetto and the granting of civil rights, were seen as allies of the French. They became the object of the peasants' hatred, fueled by traditional negative stereotypes about Jews. Several Jewish communities suffered attacks; in some places, Jews were murdered and their property was destroyed. Only in 1800 did the French succeed in quelling the uprising and restoring order. *MZ*

0218 Salvadori, Roberto G.: *La notte della rivoluzione e la notte degli orvietani: Gli ebrei di Pitigliano e i moti del "Viva Maria" (1799).* Pitigliano: Comune di Pitigliano, Assessorato alla Cultura, 1999. 83 pp.

An account of the events leading to an attack against Jews in the town of Pitigliano (Tuscany) on 28 June 1799. The town was involved in the peasants' uprising (named "Viva Maria") against the French occupation. The Jews were viewed as sympathizers with and having the protection of the French, from whom they gained civil rights. On that day, the mob killed 13 Jews. Pp. 49-66 contain original documents related to the events. Pp. 67-82 present a hymn in remembrance of the episode, composed by the Rabbi of Pitigliano, Mose Israel Urbino. *MZ*

0219 Schaechter, Elizabeth Mahler: Perspectives of Nineteenth-Century Italian Jewry. *Journal of European Studies* 31, 1 (Mar 2001) 29-69.

Challenges the view that modern political antisemitism had no repercussions in Italy. Based mainly on reactions to antisemitism in the Trieste Italian Jewish journal "Il Corriere Israelitico," discusses the reaction of the Catholic Church to Jewish emancipation. Mentions the Jesuit journal "La Civiltà Cattolica," which in the 1880s featured the topic of ritual murder. The Church encouraged anti-Jewish political activity. Pp. 54-59 discuss the appeal of Zionism to a number of Italian Jews, mainly as a reaction to antisemitism elsewhere in Europe. *YC*

Poland

0220 Blobaum, Robert: The Politics of Antisemitism in Fin-de-Siècle Warsaw. *The Journal of Modern History* 73, 2 (June 2001) 275-306.

The industrialization of Russian Poland, especially of Warsaw in 1890-1914, brought about the new politics of antisemitism which announced itself in 1912 in the form of a nationalist-led boycott of Jewish-owned shops and trade in Warsaw. The boycott was primarily a political event driven by the social position of two key groups largely centered in Warsaw: the Polish intelligentsia, radicalized by labor market competition and blocked upward social mobility, and the Polish artisan middle class, increasingly overshadowed by larger producers of mainly non-Polish ethnicity. The working class and peasantry, as well as cities other than Warsaw, ignored the boycott, which highlighted the failure of Polish liberalism. The first calls for the boycott came from the radical nationalist camp (Endecja), but it was supported by the clergy and some liberals; the left, both liberal and socialist, remained aloof. The boycott dealt a final blow to Jewish assimilation and boosted Jewish nationalism, especially Zionism. *DR*

0221 Oppenheim, Israel: The Radicalization of the Endecja's Anti-Jewish Line during and after the 1905 Revolution. *Shvut* 9 [25] (2000) 32-66. Translated from the Hebrew: "Gal-Ed" 15-16 (1997).

Russia

(see also Belarus)

0222 De Michelis, Cesare G.: *La giudeofobia in Russia: Dal "Libro del 'kahal'" ai "Protocolli dei savi di Sion": Con un'antologia di testi.* Torino: Bollati Boringhieri, 2001. 223 pp.

Presents the text of the "Protocols of the Elders of Zion" (1903) and nine other antisemitic texts – by Yakov Brafman, Mikhail Menshikov, Pavolaki Krushevan, Georgii Butmi, and Aleksei Shmakov, and three anonymous works. The introduction (pp. 13-46) states that the "Protocols" reflects the characteristics of the Russian antisemitic subculture which crystallized in the second half of the 19th century. It combines the perception of the Jews as a mysterious, esoteric people with an earlier Russian belief that the world is ruled by a superior force (i.e. God). The waves of secularism in the second half of the 19th century produced a profane version of this belief, attributing to the Jews (masters of mystery and esotericism) the role of the superior force now ruling the world. *MZ*

0223 Freiman, Naum: Yevrei v tsarskoi armii [Jews in the Tsarist Army]. *Istoki* 5 (1999) 99-122.

From the period of the campaign against Napoleon in 1812, Jews showed themselves to be loyal to the Russian government. Despite this, brutal legislation on compulsory military service for Jews was enacted in the epoch of Nicholas I. Conversion to Christianity did not always lift all the restrictions from the Jewish soldiers. The Great Reform of the 1860s alleviated the situation of the Jews in the Russian Army, but in the 1880s discrimination of Jews and even converts was introduced once more. During World War I, thousands of Jews were resettled by the Russian Army from areas close to the front line; anti-Jewish attitudes were rampant in the Army, despite the loyal service of numerous Jewish soldiers. *DR*

0224 Kelner, Viktor Y.: Voina, Rossiya, yevrei: Stenogrammy vstrech ministrov russkovo pravitelstva s predstavitelyami yevreiskoi obshchestvennosti v 1915 godu [War, Russia, Jews: Stenographic Records of Meetings between Russian Government Ministers and Representatives of the Jewish Public in 1915]. *Vestnik Yevreiskovo Universiteta* 23 (2001) 265-284.

In 1915 the High Command of the Russian Army initiated mass expulsions of Jews from the front line. The expulsions were accompanied by pogroms. In January-March 1915 political representatives of Russian and Russian-Polish Jewry (e.g. O. Gruzenberg, N. Fridman, M. Bomash) met with Russian government officials, demanding to stop the persecution of Jews on the Western front. Presents four records of conversations held by the Jewish representatives with Prime Minister Goremykin and others, as well as a summary of Russian Minister of Finance Bark's meeting with leaders of British Jewry in October 1915. English versions of these documents are found in the YIVO archives. Pp. 265-267 contain an introductory note by Kelner. *DR*

0225 Kurganov, Yefim: Filosofskaya mysl "serebryanovo veka" i yevreiskii vopros [Philosophical Thought of the "Silver Age" and the Jewish Question]. *Mezhdu vostokom i zapadom: Yevrei v russkoi i yevropeiskoi kulture* [Between East and West: Jews in Russian and European Culture], eds. I. Belobrovtseva, S. Dotsenko, V. Hazan. Tallinn: Tallinskii Pedagogicheskii Universitet, 2000. Pp. 48-79.

Contains three brief essays comparing the theological views of Rozanov on Jews and Judaism with those of Florenskii, Sergei Bulgakov, Vladimir Solovyov, and Berdyayev. Disagrees with the view that Rozanov was an inveterate antisemite. Florenskii was an antisemite; he was pessimistic regarding the future of Russian Orthodoxy due to a "Jewish plot." Rozanov derived from Florenskii's views a belief in the greater vitality and in the eternity of the Jewish people. Rozanov and Bulgakov adhered to that strain of Christian theology that maintained that God's covenant with the Jews has not ceased and that Jews continue to be God's beloved people. However, while Bulgakov believed that this fact demands of Jews that they join the Church, Rozanov did not. Both of them opposed the participation of Jews in Russian culture. Bulgakov regarded the Russian Revolution as "a Jewish pogrom against the Russians" and in 1942-43 asserted that Jews must atone for this sin against both Russia and Judaism. Rozanov saw the idealistic motives of Jewish revolutionaries. Solovyov and Berdyayev, with all their philosemitism, did not see any significance in Judaism after the coming of Christianity. *DR*

0226 Kurganov, Yefim; Mondry, Henrietta: *Vasilii Rozanov i yevrei*. Sankt-Peterburg: Akademicheskii Proyekt, 2000. 272 pp.

Pp. 5-153 contain an essay by Kurganov, "Vasilii Rozanov, yevrei i russkaya religioznaya filosofiya" ["Vasilii Rozanov, Jews and Russian Religious Philosophy"]; pp. 155-267 contain an essay by Mondry, "Vasilii Rozanov, yevrei i russkaya literatura" ["Vasilii Rozanov, Jews and Russian Literature"]. Kurganov examines Rozanov's stance toward the Jews, Judaism, and the Bible in the context of his religious and philosophical views. Argues that Rozanov was not antisemitic; he supported the idea of re-Judaizing Christianity. Believing that Jews consume Christian blood, he regarded this as a positive trait of Judaism. Juxtaposes Rozanov's views with those of Solovyov and Berdyayev. While the latter two blamed the Jews themselves for their situation and saw the solution of the "Jewish question" in the dissolution of Judaism, Rozanov insisted on the vital necessity of Judaism to exist. Presents Rozanov's criticism of Chrysostom's antisemitic rhetoric. Mondry regards Rozanov's antisemitism as part of the contemporary antisemitic tradition, feminizing the Jews and unifying the female and Jewish bodies, synthesizing an "Other" from both women and Jews. Rozanov's antisemitism increased greatly after the murder of Stolypin by a Jew and the Beilis trial: the "politicization" of the Jews destroyed Rozanov's romantic image of the "Old Testament Jew" standing aside from politics. He blamed Russian literature and the press for being pro-Jewish. *DR*

0227 Litvin, Aleksei: Generaly i yevrei (dokladnaya zapiska nachalnika Generalnovo Shtaba Ya.G. Zhilinskovo) [Generals and the Jews (a Memorandum by the Chief of the General Headquarters Ya.G. Zhilinskii)]. *Vestnik Yevreiskovo Universiteta* 22 (2000) 273-290.

The idea of a total cleansing of the tsarist army of the Jewish rank-and-file was spread among the Russian military in 1905-14. In 1912, in view of proposing such a measure to the government, the army leadership distributed a questionnaire among the highest ranking generals asking about the quality of Jews as soldiers. Presents the text of the memo compiled by General Yakov Zhilinskii based on the answers to the questionnaires, which he submitted to the war minister V. Sukhomlinov in January 1913. The memo shows that many generals wished to exclude Jews, either totally or partially, from military service; the Tsar, too, was sympathetic to this idea. Pp. 273-274 give a historical introduction. *DR*

0228 Lohr, Eric: The Russian Army and the Jews: Mass Deportation, Hostages, and Violence during World War I. *The Russian Review* 60, 3 (July 2001) 404-419.

The number of Russian Jews forced from their homes and deported to internal areas of Russia in 1914-17 is estimated as half a million to a million. Examines the stages of these actions – from spontaneous resettlement carried out by local commanders, not authorized from above, through a deliberate policy in early 1915 authorized by the chief of army headquarters General Yanushkevich, to the taking of Jewish hostages introduced by Yanushkevich's successor Alekseyev in mid-1915. This policy was opposed by the civilian administration; it was abolished at the end of 1915. Argues that the expropriation of Jewish property was a major motive for the relocations, as well as the pogroms that accompanied them; in three-quarters of the pogroms, soldiers and Cossacks were the initiators and main perpetrators. The expulsion campaign had destructive consequences for Russia; it undermined war efforts, shifted the war from an external to a civil one, caused moderates to become disillusioned with the Tsar and his regime, radicalized the Jews, and institutionalized the expropriation of private property, thereby precipitating the revolution. *DR*

0229 Mondry, Henrietta: Vasilii Rozanov on the Sexual Anomalies of Gogol' and the Jews. *Wiener Slawistischer Almanach* 48 (2001) 53-77.

Notes the shift in Rozanov's views on the Jews through the prism of his views on the complex psychology and sexual identity of Gogol. Rozanov moved from an attitude of love to hatred for Judaism, which he ultimately demonized. With his biological mysticism, Rozanov found in Gogol's portrayal of the Jew and in Judaism itself a predilection for taboos, an inherent criminality, and a tendency toward sexual perversion (e.g. incest). During the Beilis case, Rozanov expressed belief in the truth of the blood libel. He concluded that the role of Jews in Russian literature and in Russian political life was destructive; he was fixated on the Jewish body and on blood. However, after the Revolution Rozanov returned to a positive view of both the Jews and Gogol. *YC*

0230 Mondry, Henrietta: Vasilii Rozanov i yevreiskaya seksualnost: K tipologii predrassudkov [Vasilii Rozanov and Jewish Sexuality: Toward a Typology of Prejudices]. *Mezhdu vostokom i zapadom: Yevrei v russkoi i yevropeiskoi kulture* [Between East and West: Jews in Russian and European Culture], eds. I. Belobrovtseva, S. Dotsenko, V. Hazan. Tallinn: Tallinskii Pedagogicheskii Universitet, 2000. Pp. 114-127.

Rozanov adopted Solovyov's idea of the "holy corporeality" of the Jews but gave it a negative interpretation. His reflections on "femininity" of the Jewish character and body acquired an antisemitic tenor in his writings in the period of the Beilis case. Many of his ideas are borrowed from Otto Weininger. Rozanov criticized the practice of circumcision and ritual immersion, which he regarded as sexual rites. He viewed the Jews as a racial other. Rozanov's reflections on the Jewish body and sexuality is part of the "sexual science" of his time. He referred to the Jewish body in order to teach good sexuality to his fellow Russians who were "prone to asceticism," yet at the same time he demonized the Jews. *DR*

0231 Skurativskyi, Vadym L.: *Problema avtorstva "Protokolov sionskikh mudretsov"* [The Problem of the Authorship of the "Protocols of the Learned Elders of Zion"]. Kiev: Dukh i Litera, 2001. 241 pp. With an English summary (pp. 238-240).

Argues that the "Protocols" has mainly Russian origins and is of Russian authorship. The fin-de-siècle Russian conservative mythos maintained that freedom and liberalism must give way to despotism, and regarded the French Second Empire as a demonstration of this thesis. Russia was the last country where Maurice Joly's satire evoked interest. The "Protocols" was intended as an anti-liberal pamphlet, aimed not only against the Second Empire but against post-revolutionary France in general. They comprise a literary pastiche, drawing from the political and mystical literature of the time and reflecting the epoch's main intellectual vogues, including antisemitism and conspiracy theory. Suggests that the author of the "Protocols" was Matvei Golovinskii, a mediocre Russian writer. Apart from his own views and Joly's satire, the "Protocols" drew intensively from Dostoevsky's novels. *DR*

1789-1918: North America

USA

0232 Glazier, Jack: *Dispersing the Ghetto: The Relocation of Jewish Immigrants across America*. Ithaca, NY: Cornell University Press, 1998. x, 245 pp.

Presents the history of the Industrial Removal Office (IRO), established in New York in 1901 as part of the Jewish Agricultural and Industrial Aid Society. Its aim was to extend Jewish immigrant settlement beyond New York as a means of absorbing Jews into American society and, thus, preventing antisemitism. The IRO helped unemployed Jewish immigrants. Ch. 2 (pp. 62-105), "Confronting Immigration Restriction," discusses a major activity of the Office – countering economic and demographic arguments against allowing immigration of people, including Jews, from southern and eastern Europe. The anti-immigration arguments were often screens for racism and xenophobia. Partly from lack of government support, the IRO did not succeed in resettling many Jews. At the same time, anti-immigration legislation was being passed in Congress: a literacy test in 1917, a quota in 1921, and further restrictions in 1924. *YC*

0233 Rockaway, Robert Allen; Gutfeld, Arnon: Demonic Images of the Jew in the Nineteenth Century United States. *American Jewish History* 89, 4 (Dec 2001) 355-381.

Presents numerous examples of demonic representations of the Jew in 19th-century American culture. Stereotypes and antisemitic myths originating in medieval Christianity were imported, appearing in sermons, religious and secular literature, school textbooks, and the press. The press reported blood libel stories, and the Civil War heightened fear of Jewish conspiracies. The social and economic progress of American Jews led to the spread of stereotypes about Jews and business. The charge of deicide was featured, inter alia, by the American Catholic press. In the 1890s, the idea of a Jewish conspiracy to control the world appeared in the press. The immigrant Telemachus Timayenis introduced racial antisemitism into the U.S., along with blood libel charges; and disgruntled farmers were encouraged by Populists to scapegoat Jewish "financiers." There is no way to determine the effects of antisemitic propaganda; however, American Jews did not suffer from the pogroms and legal restrictions that characterized the European Jewish experience. *YC*

1919-1945: General

0234 Agosín, Marjorie: *Uncertain Travelers: Conversations with Jewish Women Immigrants to America*. Hanover, NH: University Press of New England [for] Brandeis University Press, 1999. 214 pp. On title page also: Edited and annotated by Mary G. Berg.

Contains interviews with nine Jewish women who immigrated to the U.S. Most of them experienced antisemitism in their homelands, some also in their new home. During World War II, Zezette Larsen hid in a Catholic convent in Belgium; when she returned home, she found that her parents had been arrested and deported to Auschwitz. Elena Ottolenghi Nightingale experienced Italian fascism as a child in 1938-39; her family later experienced anti-Jewish discrimination in the U.S. Matilde Salganicoff experienced antisemitism in Argentina. Renata Brailovsky witnessed the "Kristallnacht" pogrom in Breslau, then left with her family for Chile. Susan Bendor and Susan Rubin Suleiman both survived the Holocaust in Budapest. Silvia Zeldis Testa was born in Chile. Her grandparents had fled from antisemitism in Poland and Russia. She recalls experiences of antisemitism in Chilean society. Ruth Behar appreciates Cuba's acceptance of Jews fleeing Hitler; otherwise, her parents would not have survived. Each interview includes information on antisemitic tendencies in the countries discussed. *YC*

0235

בנקיר, דוד: שיקומה של גרמניה לאחר מלחמת העולם השניה והשאלה היהודית. ילקוט מורשת 71 (אפר' 2001) 11-21

[Bankier, David: The Rehabilitation of Germany after the Second World War and the Jewish Question. *Yalkut Moreshet* 71 (Apr 2001) 11-21.]

Toward the end of World War II, there were discussions between the Allies and anti-Nazi German émigrés, relating to the nature of the future democratic state. Most of the Germans did not mention the Jews at all, but some, although they opposed Nazi racism, made it clear that a return of Jews to Germany would be undesirable, since antisemitism would not immediately cease after the war and there would be much resentment if Jews were to become powerful in a democratic Germany. Among those who mentioned the Jews, some believed that they could be an official minority group and only opposed the return of the Ostjuden; some supported the return of upper-class liberal Jews only, since they would eventually assimilate; and some opposed the return of any Jews, claiming that all of them were power-hungry and that the German people would not tolerate Jews in their midst after the war. The views of the latter were affected by many decades of German antisemitism. *RE*

0236 Bauer, Yehuda: *Die dunkle Seite der Geschichte: Die Shoah in historischer Sicht: Interpretationen und Re-Interpretationen*. Trans.: Christian Wiese. Frankfurt a.M.: Juedischer Verlag im Suhrkamp Verlag, 2001. 384 pp. Originally published as "Rethinking the Holocaust" (New Haven: Yale University Press, 2001).

0237 Bauer, Yehuda: *Rethinking the Holocaust*. New Haven, CT: Yale University Press, 2001. xvi, 335 pp.

A study of aspects of the Holocaust on which the author has written in the past and on which he has done some "rethinking." Contends that the Holocaust is not inexplicable; however, its explanations must be multi-dimensional. It is unique among other genocides because of its ideological underpinning, its totality, and the degree of dehumanization of its victims. Criticizes some contemporary interpretations of the Holocaust, especially those of Zygmunt Bauman and Daniel Goldhagen. Regarding Jewish resistance, states that its general failure is to be attributed to the lack of a Jewish military tradition and to the naïveté of the Jewish resistants. Devotes a chapter to the case of Gisi Fleischmann, a leader of the Jewish Working Group in Slovakia, highlighting the role played by women in the Holocaust period. Examines Orthodox Jewish theological explanations of the Holocaust (especially those of Rabbi Menahem Mendel Schneersohn) and the problems engendered by this type of interpretation. Reflects on the Allies' failure to rescue the Jews, focusing on the "Auschwitz protocols" and the possibility of rescuing Hungarian Jewry. Discusses, also, the degree of impact of the Holocaust on the establishment of the State of Israel. Pp. 261-273 contain Bauer's speech to the Bundestag on 27 January 1998. *DR*

0238 Bédarida, François: Pour faire avancer l'histoire: Les archives du Vatican 1939-1945. *Esprit* 274 (May 2001) 15-25.

Discusses Pope Pius XII's silence during the Holocaust in light of the published Vatican documents, "Actes et documents du Saint-Siège relatifs à la Seconde Guerre mondiale." The Vatican chose diplomacy and impartiality based on a policy that emphasized the internal life of the Church rather than its relationship with the outside world. Suggests that the Pope, despite his charitable actions and an indisputable aversion to Nazism, bound to an anachronistic diplomatic practice, impotent in an era of propaganda and public opinion. Refutes allegations that a firm papal position would have made things worse for the Jews. Advocates research into when and how the policy of impartiality was adopted. Prompts the Vatican to open up the rest of its archives from this period and initiate a critical reevaluation of its past. *EO*

0239 Black, Edwin: *IBM and the Holocaust: The Strategic Alliance between Nazi Germany and America's Most Powerful Corporation*. New York: Crown, 2001. 519 pp. Published in paperback in New York: Three Rivers Press, 2002 (551 pp.).

In 1933-45 the International Business Machines Corporation greatly helped the German Nazis and their satellites throughout Europe, first to identify Jews and persons of Jewish extraction, then to expropriate, transport, exploit, and exterminate them. Through its German subsidiary, Deutsche Hollerith Maschinen Gesellschaft (Dehomag), IBM made Hitler's program of Jewish destruction a technological mission which the company pursued with success. More than 2,000 multi-machine sets were dispatched throughout Germany, and thousands more throughout Nazi-dominated Europe. Nearly every Nazi concentration camp operated a Hollerith Department, using equipment leased by IBM. IBM continued its contacts with Nazi Germany and its satellites even after the USA entered the war, and after the Nazi genocide became public knowledge. Far from being antisemitic or pro-Nazi, the IBM heads cared only for profit. Despite Nazi, and sometimes SS, custodianship over Dehomag and despite its involvement in Nazi programs, the company was not dissolved after the war. *DR*

0240 Black, Edwin: *IBM und der Holocaust: Die Verstrickung des Weltkonzerns in die Verbrechen der Nazis*. Trans.: Caecilie Plieninger, Ursel Schaefer, Heike Schlatterer. Muenchen: Propylaeen, 2001. 704, [16] pp. Originally published as "IBM and the Holocaust" (New York: Crown, 2001).

0241 Eshkoli (Wagman), Hava (Eva): The Zionist Aspect of Religious-Zionist Policy in Palestine in View of the Holocaust. *Yad Vashem Studies* 29 (2001) 281-320. Appeared in Hebrew in "Yad Vashem" 29 (2001).

Examines the wartime stance of Mizrachi-Hapoel Hamizrachi on two issues: precedence of the struggle for the establishment of a Jewish state in the Land of Israel over the rescue of European Jews, and mobilization of Jews in the Land of Israel to join the British army. The ideology of the Mizrachi movement gave precedence to the "redemption" of Jews over the "rescue" of Jews. Adherence to this principle caused the movement to advocate the concentration of all Zionist efforts on the struggle for Jewish statehood and to downplay the importance of the rescue of diaspora Jews. An ardent proponent of this view was R. Meir Berlin (Bar-Ilan). Awareness of the horror of the Holocaust, which came in November 1942, did not change the principal stance of the Mizrachi movement; however, the movement did not engage in assistance to European Jewry and rescue actions. *DR*

0242
אשכולי (וגמן), חוה: הפן הציוני במדיניות הציונית הדתית בארץ-ישראל לנוכח השואה. יד ושם 29 (2001) 219-247.

[Eshkoli (Wagman), Hava (Eva): The Zionist Aspect of Religious-Zionist Policy in Palestine in View of the Holocaust. *Yad Vashem* 29 (2001) 219-247.] Appeared in English in "Yad Vashem Studies" 29 (2001).

0243 Hacohen, Dvora: British Immigration Policy to Palestine in the 1930s: Implications for Youth Aliyah. *Middle Eastern Studies* 37, 4 (Oct 2001) 206-218.

The Arab riots which broke out in Palestine in April 1936 led to a drastic reduction in the number of immigration certificates issued by Britain. This development coincided with the rise of antisemitism in Europe, the Nazi annexations, and the beginning of World War II. The shift in British policy towards Palestine affected Youth Aliyah activities directed toward rescuing Jewish youth from Nazi-controlled areas. In particular, the British prevented the immigration to Palestine of 10,000 children from Germany after the "Kristallnacht" pogrom and rejected the request of the Polish government to allow 25,000 Polish children to enter Palestine in September 1939. It was not hostility to Jews, but concern for the status of Britain in the Middle East, that dictated limitations imposed on the immigration of Jews to Palestine, but these limitations were critical for the fate of European Jews. *DR*

0244 Heim, Susanne et al., eds.: *Fluechtlingspolitik und Fluchthilfe*. Berlin: Verlag der Buchlaeden Schwarze Risse, Rote Strasse, 1999. 187 pp. (Beitraege zur nationalsozialistischen Gesundheits- und Sozialpolitik, 15).

Partial contents: Ofer, Dalia: Die illegale Einwanderung nach Palaestina: Politische, nationale und persoenliche Aspekte (1939-1941) (9-38); Puto, Artan: Einige Plaene zur Ansiedlung deutscher Juden in Albanien [In 1934.] (39-46); Klein, Anne: Fluchthilfe aus Suedfrankreich: Das "Centre américain de secours" in Marseille, 1940/41 [Discusses, also, its difficulties in obtaining French exit and American entry visas.] (47-89); Merrienboër, Johan van: Hitlerfluechtlinge in den Niederlanden unerwuenscht: Die Politik der niederlaendischen Regierung gegenueber deutschen Fluechtlingen 1933-1940 [Appeared in Dutch in "Uitgeweken: De voorgeschiedenis van kamp Westerbork" (1989).] (91-105); Heim, Susanne: Vertreibung, Raub und Umverteilung: Die juedischen Fluechtlinge aus Deutschland und die Vermehrung des "Volksvermoegens" (107-138). *RW*

0245 Lang, Berel: "Not Enough" vs. "Plenty": Which Did Pius XII Do? *Judaism* 50, 4 (Fall 2001) 448-452.

Discusses the argument between Susan Zuccotti, author of "Under His Very Windows" (2000), and Ronald R. Rychlak, author of "Hitler, the War, and the Pope" (2000), as to whether Pius XII did "not enough" or "plenty" to save the Jews of Nazi-occupied Europe. States that particular actions or failures to act should be taken into account, and that Pius XII not only failed but refused to act. He did not threaten Nazi leaders with excommunication, although it would have been easy to do so and such an action could have influenced Nazi followers or collaborators, and he did not protest against the deportation of more than 1,000 Jews from Rome in October 1943. Even if the Pope did fear reprisal, this was not a sufficient reason for his silence and acquiescence. *DR*

0246 Laqueur, Walter Ze'ev: *Geboren in Deutschland: Der Exodus der juedischen Jugend nach 1933*. Trans.: Hans-Ulrich Seebohm. Berlin: Propylaeen, 2000. 472, [24] pp. A translation of "Generation Exodus" (Hanover, NH: University Press of New England for Brandeis University Press, 2001).

0247 Laqueur, Walter Ze'ev: *Generation Exodus: The Fate of Young Jewish Refugees from Nazi Germany*. Hanover, NH: University Press of New England for Brandeis University Press, 2001. xvii, 345, [24] pp.

Relates the multiple ways adopted by young German Jews (born mainly between 1914-28) in order to escape from Nazi Germany. Their destinations included Palestine, Great Britain, the USA, South America, and China. Dwells on the routes of escape – sometimes dangerous, and on the organizational framework of the escape – sometimes part of a political underground. Some young persons managed to survive in Germany. Many Jews, both those who escaped and those who remained in Germany, joined the anti-Nazi struggle in multiple forms – from the underground in Berlin to joining up with the Red Army. Many countries of destination were reluctant to accept the refugees, in part due to antisemitism. Other possible frameworks (e.g. communists) were also not eager to help Jews. The majority of the émigrés did not return to Germany after the war. *DR*

0248 Levy, Richard H.: Did Ben-Gurion Reverse His Position on Bombing Auschwitz? *Journal of Genocide Research* 3, 1 (2001) 89-96.

Poses the question as to whether David Ben-Gurion and other members of the Jewish Agency Executive changed their decision, expressed at a meeting in Jerusalem on 11 June 1944 (and opposed by Yitzhak Gruenbaum), not to ask the Allies to bomb locations (e.g. Auschwitz) where there were Jews. Since, less than a month later, Chaim Weizmann and Moshe Shertok (later, Sharett) did ask the British government to carry out such bombings, several Israeli scholars have claimed that Ben-Gurion and the Executive must have changed their minds. If this were so, then the

Allies would emerge as having been unresponsive to Jewish calls for help. Argues, however, that no documentation has been found that would corroborate such a reversal of Ben-Gurion's position. *YC*

0249 Madigan, Kevin: What the Vatican Knew about the Holocaust, and When. *Commentary* 112, 3 (Oct 2001) 43-52.

Positive assessments of Pope Pius XII's activities during World War II in regard to the rescue of Jews, prevalent before the mid-1960s, gave way to works deploring the Pope's silence and inactivity during the Holocaust. Shows that by the end of 1942 the Vatican had at least four reports on the Einsatzgruppen massacres in the East, provided by ecclesiastical sources, and a number of messages on deportations of Jews in Western and Central Europe. Thus, the cause of Pius XII's silence and lack of action was not the lack of information but the fact that he did not view the plight of the Jews with a sense of urgency and moral outrage. Criticizes apologetic historiography, which tends to regard Pius XII as a Righteous Gentile, but also objects to works like John Cornwell's "Hitler's Pope," which ascribes pro-Nazi views to Pius XII and depicts him as an accomplice in the Holocaust. See twelve responses to this article published in "Commentary" 113 (Jan 2002) 12-16; most of them reject Madigan's arguments and contend that Pius XII can be regarded as a Righteous Gentile, but some support Madigan's views. On pp. 16-18, Madigan responds to his critics. *DR*

0250 McMahon-Kaye, Stephanie, ed.: *The Memory of the Holocaust in the 21st Century: The Challenge for Education*. Jerusalem: Yad Vashem, International School for Holocaust Studies, [2001]. 135 pp. Cover-title. On cover also: The Second International Conference on the Holocaust and Education [Jerusalem, 1999].

Partial contents: Shalem, Motti: Thoughts on the Educational Goals of Holocaust Education (13-22); Bauer, Yehuda: The Educational Impact of the Holocaust on Contemporary Civilization: The Memory and the Challenge (23-33); Michman, Dan: Jewish Identity in Interwar Europe: Between Acculturation, Democratization and Rising Antisemitism (35-47); Eliach, Yaffa: Jewish Life in Inter-War Poland – the Shtetl of Eishyshok (49-59); Kwiet, Konrad: Perpetrators and the Final Solution [Attempts to explain how ordinary men were transformed into genocidal killers.] (61-82); Tec, Nechama: Resistance and Jewish Resistance during the Holocaust [Focuses on Poland and Belarus.] (83-97); Cesarani, David: The Holocaust Is/as News [On the continuous "newsworthiness" of the Holocaust, from the end of the war until the present day.] (109-128); Rosenak, Michael: Educational Guidelines for Teaching the Shoah in the 21st Century (129-135). *SSC*

0251 Naor, Bezalel: *Kabbalah and the Holocaust.* Spring Valley, NY: Orot, 2001. 135 pp.

Combines various kabbalistic and hasidic ideas with stories of several hasidic figures whose ideas prefigured the Holocaust, and of their descendants, whose lives were caught up in the Holocaust or affected by it. The locations discussed in relation to Nazi domination or threat include: Transnistria, Bulgaria, Poland (Radzyn), Yemen (San'a), and flight from Galicia that ended in the Siberian labor camp of Ynukcan. Includes accounts of miracles of survival perceived as Divine aid, in the cases of individuals and of two Jewish communities – Bulgaria and Yemen, the latter threatened by the Mufti of Jerusalem as an emissary of Hitler. Discusses the kabbalist rabbis Moses Simeon Pesach of Volos, who joined the partisans in Greece, and Samuel Solomon of Radzyn, who called on Jews to flee the ghetto and to resist the Nazi murderers. *YC*

0252 Neufeld, Michael J.; Berenbaum, Michael, eds.: *The Bombing of Auschwitz: Should the Allies Have Attempted It?* New York: St. Martin's Press, in association with the United States Holocaust Memorial Museum, 2000. xvii, 350, [8] pp.

A collection of articles based on papers delivered at a symposium held at the Smithsonian Institution in April 1993, jointly sponsored by the National Air and Space Museum and the United States Holocaust Memorial Museum. The papers have been revised and new ones added. Contents: Neufeld, Michael J.: Introduction to the Controversy (1-10); Weinberg, Gerhard L.: The Allies and the Holocaust [Appeared in "The Holocaust and History" (1998).] (15-26); Breitman, Richard David: Auschwitz Partially Decoded [A revised version of ch. 7 of his "Official Secrets" (1999).] (27-34); Biddle, Tami Davis: Allied Air Power: Objectives and Capabilities (35-51); Brugioni, Dino A.: The Aerial Photos of the Auschwitz-Birkenau Extermination Complex [An earlier version appeared in "Military Intelligence" (1983).] (52-57); Gilbert, Martin: The Contemporary Case for the Feasibility of Bombing Auschwitz (65-75); Riegner, Gerhart M.: The Allies and Auschwitz: A Participant's View (76-79); Kitchens, James H. III: The Bombing of Auschwitz Re-examined [Appeared in the "Journal of Military History" 58 (1994).] (80-100); Levy, Richard H.: The Bombing of Auschwitz Revisited: A Critical Analysis [Appeared in "Holocaust and Genocide Studies" 10 (1996).] (101-126); Erdheim, Stuart G.: Could the Allies Have Bombed Auschwitz-Birkenau? [Appeared in "Holocaust and Genocide Studies" 11 (1997).] (127-156); Rice, Rondall R.: Bombing Auschwitz: U.S. Fifteenth Air Force and the Military Aspects of a Possible Attack [An earlier version appeared in "War in History" 6 (1999).] (157-179); Laqueur, Walter Ze'ev: Auschwitz (186-192); Feingold, Henry L.: Bombing Auschwitz and the Politics of the Jewish

Question during World War II [Considers, also, the USSR's reluctance to bomb Auschwitz.] (193-203); Murray, Williamson: Monday-Morning Quarterbacking and the Bombing of Auschwitz (204-213); Davis, Richard G.: The Bombing of Auschwitz: Comments on a Historical Speculation (214-226); Lipstadt, Deborah Esther: The Failure to Rescue and Contemporary American Jewish Historiography of the Holocaust: Judging from a Distance (227-237). Pp. 237-280 contain documents. Pp. 281-325 contain the notes to the articles. *DR*

0253

רפל, יואל , עורך: כל הימים ההם: מבחר מחקרים. משואה 30 (2001).

[Rappel, Yoel, ed.: All Those Days. *Massuah* 30 (2001).]

תוכן חלקי: גוטמן, ישראל: הפאשיזם והאנטישמיות (18-9); כהן, אשר: האנטישמיות הפשיסטית-פלבאית [בהונגריה.] (37-19); מלצר, עמנואל: השפעת החברה הסובבת על הציונות בפולין בין שתי מלחמות העולם (52-39); זריז, רות: יהודי גרמניה בין הגירה לבריחה (71-53); לוין, דב: לפרשת היחסים בין העמים הבלטיים (93-73); שפירא, אניטה: שואה וכוח (121-95); בראון, אברהם צבי: מסר לדורות הבאים: על לקחי הארגון הלוחם של יהודי קובנא במלחמת העולם השניה (128-123); ויץ, יחיעם: הרהורים על האדם בשואה (על יסוד תצפיותיהם של ויקטור פראנקל, אלי ויזל ואחרים (164-157); ליטל, פרנקלין המלין: על משמעותה של השואה (184-165); פורת, דינה: בעיה בהיסטוריוגרפיה: יחסו של דוד בן-גוריון ליהודי אירופה בתקופת השואה (198-185); ברלס, חיים: פגישות בקושטא: עדות אישית (223-213); קרמיש, יוסף: "הנוער הציוני" ובטאונו המחתרתי בגטו וארשה "שביב" (314-307).

A collection of articles which appeared in previous issues of "Massuah." Partial contents: [Gutman, Yisrael: Fascism and Antisemitism ["Massuah" 21 (1993).] (9-18); Cohen, Asher: Plebeian Fascist Antisemitism [In Hungary. "Massuah" 6 (1978).] (19-37); Melzer, Emanuel: The Influence of the Surrounding Society on Zionism in Poland between the Two World Wars ["Massuah" 16 (1988).] (39-52); Zariz, Ruth: German Jews between Emigration and Escape ["Massuah" 17 (1989).] (53-71); Levin, Dov: On Relations between the Baltic Peoples [Their relations with the Jews. "Massuah" 17 (1989).] (73-93); Shapira, Anita: Holocaust and Power ["Massuah" 20 (1992).] (95-121); Bar-On, Abraham Zvie: A Message to Future Generations: On the Lessons of the Fighting Organization of the Jews of Kovno in the Second World War ["Massuah" 8 (1980).] (123-128); Weitz, Yehiam: Thoughts on Man in the Holocaust (Based on the Observations of Viktor Frankl, Elie Wiesel and Others) ["Massuah" 7 (1979).] (157-164); Littell, Franklin Hamlin: The Meaning of the Holocaust ["Massuah" 6 (1978).] (165-184); Porat, Dina: A Historiographical Problem: David Ben-Gurion's Attitude toward European Jewry during the Holocaust ["Massuah" 19 (1991).] (185-198); Barlas, Chaim: Meetings in Constantinople: A Personal Testimony ["Massuah" 4 (1976).] (213-223); Kermish, Joseph: "Hanoar Hazioni" and Its

Clandestine Journal in the Warsaw Ghetto "Shaviv" ["Massuah" 8 (1980).] (307-314).] *SSC*

0254 Riegner, Gerhart M.: *Niemals verzweifeln: Sechzig Jahre fuer das juedische Volk und die Menschenrechte.* Trans.: Michael von Killisch-Horn. Gerlingen: Bleicher, 2001. 638 pp. Originally published as "Ne jamais désespérer: Soixante années au service du peuple juif et des droits de l'homme" (Paris: Cerf, 1998).

0255

רובינסון, רינת-יה: התנועה הרביזיוניסטית נוכח התגברות האנטישמיות באירופה ועליית הנאצים לשלטון 1933-1940. דיס' – האוניברסיטה העברית בירושלים, 2001. 346, LXIII, [50] עמ'.

[Robinson, Rinat-ya: *The Revisionist Movement from 1933 through 1940: The Impact of European Antisemitism and the Rise of Nazism in Germany.* Diss. – Hebrew University of Jerusalem, 2001. 346, lxiii, [50] pp.]

Surveys the establishment, goals, and activities of the Revisionist Zionist movement in the 1930s. The movement was led by Zeev Jabotinsky and included the Tsohar political faction, the Betar youth movement, and the Irgun Tsva'i Le'umi (IZL) underground organization. Discusses, inter alia, Jabotinsky's attitude toward antisemitism in general and Nazi antisemitism in particular, and various approaches to the latter within the movement; support for the economic boycott of Germany; opposition to the Transfer Agreement between Germany and the Jewish Agency; the fate of the movement in Germany; the unfruitful "policy of alliances," which included a petition campaign and attempts to influence the League of Nations; and the rescue of Jews through illegal immigration. *RE*

0256 Stone, Dan: Recent Trends in Holocaust Historiography. *The Journal of Holocaust Education* 10, 3 (Win 2001) 1-24.

Holocaust historiography is fragmented into many specialist sub-fields, but one basic split is striking – that between empirical historians working largely on the decision-making process and "mechanics" of the Final Solution, and historiographers and other scholars who write about "representations of the Holocaust." In the first group, the debate between intentionalists and functionalists is ongoing, but a moderate synthetical viewpoint has evolved. Criticizes the excessive concentration of recent historiography on the decisions of the perpetrators, and especially on the timing of the Final Solution. The theoretical approaches of the second group, far removed as they are from the history written by empiricists, have impacted on the work of the latter. Empirical historiography has also engaged in theorizing about the nature of historical writing. The interaction of these two aspects of Holocaust studies can be fruitful. *DR*

0257 *Studia Judaica* 10 (2001).

Pt. II (pp. 22-203) contains papers delivered at an international conference on "The Resistance of the Jews against the Totalitarian Forces in the Time of the Holocaust," Cluj-Napoca, October-November 2000. Contents: Carmilly-Weinberger, Moshe: Communal and Spiritual Resistance of the Transylvanian Jews between 1933-1944 (22-42); Paucker, Arnold: German/Austrian Jewish Resistance – European Jewish Resistance: A Comparison (43-57); Lustiger, Arno: Some Problems and Aspects in the Historiography of the Jewish Resistance (58-70); Steinberg, Lucien: Reflections on Jewish Resistance in France 1940-1944 (71-91); Ben-Efraim, Yitzhak (Menu): From a Land under Siege to a People Besieged: The Israeli Parachutists in Romania in World War II: A Personal Testimony (92-109); Don, Yehuda: Why Was There No Massive Jewish Resistance in Hungary? (110-123); Blatman, Daniel: The Jewish Resistance through the Prism of the Underground Press in the Warsaw Ghetto (124-136); Radosav, Maria: Defying Death: The Resistance of Jewish Women during the Holocaust and Their Social Reintegration after the Holocaust [Based on interviews with five women survivors who lived in northern Transylvania when the deportations began.] (137-160); Rotman, Liviu: A Specific Form of Jewish Resistance: The Parallel Space [On Jewish institutions which operated parallel to the official reality, with the example of Romania.] (161-167); Vago, Raphael: The Concepts of Resistance and Heroism in Israeli Historiography and Public Opinion [From the Holocaust period to the present.] (168-179); Mudure, Mihaela: The Power of Words and Resistance to the Holocaust [Discusses discrete references to the Holocaust in the works of Isaac Bashevis Singer.] (180-192); Ursuţiu, Claudia: Jewish Resistance to the Anti-Semitic Policy of the Goga-Cuza Government (193-203). *SSC*

0258 Wistrich, Robert Solomon: *Hitler and the Holocaust*. London: Weidenfeld & Nicolson, 2001. xii, 322 pp.

Discusses major aspects of the Holocaust, emphasizing the centrality of antisemitism, both in terms of Europe's long history of Christian anti-Judaism and of Hitler's adoption of some aspects of Christianity and perversion of others. Hitler was also influenced by Austrian racist antisemitism. He exploited Germany's post-World War I national traumas resulting from defeat and fear of the Bolshevik threat, both of which he linked to the myth of Jewish power. Hitler's Final Solution was made possible by the initial success in his war against the USSR, which involved the brutal mass execution of Jews. Details the failure of Pope Pius XII to speak out against the Holocaust and of Christian states to protect their Jews. The extent of collaboration in the mass murder of Jews across Europe challenges Goldhagen's thesis of a uniquely German eliminationist antisemitism. Antisemitism is cited as one of the reasons for the failure of the USA and England to rescue Jews, provide them with a haven, or bomb Auschwitz. Disagrees with Zygmunt Bauman's theory that modernity is central to the Holocaust, stressing that apocalytic antisemitic ideology was a far more important factor than economic rationality in determining Nazi policy toward the Jews. *YC*

1919-1945: Europe

General

0259 Albrich, Thomas, ed.: *"Wir lebten wie sie...": Juedische Lebensgeschichten aus Tirol und Vorarlberg*. Innsbruck: Haymon, 1999. 383 pp.

A collection of essays resulting from a research project at the University of Innsbruck, "Biographische Datenbank zur juedischen Bevoelkerung in Tirol und Vorarlberg im 19. und 20. Jahrhundert." Several of them relate the experiences of Jews during the Holocaust. *ID*

0260 Altman, Ilya A.; Agmon, Pinchas, eds.: *Kholokost, soprotivleniye, vozrozhdeniye: Yevreiskii narod v gody Vtoroi mirovoi voiny i poslevoyennyi period 1939-1948* [Holocaust, Resistance, Rebirth: The Jewish People during and after the Period of the Second World War 1939-1948]. Moskva: Mezhregionalnyi fond "Kholokost," 2000. 342 pp.

A textbook on the history of the Holocaust and its aftermath, intended mainly for teachers. Each chapter is followed by documentary material, questions for the reader, suggested topics for students' work, etc. *DR*

0261 Altman, Ilya A., ed.: *Prepodavaniye temy Kholokosta v XXI veke* [Teaching the Holocaust in the 21st Century]. Moskva: Fond "Kholokost," 2000. 223 pp.

Presents 26 brief papers by historians and educators from Russia, Israel, Great Britain, Belarus, and other countries, delivered at a conference held in Moscow in 1998. They deal with such issues as Western and Israeli experiences in Holocaust education, Holocaust education in the CIS, new methods of Holocaust education, and textbooks. Contains, also, ten lesson plans dealing with the Holocaust, or with World War II and the Nazi occupation as background to the Holocaust. Pp. 208-220 gives a bibliographical list of books in Russian dealing with the Holocaust, its commemoration, and teaching its history. *DR*

0262 Anissimov, Myriam: *Sa majesté la mort*. Paris: Seuil, 1999. 279 pp.

Recounts the experiences of her family (originally from Szydłowiec, Poland) during the war, based on a dossier of recently-discovered personal letters written by her uncle, Israël Frydman. Anissimov's paternal grandparents were murdered in Szydłowiec in 1942; most of the other family members were murdered in Treblinka. Frydman was sent to the HASAG factory in Skarżysko-Kamienna and later transferred to the labor camp in Częstochowa. He and his wife survived the war. Anissimov's parents were living in Lyon when the war broke out. Her father escaped from the Fort Chapoly internment camp in 1940. The family escaped to Switzerland, but Anissimov's father was arrested as an illegal alien and returned to France; he survived the war. Her mother was permitted to remain in Switzerland (where Myriam was born in 1943); they lived in immigrant work camps and returned to France after the war. *TS*

0263 Barnett, Victoria J.: *Bystanders: Conscience and Complicity during the Holocaust*. Westport, CT: Greenwood Press, 1999. xviii, 185 pp.

Defines a bystander as a passive spectator of the Holocaust, who stands closer to the perpetrator than to the victim. The behavior of bystanders has both historical and ethical significance for any understanding of the Holocaust, which would have been impossible without the bystanders' passivity. Focuses on Germany, but states that the conclusions drawn are relevant for other parts of Nazi Europe as well. With the incremental development of their anti-Jewish policy, the Nazis succeeded in getting all Germans involved in the Holocaust. Indifference to the victims of the Holocaust was not only a product of prejudice; it was conditioned by Nazi totalitarianism, modernity, and the average person's wish to conform. Contrasts rescuers (such as the people of Le Chambon-sur-Lignon) with the bystanders. Concludes that the ethical person, as opposed to the passive bystander, is one who recognizes his responsibility for all his neighbors and is not self-centered. *DR*

0264 Bartov, Omer: *Mirrors of Destruction: War, Genocide, and Modern Identity*. Oxford: Oxford University Press, 2000. viii, 302 pp.

Reflects on the interrelationship between war, genocide, and identity in the 20th century. Contends that World War I provided a pattern for the industrial killing that reached its apex in the Nazi Holocaust. Comparing France and Germany, both of which looked for an "elusive enemy" to help them cope with the trauma of World War I, notes that Germany found it in the Jews. This perception allowed the Germans to see themselves as victims and their actions against the Jews as self-defense. The search for "elusive enemies" (to blame for perpetrating the genocide) and "elusive victims" (e.g. the French or German people) continued after the war.

Compares the French, German, and Jewish postwar discourses of the Holocaust. Presenting themselves as the main victims of Nazism, the French and Germans have tended to neglect Jews as the main victims. Criticizes writings by Hannah Arendt, Ka-Tzetnik, and some other postwar writers on the Holocaust and its roots. *DR*

0265 Bartov, Omer; Mack, Phyllis, eds.: *In God's Name: Genocide and Religion in the Twentieth Century*. New York: Berghahn Books, 2001. vii, 401 pp.

A collection of papers dealing with, in addition to the Holocaust, the Armenian genocide and the genocides in Rwanda, East Africa, and Bosnia. Partial contents: Ericksen, Robert P.: Genocide, Religion, and Gerhard Kittel: Protestant Theologians Face the Third Reich (62-78); Heschel, Susannah: When Jesus Was an Aryan: The Protestant Church and Antisemitic Propaganda [An earlier version appeared in "Betrayal: German Churches and the Holocaust" (1999).] (79-105); Griech-Polelle, Beth A.: A Pure Conscience Is Good Enough: Bishop von Galen and Resistance to Nazism [Von Galen protested against euthanasia but not the persecution or genocide of the Jews.] (106-122); Bergen, Doris L.: Between God and Hitler: German Military Chaplains and the Crimes of the Third Reich (123-138); Sheetz-Nguyen, Jessica A.: Transcending Boundaries: Hungarian Roman Catholic Religious Women and the "Persecuted Ones" [On Margit Slachta and her Society of the Sisters of Social Service, which mounted rescue and relief efforts between 1939-45 for Jewish families.] (222-242); Kellenbach, Katharina von: Denial and Defiance in the Work of Rabbi Regina Jonas [Jonas (1902-1944) exemplified spiritual resistance to the Nazis; she died in Auschwitz.] (243-258); Vermes, Gábor: A Personal Account [Memoir of a Jew who as a boy in Hungary was hidden by a Lutheran pastor; Vermes' personal prayers helped him survive.] (259-263); Greenberg, Gershon: Orthodox Jewish Thought in the Wake of the Holocaust: "Tamim Pa'alo" of 1947 [On Hayim Yisrael Tsimerman's religious treatise published in Palestine in 1947 in reaction to the Holocaust.] (316-341). *YC*

0266 Baumeister, Stefan: *Zur Organisation und Realisation der Schoah: Rechtliche, institutionelle, organisatorische und verwaltungstechnische Voraussetzungen des Massenmords an den europaeischen Juden*. Konstanz: Hartung-Gorre, 2001. iii, 528 pp.

Describes the main bodies involved in the Holocaust: their functions, interaction, and conflicts of interest. Discusses the total subjugation of the legal and judicial system to the Fuehrer's ideology, as exemplified in the special legislation concerning the Jews and the increasing replacement of legislation by ministerial decrees or the Fuehrer's orally-expressed wishes; the SD's assumption of hegemony as experts on the "Jewish question," shaping policy for the removal of Jews from Germany and later from all of Europe by more and more radical means; and the role of the Einsatzgruppen, together with police batallions and local militia, in the massacre of the Jews of Eastern Europe. Also discusses contributions of the bureaucracy and of economists and demographers; the Jewish councils; the economic exploitation of the Jews by expropriation and forced labor; what ordinary Germans knew, and their fear, nourished by propaganda, of Jewish retribution after Germany's defeat. Analyzes the interaction of local and centralized decision-making, emphasizing that the radicalization of the Final Solution was not a mere product of circumstances but a result of the ideological intention that created those circumstances. *RW*

0267 Baumel, Judith Tydor: Women's Agency and Survival Strategies during the Holocaust. *Women's Studies International Forum* 22, 3 (1999) 329-347.

Focuses on ways in which Jewish women survived during the Holocaust by communal cooperation and the development of leadership. Discusses these phenomena against the background of gender roles in the Jewish community in Europe prior to Nazism, in Nazi Germany in 1933-39, in the communal ghetto framework in 1939-43, in Theresienstadt as a transition between ghetto and concentration camp, and in concentration camps in 1942-45. Concludes that women's survival strategies were different from men's, often reflecting radical changes in society. Examples were women's adoption of alternative female leadership roles and their functions in the compulsory single-sex frameworks created by the Nazis. Exemplified by various women's networks, the communal bond is highlighted as a means of survival, albeit often temporary, in the face of ultimate trauma. *YC*

0268 Behrens, Christian Guenther: *Holocaust, nicht im Namen, nicht mit Wissen der weit ueberwiegenden Mehrheit des Volkes*. Berlin: Christian G. Behrens co-tangens Verlag, 2001. 144 pp.

Behrens, a former soldier in the Wehrmacht, asserts that the Nazis kept the secret of the extermination of the Jews so well that no one but the actual perpetrators could have known about it. This was true of the mass of ordinary Germans and the citizens of the occupied countries who would surely have risen to prevent the deportations, and of the Jews themselves, who would have put up greater resistance. States that disregard of these facts leads to accusations of collective guilt. Care must be taken when teaching the Holocaust to young people not to overwhelm them with its horrors without proper preparation; this only leads to denial. Neither must they be made to feel guilty or ashamed, or to see their grandparents as guilty. Public discourse and the inscriptions on Holocaust memorials must avoid implying collective guilt and explain the historical facts as they were. *RW*

0269 Beker, Avi, ed.: *The Plunder of Jewish Property during the Holocaust: Confronting European History*. Basingstoke, Hants: Palgrave, 2001. xii, 355 pp.

A collection of articles dealing with the plunder of Jewish property during the Holocaust and problems of its restitution, as well as with the role of neutrals in the Holocaust and their relationship with Nazi Germany. *DR*

0270 Benz, Wolfgang: Illegale Einwanderung nach Palaestina. *Exilforschung* 19 (2001) 128-144.

Nazi persecution forced many Jews, not originally Zionists, to try to immigrate to Palestine in spite of British restrictions. Tells the story of the illegal transports from Central Europe organized by the Jewish Agency; the ships that were seized by the British, who deported the refugees to Cyprus or Mauritius; the ships that sank; and the Kladovo transport that was delayed until it fell prey to the Nazis after their invasion of Yugoslavia. *RW*

0271 Berler, Willy: *Itinéraire dans les ténèbres: Monowitz, Auschwitz, Gross-Rosen, Buchenwald*. Ed.: Ruth Fivaz-Silbermann. Paris: L'Harmattan, 1999. 293 pp.

Memoirs of a Jew born in 1918 in Czernowitz. He immigrated to Palestine in 1936, and then to Liège in Belgium in 1938. He went into hiding in 1941, but was denounced and deported to Auschwitz in 1943. At first, Berler worked at the Buna factory; he was transferred to the infirmary as an assistant, and later worked at the SS Institute of Hygiene at the nearby Rajsko camp. In 1945, after a death march to Gross-Rosen, Berler was sent to Buchenwald, where he was liberated. *TS*

0272 Bernhard-Ithai, Lilli: *Erinnerung verpflichtet: Von Berlin ueber Bruessel nach Lyon in die Schweiz und durch Gurs nach Auschwitz: Juedische Schicksale 1933-1945*. Konstanz: Hartung-Gorre, 1999. 83 pp.

Memoirs of a Jew born in 1920 in Berlin. In 1933 she and her family (the Bernhards) immigrated to Belgium; in 1939 her parents were arrested, together with their elder son, and interned in Gurs and Saint-Cyprien. In 1942 her father died of a stroke, in Drancy, while her mother and brother were deported to Auschwitz. Lilli fled with her younger brother from Belgium to France in August 1942, and in December they crossed into Switzerland. In Lausanne, she worked for the Youth Aliyah home and as a representative of the Hashomer Hatzair. She married in 1944 and immigrated to Eretz-Israel with her husband in June 1945. Her elder brother was interned in Blechhammer and in 1945 survived a death march to Gross-Rosen and Buchenwald. Pp. 47-72 contain a report on Gurs written ca. 1949 by Elisabeth Bernhard, a cousin. *ID*

0273 Bernhard-Ithai, Lilli: *Una piccola tessera: Il dovere di ricordare*. Trans.: Cristina Tabbia. Trieste: Edizioni E, 2001. 102 pp. Originally published as "Erinnerung verpflichtet: Von Berlin ueber Bruessel nach Lyon in die Schweiz und durch Gurs nach Auschwitz: Juedische Schicksale 1933-1945" (Konstanz: Hartung-Gorre, 1999).

0274 Birenbaum, Halina: *Wołanie o pamięć* [A Call for Memory]. Oświęcim: Wydawnictwo Państwowego Muzeum Oświęcim-Brzezinka, 1999. 189 pp.

Recounts the experiences of twenty Holocaust survivors from Poland as told to Halina Birenbaum. Pp. 148-181 contain an account by Jaco Poliker (the father of Israeli singer Yehuda Poliker) who was deported from Salonika to Auschwitz. *HV*

0275

בשביל הזיכרון 40 (ינ'-פבר' 2001).

[*Bishvil ha-Zikaron* 40 (Jan-Feb 2001).]

חוברת זו מוקדשת לנושא "חקיקה יהודית בשנות ה-30 וה-40". התוכן: גרוכמן, לותר: "חוק הגנת הדם" והמערכת המשפטית: על חוקי נירנברג ועל השפעתם (4-12); נצר, שלמה: בין חקיקה אנטי-יהודית להשתלבות: פולנים ויהודים, 1918-1939 (13-21); נתנזון, אפרים: אנטישמיות וחקיקה יהודית ברומניה, 1919-1940 (22-30); ראובני, שרי: אופיים של החוקים היהודיים ויישומם בהונגריה, 1938-1944 (31-36); ברוך, מרק-אוליביה, עורך: כאשר המנגנון הממשלתי הצרפתי היה אנטישמי: סטנוגרמה של הישיבה הבין-משרדית מה-16 בדצמבר 1940 (37-45); בלברג, אפרת: השפעות הגומלין בין החקיקה האנטישמית לבין דעת הקהל (46-54).

This issue is entitled "Jewish Legislation in the 1930s and 1940s." Contents: [Gruchmann, Lothar: "The Law for Protection of Blood" and the Judicial System: On the Nuremberg Laws and Their Influence [Appeared in German in "Vierteljahrshefte fuer Zeitgeschichte" 31 (1983).] (4-12); Netzer, Shlomo: Between Anti-Jewish Legislation and Integration: Poles and Jews, 1918-1939 (13-21); Nathanson, Efraim: Antisemitism and Jewish Legislation in Romania, 1919-1940 (22-30); Reuveni, Sari: The Character of the Jewish Laws and Their Implementation in Hungary, 1938-1944 (31-36); Baruch, Marc-Olivier, ed.: When the French Government Was Antisemitic: A Stenogram of the Inter-Departmental Meeting of 16 December 1940 (37-45); Balberg,

Efrat: Mutual Effects between the Antisemitic Legislation and Public Opinion (46-54).] *SSC*

0276

בשביל הזיכרון 42 (יוני-יולי 2001).

[*Bishvil ha-Zikaron* 42 (June-July 2001).]

חוברת זו מוקדשת לנושא "הפתרון הסופי – מגמות חדשות במחקר." התוכן: מכמן, דן: הפתרון הסופי של שאלת היהודים – התגבשותו ויישומו: מצב המחקר (4-21); גרלך, כריסטיאן: ועידת ואנזה, גורל יהודי גרמניה והחלטתו העקרונית של היטלר להשמיד את כל יהודי אירופה (22-40); בראונינג, כריסטופר רוברט: המדיניות הנאצית: החלטות לקראת הפתרון הסופי (41-55); בלברג, אפרת: החלטות, מבצעים ושגרת היומיום של הרצח (56-62).

This issue is entitled "The Final Solution – New Directions in Research." Contents: [Michman, Dan: The Final Solution of the Jewish Question – Its Development and Implementation: The State of Research (4-21); Gerlach, Christian: The Wannsee Conference, the Fate of German Jews, and Hitler's Decision in Principle to Exterminate All European Jews (22-40); Browning, Christopher Robert: Nazi Policy: Decisions for the Final Solution [Translation of a chapter from his book "Nazi Policy, Jewish Workers, German Killers" (2000).] (41-55); Balberg, Efrat: Decisions, Operations and the Routinization of Mass Murder [Guidelines for teaching the subject.] (56-62).] *SSC*

0277 Bloxham, Donald: A Survey of Jewish Slave Labour in the Nazi System. *The Journal of Holocaust Education* 10, 3 (Win 2001) 25-59.

Examines stages and trends in the Nazi use of the Jewish work force from 1939-45. The exigencies of the economy and the course of the war were important factors which affected the use of slave labor during this period. Ideological considerations, those of the Final Solution, contested with practical ones. The mass murder of Jews went side by side with their exploitation. The use of Jewish labor began in concentration camps even before the war and was intensified after the conquest of Poland, where it was implemented in labor camps and ghettos. In 1942-43 the exploitation of Jewish labor was concentrated in hundreds of camps run by the SS. A large part of the Jewish work force was used in the armament industry. In 1943-44 slave labor began to be used increasingly by private industries; thousands of slave laborers perished due to the inhuman conditions in private enterprises. *DR*

0278 Burleigh, Michael: *The Third Reich: A New History*. New York: Hill and Wang, 2000. xxv, 965 pp.

A history of Nazi Germany, viewing it as a church-state based on a quasi-religious ideology, and as one of the totalitarian regimes of the 20th century. Nazi totalitarianism was similar to that of the Soviet regime, from which it borrowed many elements. Racism was a cornerstone of Nazi ideology; the Jews were seen as the most dangerous of Germany's racial enemies. Ordinary Germans displayed opportunism in their support for Nazi actions, including those against Jews; opposition was sporadic and weak. Ch. 4 (pp. 279-342) deals with Nazi anti-Jewish policy in 1933-39 and focuses on the "Kristallnacht" pogrom, which was largely a spontaneous popular action. Ch. 8 (pp. 569-662) deals with the wartime anti-Jewish policy. In 1939-41 the main goal was resettlement. Only with the invasion of the USSR, did the Einsatzgruppen receive orders to carry out systematic murder. The murderers were not motivated by antisemitism, but by their conviction that they performed necessary work for the Reich. The false belief that the Jews were communists, and vice versa, provided them with a pretext for the murders and helped mobilize helpers from the East European non-Jewish population. Dwells, also, on Romania's role in the Holocaust. *DR*

0279 Chargüéraud, Marc-André: *L'étoile jaune et la Croix-Rouge: Le Comité international de la Croix-Rouge et l'Holocauste, 1939-1945*. Genève: Labor et Fides, 1999. 140 pp.

A concise history of the activities of the International Committee of the Red Cross during World War II. Aims to give a balanced answer to the question of the organization's share in the responsibility for the extermination of the Jews of Europe. Lists attempts by the ICRC to come to the aid of the victims, while shedding light on ambiguities among the Allies concerning human rights. Concludes that although the accomplishments of the ICRC were modest compared to the scope of the disaster, the dedication of its representatives was exemplary. *EO*

0280 Cling, Maurice: *Vous qui entrez ici...: Un enfant à Auschwitz*. Paris: Graphein, 1999. 239 pp.

Memoirs of a Jew born in Paris in 1929, based on notes he made upon his return from the camps. He was deported to Drancy with his parents and older brother Willy in May 1944, and from there to Auschwitz, where his parents were killed upon arrival; his brother died in the autumn of 1944. In January 1945 Cling was deported to Dachau, from there evacuated twice, and finally liberated by American troops in April 1945. After the war he returned to Paris. *EO*

0281 Cohen, Israel I.: *Destined to Survive: Uplifting Stories from the Worst of Times*. Portions trans. from Yiddish: Shmuel Yaakov Klein. Ed.: Gitta Cohen Shafran. Brooklyn, NY: Mesorah, 2001. 317 pp.

Memoirs of a Jew born ca. 1925 in Łódź to a hasidic family. Relates his experiences in the Holocaust, stressing his efforts to observe religious practices whenever possible. He and his family were interned for four years in the Łódź ghetto, where his father died of starvation. In summer 1944 the family was deported to Auschwitz, where his mother was immediately gassed. Cohen was sent to the Kaufering labor camp, where he was liberated. In 1946 his younger sister, Mirel, was killed by antisemites in Łódź; his elder sister, Gittel, perished in the Holocaust. After the war Cohen lived in Switzerland; he married and in 1951 immigrated to Canada. *YC*

0282 Dantzig, Emmy A. van: *Zeg nooit dat je joods bent: Vlucht naar Zwitserland 1942-1944* [Never Say That You Are Jewish: Escape to Switzerland 1942-1944]. Schoorl: Conserve, 2000. 131 pp.

Relates the story how her grandparents, Emmy and Joop van Dantzig, along with their son and daughter (aged 13 and 15), escaped from the Netherlands to Switzerland during the war, based on a diary written by the grandmother. They were in flight for more than six months. They left Arnhem on 18 August 1942, went through Brussels, then through occupied France to the Free Zone, and over the mountains to Switzerland. On 31 August 1944, Emmy died of cancer. Joop and the children returned to the Netherlands right after the war. *MF*

0283 Douer, Alisa, ed.: *Kladovo – eine Flucht nach Palaestina = Kladovo – Escape to Palestine.* English trans.: Karin Hanta-Davis. Wien: Mandelbaum, 2001. 124 pp. In German and English.

A catalogue accompanying an exhibition held at the Juedisches Museum Wien, July-November 2001. Pp. 10-22, "Die Geschichte des Kladovo-Transports 1939-1942 = The Story of the Kladovo Transport, 1939-1942," by Gabriele Anderl and Walter Manoschek, relate how ca. 800 Jews who left Vienna for Palestine in November 1939 by way of the Danube were stopped near Kladovo, Serbia; all efforts to continue the transport failed. In early March 1941, 200 adolescents who obtained certificates were able to immigrate to Palestine. A few days later the German army invaded Yugoslavia and all the remaining Jews of the Kladovo transport were murdered. Pp. 23-109 contain the photographs shown in the exhibition, brought to Israel by one of the adolescents, Ehud Nahir, in 1941 and preserved in an album. P. 113 lists the names of the survivors, and pp. 116-122 contain a list of the victims. *ID*

0284 Dubossarskii, Lev: *Stranitsy zhizni* [Pages of a Life]. Tel Aviv: [The Author], 1999. 112 pp.

Memoirs of a Jew born in 1914 in Belaya Tserkov, Ukraine. During the Civil War in 1919-20, Jews of his native town fell victim to looting and pogroms perpetrated by all the armies (including the Reds) that passed through the town. In 1939 Dubossarskii was drafted into the Army and fought in the war from 1941-45. On pp. 93-109 he recounts stories of Jews who survived the German occupation in Ukraine and Poland, and the story of his sister who perished with her family in Belaya Tserkov. In the 1990s he settled in Israel. *DR*

0285 Eisenstadt, Oona: Making Room for the Hebrew: Luther, Dialectics, and the Shoah. *Journal of the American Academy of Religion* 69, 3 (Sept 2001) 551-575.

Discusses two "works of art," Franz Rosenzweig's essay "Scripture and Luther" (1926) and Peter Kien and Victor Ullmann's opera "The Emperor of Atlantis." The former implicates German culture, in terms of the founding document of its language and culture, Luther's translation of the Bible, for having no room for Hebrew and, therefore, for the Jews. Rosenzweig had a premonition of a coming rupture of history, but the German translation of the Hebrew Bible that he and Buber were working on eventually became a gravestone for European Jewry, after the latter's destruction by Germans who had gone mad along with their language. "The Emperor of Atlantis" was written for performance in Theresienstadt, but was banned. Both of the authors were killed in Auschwitz, but their work was saved. Death and war are central themes in the opera, which criticizes both modernity and Christianity, and specifically Luther's Christianity for its triumphalism and demonizing of non-Christians. *YC*

0286 Elam, Shraga: *Hitlers Faelscher: Wie juedische, amerikanische und schweizer Agenten der SS beim Falschgeldwaschen halfen*. Wien: Ueberreuter, 2000. 208 pp.

Traces the SS's use of the counterfeit British pounds produced in Sachsenhausen, as well as of the booty from occupied Hungary, to prepare for its survival after the fall of Hitler. Among the agents who disseminated and laundered the counterfeit money were Jews, prominent among them Jaac van Harten, who was active in Hungary and northern Italy. Notes that he used some of the funds to free Jews, and after the war he lent money to Aliyah Bet. At the same time, he helped former SS-men escape together with their booty; the routes of SS and Jews often crossed. Later, to rehabilitate himself, he probably exaggerated his part in the rescue of Jews. Also chronicles SS attempts to "sell" Jews to the Allies, including the Becher-Kasztner affair – another survival tactic of the SS leadership. *RW*

0287

אנציקלופדיה "שמע ישראל": לתיעוד ולהנצחת מעשים של מסירות-נפש בשנות הזעם תרצ"ט-תש"ה (1945-1939) בגיטאות, במחנות העבודה, במחנות הריכוז ובמחנות ההשמדה. כר' א-ב. בני- ברק: מכון "שמע ישראל" שע"י מרכז קאליב העולמי, תשנ"ז-תש"ס. 337; 373, [5] עמ'. בשער גם: יוזם, מייסד ונשיא המפעל כ"ק אדמו"ר מקאליב.

[*Encyclopedia "Shema Yisrael": To Document and Commemorate Acts of Self-Sacrifice in the Years of Wrath 1939-1945 in the Ghettos, in the Labor Camps and in the Extermination Camps. Vol. 1-2*. Bnei-Brak: "Shema Yisrael" Institute at the Kaliv World Center, 1997-2000. 337; 373, 5 pp.]

Presents brief excerpts from various publications, describing acts of spiritual resistance by Orthodox Jews in ghettos and Nazi camps during the Holocaust. Many Jews attempted to uphold religious observance, sometimes risking their lives. Includes photographs taken in various ghettos and camps. Pp. 257-291 in Vol. 1 and pp. 275-309 in vol. 2 contain brief biographies, with photographs, of rabbis who perished or suffered in the Holocaust. *RE*

0288 Fastag, Miriam: *Dr. Josef Mengele's 301st: A Memoir of Darkness, Faith and Survival*. Trans.: Ruth Finkelstein. Brooklyn, NY: [The Author], 2000. 238 pp. Translated from the Yiddish manuscript.

Memoirs of an Orthodox Jew (née Feder) who was born ca. 1920 in the village of Porąbka, Poland, and later moved to Sosnowiec. Her father died ca. 1930. Two of her brothers immigrated to Argentina and one sister to Eretz-Israel before the war. Pp. 35-132 describe her experiences in the Holocaust. Her mother and the other members of her family (they were 12 siblings) were killed in 1942. Fastag was sent to the Gleiwitz labor camp. In January 1945 she was transported to Ravensbrueck, then to a camp called Rechlin(?), then again to Ravensbrueck. In Rechlin she was picked by a Nazi officer, whom she identifies as Mengele, to join a group of prisoners who were sent to Sweden due to efforts of the Swedish government. The rest of the book relates her activities in Lidingo, Sweden, where she helped Jewish refugee girls maintain their Jewish identity. She immigrated to the U.S. in 1947. *DR*

0289 Frei, Norbert et al., eds.: *Standort- und Kommandanturbefehle des Konzentrationslagers Auschwitz 1940-1945*. Muenchen: K.G. Saur, 2000. xii, 604 pp.

A collection of documents dated between June 1940-January 1945, written by SS officers in Auschwitz who numbered ca. 7,000 employees, among them 200 women, during the period covered. The documents demonstrate the course of the daily life of the SS, ruled by orders and commands which seem to be insignificant at first sight, but point to the bureaucratic organization of the murder of the Jews, and thus to the systematic policy of annihilation. The introduction (xii pp.) emphasizes the extent of exploitation of the prisoners' property which was much more important in Auschwitz than in other camps; the belongings were stockpiled, assorted, and stored in 29 barracks. Notes that while the instructions of Rudolf Hoess to his subordinates (regarding their own personal lives) were characterized by harshness and cruelty, they were often not fully implemented. The orders rarely refer directly to the Final Solution, but between the lines one can find revealing information on its process. *ID*

0290

פריעדמאן, הערשל, עורך: ספר מאפילה לאור גדול: נס פון הצלה פון רבי'ן זי"ע, כ"א כסלו תש"ה לפ"ק. קרית יואל [ניו-יארק]: "די אידישע פענע", 2001. 572 עמ'.

[Friedman, Hershel, ed.: *From Darkness to a Great Light: The Miracle of Rescue of the Rebbe, 21 Kislev 5705*. Kiryas Joel, NY: "Di Iddishe Pene," 2001. 572 pp.]

Describes in detail the rescue of Rabbi Joel Teitelbaum, the hasidic Rebbe of Szatmár (Satu-Mare), Romania. The town was annexed to Hungary in 1940 and occupied by the Germans in 1944. In April 1944 Teitelbaum fled in an ambulance to Klausenburg (Cluj) in Romania, where he was interned in the newly established ghetto. The Rebbe and his entourage were then transported on the "Kasztner train" from Budapest through Bergen-Belsen to Switzerland. Attributes the rescue of this group of Jews not to the efforts of the Zionist Rudolf Kasztner, who arranged it, but to the American Orthodox Vaad ha-Hatzalah (which provided the money that was paid to the Germans) and its representative in Switzerland, Jitschak Sternbuch. *RE*

0291 Gál, Fedor; Skálová, Dita; Zádorová, Marketa: *Trojí svědectví* [Three Testimonies]. Praha: G plus G, 2000. 141 pp.

Presents three testimonies of Holocaust survivors. Pp. 7-49, by Skálová (née Picková), relate the story of her deportation to the Terezín ghetto, from where she followed her husband to Auschwitz at the end of 1944. After being sent to the Birnbaum camp, she fled with two other prisoners from a death march and spent the last weeks of the war in the Els and Katowice refugee camps established by the Russians. On pp. 51-75, Gál, who was born at the end of the war, presents testimonies which he collected about the fate of his parents and other relatives from Německá L'upč in eastern Slovakia during the Shoah. He combines fragments of intimate personal memories with historical remarks and explanations. Pp. 77-138, by Zádorová (born in Mukachevo, which was under Hungarian rule during the war years), describe her experience of deportation to Auschwitz in the form of a fictional narrative. *JT*

0292 Gastfriend, Edward: *My Father's Testament: Memoir of a Jewish Teenager, 1938-1945.* Ed. with an afterword by Bjoern Krondorfer. Philadelphia: Temple University Press, 2000. xvi, 187 pp.

Memoirs of a Jew born in 1926 in Sosnowiec as Eliezer Gastfreund, the youngest son of a hasidic family of eight children. One brother immigrated to Palestine before the war; two brothers fled to the Soviet Union; one brother was arrested as a Polish POW; one sister married and went to Bedzin; another sister was arrested. The last three perished in the war. Between 1940-42 Gastfriend participated in resistance actions. In August 1942 his parents and remaining sister were deported (and perished). Gastfriend posed as a non-Jew, moving from place to place in Sosnowiec. In summer 1943 he was arrested by the Jewish police and sent to the Seibersdorf (Zebrzydowice) labor camp in Upper Silesia. In April 1944 all of the inmates were transferred to Blechhammer; Gastfriend was sent for forced labor at the construction plant Oberschlesische Hydrierwerke. In January 1945 he was taken on a death march to Gross-Rosen, and from there transported to Buchenwald and then to the Langenstein concentration camp, where he remained from February to April 1945. He was liberated when he escaped during the death march from Langenstein. After the war, Gastfriend immigrated to the USA. Pp. 173-187 contain an afterword by Bjoern Krondorfer. *DR*

0293

גטניו-אוסמו, נטה (אפטיחיה מזל): מקורפו לבירקנאו ולירושלים: סיפורה של ילידת קרקירה. תל-אביב: עקד, 1999. 96 עמ'.

[Gattegno-Osmo, Nata (Evtihia Mazal): *From Corfu to Birkenau and Jerusalem: The Story of a Woman Born in Kerkyra.* Tel-Aviv: Eked, 1999. 96 pp.]

Memoirs of a Jew born in Corfu in 1923. Pp. 48-69 describe her experiences in the Holocaust, including the German occupation of the island in fall 1943. Gattegno-Osmo was deported to Auschwitz in June 1944, together with her parents and one of her sisters, who perished in the camp. In December 1944 she was sent to Bergen-Belsen, and from there on a death transport to Mauthausen, where she was liberated. She returned to Corfu for a short time, where she found two sisters who had survived in hiding, and then she immigrated to Eretz-Israel. *LFo*

0294 Gilbert, Martin: *Never Again: A History of the Holocaust.* London: HarperCollinsIllustrated, in association with the Imperial War Museum, 2000. 192 pp.

An illustrated history of the Holocaust, including many photographs, maps, informative side-bars, and excerpts from documents, e.g. diaries and memoirs. *YC*

0295 Grohs-Martin, Silvia: *Silvie*. New York: Welcome Rain, 2000. 373, [16] pp.

Memoirs of a Jewish actress, Silvie Grohs, who was born in Vienna in 1918. In March 1938 she left Austria to work abroad. She settled in Amsterdam, where she was caught by the German occupation in 1940. Grohs and other Jewish refugee actors played in the Schouwburg theater. In mid-1942 the Germans turned the theater into a detention center for Jews. Grohs fled to Brussels. In 1943 she was sent to the Malines detention camp, from there to Auschwitz, and in September 1943 to Ravensbrueck. In spring 1945 she was among 700 women transferred to Sweden due to the efforts of the Swedish Red Cross. Just when the trucks were about to pass the last German stronghold it was air-bombed by the British; caught in the attack, 240 of the women prisoners were killed and 460 survived. Grohs was wounded. Later, she married and immigrated to the USA. *DR*

0296 Gryn, Hugo; Gryn, Naomi: *Chasing Shadows*. London: Viking, 2000. xxxvi, 264, [24] pp. Appeared also in paperback: London: Penguin Books, 2001 (xxxiii, 270 pp.).

The unfinished memoirs of Hugo Gryn (1930-1996), born in Berehove, Ruthenia, completed by his daughter. Ch. 11-13 (pp. 123-229), written by Gryn in the USA in 1951, describe his experiences in the Holocaust from the viewpoint of a teenager. In April 1944 Gryn and his family, with the other Jews of Berehove, were interned in a nearby brick factory. In June 1944 they were deported to Auschwitz. Gryn and his father were then sent to the Lieberose concentration camp in Upper Silesia. Naomi Gryn pieces together the remaining account from various writings and interviews, describing a death march which ended at Gunskirchen. Gryn's father succumbed there, shortly after the liberation; his brother perished at Auschwitz, but his mother survived. Gryn moved to England with a group of child survivors in 1946, and later became a well-known rabbi. *YC*

0297 Gutterman, Bella: Jews in the Service of Organisation Todt in the Occupied Soviet Territories, October 1941-March 1942. *Yad Vashem Studies* 29 (2001) 65-107. Appeared in Hebrew in "Yad Vashem" 29 (2001).

In 1941-42, 350 Jews from four labor camps under the auspices of Organisation Schmelt in Upper Silesia were disguised as employees of Organisation Todt and sent to the eastern front for work on rehabilitating the Soviet railway system. Based mainly on survivors' accounts, reconstructs the story of this work detail. It is difficult to ascertain which German office initiated the deployment of Jewish forced laborers for railway work near the frontline. Nonetheless, in the fall of 1941, the Jewish slave laborers were collected in the Gross-Masselwitz camp near Breslau and, after receiving Todt uniforms, were sent to Sebezh (Pskov region, Russia). Many of them died due to the harsh conditions: the hard work, frost, and typhus. In March 1942, the survivors were returned to Silesia. Subsequently, most of them perished in Gross-Rosen and Auschwitz. *DR*

0298

גוטרמן, בלה: יהודים בשורות ארגון טוט במסע לכיבוש ברית-המועצות (אוקטובר 1941-מרס 1942). יד ושם 29 (2001) 53-83.

[Gutterman, Bella: Jews in the Service of Organisation Todt in the Campaign to Occupy the Soviet Union (October 1941-March 1942). *Yad Vashem* 29 (2001) 53-83.] Appeared in English in "Yad Vashem Studies" 29 (2001).

0299 Győri, György: *Megmenekülėseim: Hosszú levél unokáimnak* [My Survival: A Long Letter to My Grandchildren]. Budapest: Belvárosi Könyvkiadó, 2001. 288 pp.

Memoirs of a Jew born in 1924 in Budapest. Describes his involvement in the underground movement as a member of the Hashomer Hatzair, forging documents and saving persecuted Jews. He was captured in 1944 on a mission to find an escape route. He was deported by the Nazis to Auschwitz and was later transferred to Mauthausen where he was liberated. Most of the book deals with his experiences in the concentration camps. *AV*

0300 Hagen, William W.: Murder in the East: German-Jewish Liberal Reactions to Anti-Jewish Violence in Poland and Other East European Lands, 1918-1920. *Central European History* 34, 1 (2001) 1-30.

Traces the reactions of liberal Jews in Germany, mainly through the Centralverein's periodicals "Allgemeine Zeitung des Judenthums" and "Im deutschen Reich," to the upsurge of antisemitism and pogroms in Eastern Europe, mainly in Poland but also in Romania, Hungary, Ukraine, etc. The CV press delivered insightful interpretations of the pogrom in Lvov in November 1918, as well as pogroms in Pinsk and other Polish cities. The violence was interpreted as a modern phenomenon of East European political life, associated with the rise of nationalism, anti-German and anti-Bolshevik moods, etc. The CV was disappointed by Poland and other new states, whose soldiers and officers perpetrated the pogroms, and by their governments' and Western powers' unwillingness to stop the violence. Despite this, German Jewish liberals remained loyal to their principles, believing that relations between the new states and the Jews would improve and that Jews should identify firmly with the countries in which they lived. *DR*

0301 Hardman, Anna: *Women and the Holocaust.* London: Holocaust Educational Trust, 2000. ii, 36 pp. (Holocaust Educational Trust Research Papers, 1, 3).

Discusses research from the 1970s to the present that has combined the perspectives of women's studies and Holocaust studies, basically with an overt feminist agenda. Deals with both secondary literature and memoirs of women who were prisoners in Auschwitz between 1942-45. Stresses the value of the woman's perspective to complement dominant male views of the Holocaust, but emphasizes the need for critical methodologies to overcome simplistic overviews of women's experiences, which indeed had gender-specific elements, both biological and social. Warns against accepting memoirs such as Fania Fenelon's as unmediated testimony, both in regard to sexuality in the camp and in regard to gender stereotyping in general. Contrasts the oral testimony of Anita Lasker-Wallfisch with Fenelon's account of the women's orchestra in Auschwitz. *YC*

0302 Harris, Mark Jonathan; Oppenheimer, Deborah: *Into the Arms of Strangers: Stories of the Kindertransport.* New York: Bloomsbury, 2000. xiii, 292 pp.

A collection of excerpts from narratives (testimonies, transcripts of videotaped interviews, letters, etc.) of former children of the Kindertransport from Germany, Austria, and Czechoslovakia to Great Britain between December 1938-August 1939. Some of the interviews took place in 1999, during the 60-year reunion of the children and their rescuers, held in London. The accounts in the book are by former "Kinder," parents or foster parents, and those who organized the transports. The excerpts are divided into chapters dealing with first encounters with antisemitism in pre-Nazi and Nazi Germany and Austria, the "Kristallnacht" pogrom, departure from Germany, reception in Great Britain, and wartime and postwar experiences (including the internment of some of the children in 1940). One of the interviewees failed to depart from Germany; she was sent to Theresienstadt, Auschwitz, and Bergen-Belsen, where she was liberated. *DR*

0303 Heinemann, Isabel: "Another Type of Perpetrator": The SS Racial Experts and Forced Population Movements in the Occupied Regions. *Holocaust and Genocide Studies* 15, 3 (Win 2001) 387-411.

Discusses the activities of the Rasse- und Siedlungshauptamt (RuSHA) – the Race and Settlement Main Office – of the SS. Having started with racial screening of prospective members of the SS and their spouses, this office became instrumental in the Nazi policy of "re-Germanization" of the occupied Polish and Soviet territories. Ca. 500 "racial experts" of RuSHA not only prepared large expulsions of "racially unfit" Slavs from areas chosen for German colonization, but were also involved in the persecution of Jews and "Mischlinge." For example, these experts opposed the treatment of "Mischlinge of the second degree" as Germans. They provided information for tracing Jews and persons with Jewish ancestry throughout Europe. The RuSHA experts' proposals and measures were always radical and spelled imminent death for victims in camps and elsewhere. *DR*

0304 *The Journal of Holocaust Education* 9, 2-3 [Fall-Win 2000] (2001).

A special issue, entitled "'Bystanders' to the Holocaust: A Re-evaluation," edited by David Cesarani and Paul A. Levine, based on papers delivered at an international colloquium held in Uppsala, Sweden, in September 1999. The papers focus on the response of "neutral" or Allied nations to the Holocaust. *SSC*

0305 Kádár, Gábor; Vági, Zoltán: *Aranyvonat: Fejezetek a zsidó vagyon történetéből* [The Hungarian Loot Train: Chapters from the Story of Jewish Property]. Budapest: Osiris, 2001. 304 pp.

Describes the demographic and economic situation of Hungarian Jews before March 1944. Relates how Jews were robbed of their property before and after the ghettoization. Tracks the route of the most precious valuables – gold, jewels, foreign currencies, paintings and carpets – which were loaded onto a train in March and sent to Austria. After 42 days, the train was handed over to the American army on Austrian soil. In 1946, the Inter-Governmental Committee on Refugees became responsible for the treasures, which were used to help survivors in DP-camps and elsewhere. Pp. 133-222 portray the senior SS-officer Kurt Becher, who served in Hungary from March 1944. He seized all the property of leading Hungarian Jewish industrialist Manfred Weiss for the SS, which enraged the Hungarian government. Becher held talks with Kasztner and saved a lot of Jews in return for their money. He also helped in the handover of Bergen-Belsen. He assumed the role of the rescuer to save himself and to become rich, in both of which he succeeded. *AV*

0306 Karsai, László: *Holokauszt* [Holocaust]. Budapest: Pannonica, 2001. 426 pp.

A general history of Nazism and the Holocaust. Relates the prewar policy of Hitler and Nazism towards the Jews; then describes the fate of the Jews during the war, country by country. Includes brief chapters on the actions of the Einsatzgruppen and on the concentration camps. Pp. 209-269 focus on the Holocaust in Hungary. *AV*

0307 Katzenelson, Itzhak: *Skazaniye ob istreblyonnom yevreiskom narode* [The Song of the Murdered Jewish People]. Trans.: Ye.G. Etkind. Ed.: Arno Lustiger. Moskva: Yazyki Russkoi Kultury, 2000. 239 pp. Translation of "Dos lied vunem ojsgehargetn jidischn volk" (Koeln: Kiepenheuer & Witsch, 1994), with the addition of the article by Lustiger.

Presents the text of Katzenelson's epic poem, in Yiddish (transcribed in Latin letters) and Russian, with commentary by Lustiger (pp. 191-200). Pp. 201-239 contain two articles. Pp. 201-212, "Yevreiskoye soprotivleniye natsistskomu rezhimu" ("Jewish Resistance to the Nazi Regime"), by Lustiger, shows that despite the allegations of some historians, Jewish resistance existed; illustrates this with the story of the Warsaw ghetto uprising. Pp. 213-239, "Itskhak Katsenelson, yevrei" ("Yitzhak Katzenelson, the Jew"), by Wolf Biermann, relates how the epic poem was written by Katzenelson, reflects on the poem and its author, and on Jewish passivity and Jewish resistance to the Nazi genocide. *DR*

0308 Klarsfeld, Serge, ed.: *Hommage à trois scientifiques juifs déportés de France: Ayzyk Gorny, mathématicien..., Vladislas Zeitline, physicien..., Xénia Gourvitch, chimiste....* Paris: Association "Les Fils et Filles des Déportés Juifs de France," 1999. 92 pp.

A memorial book for Ayzyk Gorny, Vladislas Zeitline, and Xenia Gourvitch, who were deported to Auschwitz, where Gorny and Zeitline were killed. Consists of letters, journals, and photographs kept by Fenia Berz, a friend of the three scientists. Gorny, born in Kamienec in 1908, immigrated to France in 1926, received his doctorate from the Sorbonne in 1940, and was deported via Drancy to Auschwitz in September 1942 after a year in the concentration camp at Le Vernet, although he had a visa for entering the United States. Zeitline, born in Tiflis in 1907, was a deaf-and-dumb physicist and a pioneer of television. He immigrated to France in the 1930s, was arrested in the Vel d'hiv roundup in July 1942, and deported a few days later. Gourvitch, who was born in Berlin in 1914 and settled in France in the mid-1920s, was arrested in a roundup on 6 March 1944 and deported to Auschwitz. She wrote a reconstructed journal of her experiences in the summer of 1945, published here for the first time (pp. 55-92). *EO*

0309 Knopp, Guido et al.: *Holokaust*. Eds.: Alexander Berkel, Silke Gampper, Christine Kisler. Muenchen: C. Bertelsmann, 2000. 416 pp.

A historical survey of the Holocaust, based on research undertaken for the A&E History Channel six-part documentary "Holokaust" (2001). Includes photographs from the series. Contents: Knopp, Guido; Bassewitz, Vanessa von: Menschenjagd (23-87); Knopp, Guido; Muellner, Joerg: Entscheidung (89-146); Knopp, Guido; Hartl, Peter: Ghetto (147-213); Knopp, Guido; Liechtenstein, Michaela: Mordfabrik (215-276); Knopp, Guido; Deick, Christian: Widerstand (277-338); Knopp, Guido; Dreykluft, Friederike: Befreiung (339-401). *ID*

0310 Kolin, Andrew: *One Family: Before and during the Holocaust.* Lanham, MD: University Press of America, 2000. 189 pp.

Based on archival documents and interviews with relatives and their friends, Kolin traces his family history from his great-grandparents to his parents, who were cousins. The family Kolniczanski stemmed from Kolnica, northeastern Poland. In the early 20th century one branch of the family settled in Warsaw, while another branch, to which his mother was born, immigrated to Paris and changed their surname to Kolnitchanski. In 1934, Motke Kolniczanski from Warsaw, Kolin's paternal grandfather, immigrated to the USA and changed his surname to Kolin; among his motives were growing antisemitism and fear of war. The Kolniczanskis, except for some family members who fled to the USSR, perished under the Nazi occupation, some of them as resistance fighters. Part of the Kolnitchanskis from Paris survived (Kolin's mother, Helen, by fleeing to Grenoble), while others were deported via Drancy to Auschwitz and perished. *DR*

0311 Kovner, Abba: *Scrolls of Testimony.* Trans.: Eddie Levenston. Philadelphia: Jewish Publication Society, 2001. xxxviii, 191 pp. Originally published as "Megillot ha-edut" (Jerusalem: Bialik Institute, 1993).

0312

קוז'ניצקי, חיים: נעורים בתופת. תרגום ועריכה: יוסף רב. תל-אביב: אבישלום, 1999. 207 עמ'.

[Kozienicki, Chaim: *Adolescence in Hell.* Trans. & ed.: Yosef Rab. Tel-Aviv: Avishalom, 1999. 207 pp.] Translated from the Polish manuscript.

Memoirs of a Jew from Łódź (b. 1928), part of which was written in 1947. Kozienicki, along with his parents and brother, was interned in the Łódź ghetto in May 1940. Describes life in the ghetto, and the activities of the Hanoar Hazioni youth movement, of which he was a member. In September 1944 he was deported to Auschwitz, then to Stutthof and to the Stolp labor camp. He was sent on a death march to the Baltic coast and then transported on a boat which was abandoned at sea by the Nazis. The prisoners managed to return to shore, but many of them were killed just before the liberation. His parents perished in the Holocaust, but his brother survived. In 1947 Kozienicki immigrated to Eretz-Israel. *RE*

0313 Kranzler, David H.: *The Man Who Stopped the Trains to Auschwitz: George Mantello, El Salvador, and Switzerland's Finest Hour*. Syracuse, NY: Syracuse University Press, 2000. xxv, 341 pp.

A biography of Mantello, born in 1901 to the Orthodox Jewish family of Mandl in Transylvania. As a businessman, he became honorary consul of El Salvador for several East European countries; in 1942-45 he served as first secretary of the El Salvador consulate in Geneva. With the approval of the consul general, Mantello provided thousands of Jews in Hungary and Transylvania with protective papers. In addition, during 1941-44 he gathered information on the Nazi Final Solution, including operations at Auschwitz. In 1944, using this data, he initiated two press campaigns in Switzerland which alerted that country and the entire world to the Nazi genocide of Jews. The first press campaign forced Horthy to stop the deportations. Dwells, also, on the activities of Florin Manoliu, a Romanian businessman who helped Mantello. During the war, Mantello was arrested several times by Swiss authorities on superficial pretexts. After an inquiry into his wartime activities in the 1950s, Mantello was forgotten and recalled only shortly before his death in 1992. *DR*

0314 Land-Weber, Ellen: *To Save a Life: Stories of Holocaust Rescue*. Urbana: University of Illinois Press, 2000. ix, 331 pp.

Land-Weber participated in the Altruistic Personality Project carried out by Samuel and Pearl Oliner; in the early 1980s she interviewed fifty rescuers for the project. Between 1985-88 she interviewed seven of the rescuers, as well as some of the Jews they had rescued. This book contains accounts by rescuers – Tina Strobos, Bert Bochove, and Mirjam Waterman Pinkhof from the Netherlands; Anna Chlup (who spoke for herself and her husband Jaruslav) from Czechoslovakia; Barbara Szymanska Makuch and her sister Halina Szymanska Ogrodzinska, and John Damski from Poland – as well as by ten people whom they rescued. Pinkhof, herself a Jew and a member of a Zionist youth movement in Loosdrecht, participated in the rescue network established by Joop Westerweel and Joachim Simon. Makuch was a courier for Żegota. Herman Feder, rescued by the Chlups, describes his experiences in Sachsenhausen. After the war these rescuers immigrated to the USA and Canada. *DR*

0315 Langerbein, Helmut Paul: *Profiles of Mass Murder: The Einsatzgruppen Officers*. Diss. – University of California, Santa Cruz, 2000. 272 pp. Unseen.

This study attempts to answer the question: how could so many Germans willingly contribute to the torture and murder of the Jews? Examines the behavior and the motivation of the perpetrators most directly involved in the mass shootings of the Soviet Jews, the officers in the Einsatzgruppen. Based on records from German postwar trials, analyzes the careers of individual officers and their actions before, during, and after their time in the Einsatzgruppen. Describes the various types of killers, and concludes that not all of the officers responsible for the massacres were willing tools of the Nazi leadership, driven by blind antisemitic fanaticism. Rather, human beings faced with the situation in the killing fields responded differently. Most of the officers were motivated by career considerations, submission to orders, peer pressure, human weakness, and other factors.

0316

לנגלבן, לאה: נשים יהודיות מהונגריה באושוויץ-בירקנאו ב-1944. הגיגי גבעה 7 (1999) 93-118.

[Langleben, Leah: Jewish Women from Hungary in Auschwitz-Birkenau in 1944. *Hagigei Giv`ah* 7 (1999) 93-118.]

Analyzes 16 testimonies of Jewish Orthodox women, survivors of Auschwitz, who were deported in 1944 from Hungary and territories annexed to Hungary. The texts are part of 46 testimonies documented by Moshe Davis, in conjunction with Yad Vashem, during group therapy sessions conducted by him with Holocaust survivors who founded the moshav Nir Galim in Israel and attended discussions and lectures about the Holocaust. Presents data on the background of the interviewees, e.g. age, origins (family and community), and socioeconomic status. Focuses on their interpretations of their ordeal, their survival strategies, and the impact of their cultural, spiritual, and religious origins on their coping mechanisms in the concentration camp. States that their hope to be reunited with their families, supportive social connections in the camp, and their religious faith probably had a great impact on their survival. *LFo*

0317 Levin, Dov: How the Jewish Police in the Kovno Ghetto Saw Itself. *Yad Vashem Studies* 29 (2001) 183-240. Appeared in Hebrew in "Yad Vashem" 29 (2001).

Presents an English translation of excerpts (parts of chs. 3, 6, and 11) from a Yiddish manuscript entitled "Di geshikhte fun der Vilijampoler idisher geto politsai" ("The History of the Jewish Police in Vilijampole"), apparently written in 1942-43 in Kovno. Vilijampole was Kovno's Jewish district, where the ghetto was established. The excerpts deal with the behavior of the Jewish police during the "great action" – the deportation of 10,000 Jews to the Ninth Fort for execution; organizational changes in the police after the "action"; subsequent small "actions"; and some internal matters of the Jewish police. They show the police as a body identifying with the ghetto population, not with the Germans, sometimes daring rescue attempts, and even tinged with Jewish nationalist spirit. Chs. 6 and 11 of the manuscript were published in Yiddish in "YIVO Bleter" 3 (1997). Pp. 183-202 contain an introduction describing the history of the Kovno ghetto and of its Jewish administration. *DR*

0318 לוין, דב: המשטרה היהודית בגטו קובנה בעיני עצמה. יד ושם 29 (2001) 143-185.

[Levin, Dov: How the Jewish Police in the Kovno Ghetto Saw Itself. *Yad Vashem* 29 (2001) 143-185.] Appeared in English in "Yad Vashem Studies" 29 (2001).

0319 Levine, Allan Gerald: *Fugitives of the Forest*. Toronto: Stoddart, 1998. xxxi, 400 pp.

Describes the escape of thousands of Jews to the forests in Nazi-occupied Poland and the USSR between 1941-44. Relates the problems they faced, as well as Jewish participation in partisan warfare, Soviet or Polish. The relations of the Jews with their non-Jewish comrades in the partisan units, and with the surrounding non-Jewish population, were complicated. Many partisans were reluctant to accept non-combatants into their units and held Jewish fighting abilities in low esteem; antisemitism was widespread. Dwells on the two largest Jewish partisan "family camps" in Belorussia – the camp of the Bielski brothers and of Shalom Zorin. *DR*

0320 Liekis, Sarunas; Milyakova, Lidiya V.; Polonsky, Antony B.: Three Documents on Anti-Jewish Violence in the Eastern Kresy during the Polish-Soviet Conflict. *Polin* 14 (2001) 116-149.

Following an introduction, presents three documents (on pp. 121-149) describing anti-Jewish violence by Polish troops in Lida and Vilna in 1919, when Jews were widely identified with the Bolsheviks and many were massacred. Document 1 is the report of a local Jewish committee that investigated events after the Polish Legion captured Lida, including the murder of innocent Jews and looting of their property. The pogrom was attributed to rumors based on slander and prejudice. Jews were also subjected to insult and humiliation, forced labor, requisitions, and arrest. Document 2 is a list of 38 Jews killed in Lida, giving their ages and occupations and the circumstances of death. Document 3 is a report by Jews of Vilna to the Polish government about deportations, murders, atrocities, and looting suffered by Jews in Vilna and the surrounding areas. The Polish press is blamed for accusing all Jews of being Bolsheviks and, hence, making them targets for abuse and destruction. *YC*

0321 Longerich, Peter: *The Wannsee Conference in the Development of the 'Final Solution'*. London: Holocaust Educational Trust, 2000. ii, 29 pp. (Holocaust Educational Trust Research Papers, 1, 2).

The Wannsee Conference of 20 January 1942 did not launch the genocide of European Jewry – it had been under way in the Soviet territories and the Balkans since summer 1941. The annihilation of the Jews was a central aim of the Nazi regime and is not to be regarded as a mere side-effect of historical developments in this period. In fact, it began in 1939 with the implementation of "Jewish reservation" projects, and it accelerated in 1941-42. The Wannsee Conference coincided with the USA's entry into the war and the Soviet counter-offensive from Moscow, two events which put an end to German hopes for a quick victory. Thus, Wannsee shifted the Final Solution from the postwar plan to wartime, and also decided on the method – annihilation through work, which had been tried before 1942 as well as outlined in a pan-European deportation program. The order for the total annihilation of all Jews was issued only in May 1942; the planned "final solution" in the form of the systematic annihilation of the Jews in the Generalgouvernement and in the Soviet territories reached its full scope in the summer of 1942. *DR*

0322 Margalit, Ruth; Margalit, Benjamin: *"Zerbrochene Kindheit": Lebenserinnerungen*. Eds.: Micheline Prueter-Mueller, Gertrud Wagner. Detmold: Gesellschaft fuer Christlich-Juedische Zusammenarbeit in Lippe, 2001. 114, [32] pp. (Panu derech, 20).

Presents, separately, the memoirs of Ruth and Benjamin Margalit. Ruth (née Ehrmann) was born in 1931 in Detmold. The family was considered a mixed marriage, although her mother had converted to Judaism. In 1943 Ruth's father and her two brothers were sent to Bielefeld for forced labor. At the end of 1944 all of the family, except the mother, were deported to Theresienstadt, where they were liberated. In 1949 the family immigrated to Israel. Benjamin Margalit (né Margulies) was born in 1923 in Vienna; in March 1940 he immigrated to Palestine. His parents were deported in May 1942 to the Minsk ghetto and they were murdered in Maly Trostinets. *ID*

0323 Maron, Monika: *Pawels Briefe: Eine Familiengeschichte*. Frankfurt a.M.: S. Fischer, 1999. 205 pp.

Maron, whose Jewish grandfather Pawel had converted to Christianity in his youth, relates the history of the family by means of excerpts from his correspondence. Pawel Iglarz was born in 1879 in Ostrów; he immigrated to Berlin in 1905, after his marriage to a Polish woman. In November 1938 he was expelled from Germany. During his stay in a camp near the border he applied for refuge in England, but was refused. His wife, who could have chosen divorce or expulsion, went to live with him in Kurów near Łódź. From this time they corresponded with their four grown children who remained in Germany. In spring 1942 Iglarz was sent to the Bełchatów ghetto, from where he was presumably deported to Chełmno and gassed. *ID*

0324

מאיר, מנחם; ריימס, פרדריק: האם העצים פורחים אצלכם? זיכרונות ומחשבות של שני אחים. ירושלים: יד ושם, 2001. 170, [16] עמ'.

[Mayer, Menachem; Raymes, Frederick: *Are the Trees in Bloom Over There? Memories and Thoughts of Two Brothers*. Jerusalem: Yad Vashem, 2001. 170, 16 pp.]

Memoirs of two brothers from Hoffenheim, Germany, born as Heinz and Manfred Mayer. Raymes was born in 1929 and Mayer in 1932. During the "Kristallnacht" pogrom the synagogue, in which they lived, was destroyed; their father was interned in Dachau. In October 1940 the family was deported to Gurs, from which the brothers (and 46 other children) were rescued through the efforts of Andrée Salomon (of the OSE) and of the American Friends Service Committee. They were taken to a children's home, but were separated in 1943. Mayer was sent to a children's home in Toulouse and was later smuggled into Switzerland; he immigrated to Israel in 1948. Raymes, through rescue efforts of the Eclaireurs Israélites de France, was sent to Moissac. He worked for the group's underground movement. He immigrated to the U.S. in 1946. Pp. 71-104 contain letters to the brothers from their parents and other relatives, written between March 1941-September 1942. Their parents were deported in summer 1942 to Auschwitz, where they perished. *RE*

0325 Meschenmoser, Alfred: *Die Judenverfolgung im Dritten Reich: Eine philatelistisch-zeitgeschichtliche Darstellung. Vol. 1-3*. Schwalmtal: Phil-Creativ, 2001. 148; 69; 124 pp.

A history of the Nazi persecution of the Jews in Germany based on philatelic sources – postcards, stamps, etc. Vol. 1, covering the period 1933-37, surveys the Nazi seizure of power, the boycott against Jewish enterprises, the exclusion of Jews from professional and cultural life, and the Aryanization of Jewish businesses. Discusses, also, Jewish reactions (e.g. reorganization of communal institutions, welfare activities), Nazi legislation, and Jewish emigration. Vol. 2, on 1938-39, discusses the dispossession of the Jews in all spheres and analyzes Jewish emigration as a problem from the international and the German point of view. Vol. 3, on 1940-45, discusses forced labor, deportations, German policies in the Protectorate, Poland, and the USSR, and the ghettoization and the concentration camps in all of Europe. *ID*

0326 Messerschmidt, Manfred: Ideologie und Befehlsgehorsam im Vernichtungskrieg. *Zeitschrift fuer Geschichtswissenschaft* 49, 10 (2001) 905-926.

Assigns responsibility for the massacres of Jews and suspected partisans, and of the general population in the USSR and the Balkans, to the German military commanders. These were part of the conservative elite which shared an ideology very close to that of the Nazis: the hegemony of Germany and its mission to impose a new order in Europe, wiping out the Jewish-Bolshevist arch-enemy, the foe of all that is cultured, pure, and noble. Historians propagated this ideology, and jurists backed it asserting that in this conflict Germany had the right to be a law unto itself. Soldiers, already infused with this ideology by Nazi education, were further indoctrinated by their commanders. Thus the commanders issued, and the soldiers obeyed without question, orders to cooperate with the SD in massacres, or (especially in the Balkans) themselves to take and shoot hostages, first and foremost Jews. Gives many examples, up to the very end of the war, with only a few instances of commanders who tried to protest. *RW*

0327 Michman, Dan: *Pour une historiographie de la Shoah: Conceptualisations, terminologie, définitions et problèmes fondamentaux.* Trans.: Nelly Hansson. Paris: In Press, 2001. 536 pp. Originally published as "Ha-Shoah ve-hikrah: Hamsagah, minuah, ve-sugyot yesod" (Tel-Aviv: Moreshet, with Beit Lohamei Haghetaot and Yad Vashem, 1998).

A collection of articles and lectures, some of them published previously, including some material not published in the Hebrew version. Partial contents: La Shoah vu par les historiens: Questions de conceptualisation, de périodisation et d'interprétation (17-66); La dimension juive de la Shoah: Le contexte de l'histoire juive de ces derniers siècles (67-112); Quel rapport entre fascisme et nazisme? Le fascisme hollandais avant et pendant la Shoah: Un cas d'étude (115-139); "Autorités" juives ("Headships") sous le nazisme: Naissance et mise en application d'une conception administrative (157-180); Clarifier la notion de "sauvetage pendant la Shoah" (183-219); Zeev Jabotinsky – le "plan d'évacuation" et la question de la prévision de la Shoah (221-234); La résistance juive à la Shoah et sa signification: Observations théoriques (237-284); La vie religieuse juive sous le nazisme: Attitudes nazies et problèmes juifs (287-313); La vie quotidienne du juif religieux dans les conditions créées par la Shoah (315-335); Croire pendant la Shoah: Courants et orientations (337-355); La recherche historique sur la Shoah: Histoire et contexte (403-433); Sujet singulier, voix plurielles: Langue et culture dans la recherche sur la Shoah (435-478); L'étude de la question du "sionisme face à la Shoah": Problèmes, polémiques et terminologie de base (479-516). *SSC*

0328 Nolte, Ernst: Der Nationalsozialismus als Vernichtungsideologie und -realitaet vor dem Hintergrund des Bolschewismus. *Studia nad Faszyzmem i Zbrodniami Hitlerowskimi* 24 (2001) 33-70.

Based on a lecture held at the Freie Universitaet Berlin in the winter semester 1998/99, his final lecture before retirement. Analyzes the similarities and differences between Soviet communism and Nazism and the fight to the death between them. Argues that Hitler and other Nazi leaders genuinely feared the spread of communist terror over all of Europe, and that they used the terms "Bolshevists" and "Jews" synonymously. They held that communist brutality could only be overcome by equal or greater brutality. But whereas the Soviets purged their country of certain social classes, the Nazis purged themselves of the Jews, a "foreign" race identified with Bolshevism. Finds a large element of truth in this identification, since many Jews were socialists, and Jews were prominent in both the Russian Revolution and the communist revolutions of 1919 in Germany; but adds that this view ignores the ideological pluralism among Jews. Asserts that Jews were certainly enemies of the Nazis, who were thus justified in interning them like other enemies in wartime, but that the enmity was caused by the Nazis themselves. Also suggests that in destroying the Jews the Nazis attempted to destroy modernism. Calls for greater freedom of scholarship in Germany, permitting understanding of the Holocaust in perspective. *RW*

0329

אודלר, מיכל: עדותה של רוזה אייזקס ילידת יאנינה. לא נשכח! 16 (2001) 14-26.

[Odler, Michal: The Testimony of Roza Izaks from Ioannina. *Lo Nishkach!* 16 (2001) 14-26.]

Memoirs of Izaks (1919-1999), who was born in Ioannina, as told to Odler, her granddaughter. Describes the German occupation of Ioannina in 1942 and the deportation of the Jews in March 1944. Izaks and her parents were deported to Auschwitz; her parents were killed immediately, while she was sent to work. After the liberation Izaks returned to Ioannina; she immigrated to the U.S. in 1951 and to Israel in 1976. *RE*

0330 Ondřichová, Lucie: *Fredy Hirsch: Von Aachen ueber Duesseldorf und Frankfurt am Main durch Theresienstadt nach Auschwitz-Birkenau: Eine juedische Biographie 1916-1944*. Trans.: Astrid Prackatzsch. Konstanz: Hartung-Gorre, 2000. 116 pp. Translated from the Czech: "Příběh Fredyho Hirsche" (Praha: Sefer, 2001).

A biography of Alfred (Fredy) Hirsch, born in 1916 in Aachen. He was active there as a leader of the Jewish scout movement from 1933, and later in Duesseldorf and Frankfurt. In 1935 he became a leader of

Maccabi Hatzair in Czechoslovakia (first running a camp in Slovakia; from 1936-39 he was based in Brno). In 1939 he moved to Prague, where he organized illegal transports to Palestine in the framework of the Hehalutz movement. Between 1942-43 Hirsch was interned in Theresienstadt, where he organized educational programs for children. In September 1943, Hirsch was deported to Auschwitz, along with a group of 274 children. He managed to obtain a special block for them and to carry on with his educational and cultural programs. On 8 March 1944 almost all of the children were gassed; Hirsch committed suicide several hours before the children were murdered. *ID*

0331 Ondřichová, Lucie: *Příběh Fredyho Hirsche* [The Story of Fredy Hirsch]. Praha: Sefer – Institut Terezínské iniciativy, 2001. 113 pp. See the abstract for the German translation: "Fredy Hirsch" (Konstanz: Hartung-Gorre, 2000).

0332

אוסטרובר, חיה: ההומור כמנגנון הגנה בשואה. דיס' – אוניברסיטת תל-אביב, 2000. [358] עמ'.

[Ostrower, Chaya: *Humor as a Defense Mechanism in the Holocaust.* Diss. – Tel-Aviv University, 2000. [358] pp.]

Based on interviews with 55 survivors, analyzes types and functions of humor in the Holocaust. Humor was a stress reducer in the horrible circumstances of the Holocaust. It served mainly as a defense mechanism, through self-humor and gallows humor. Also discusses aggressive (jokes about prominent Jews holding official jobs), sexual, social, and intellectual humor. Findings show that survivors who had previously had a sense of humor maintained it during and after the Holocaust. Humor reduced the horrors and facilitated coping with them. In addition to humorous utterances, discusses humorous songs and cabarets, and presents examples of humorous paintings and drawings. *RE*

0333 Pick, Miloš: *Verstehen und nicht vergessen: Durch Theresienstadt, Auschwitz und Buchenwald-Meuselwitz: Juedische Schicksale in Boehmen 1939-1945.* Trans.: Liselotte Teltscherová. Konstanz: Hartung-Gorre, 2000. 90 pp. Originally published in Czech as "Přes Acherón zpátky to šlo hůř: Opožděná zpověď' vnoučatům – Šimůnkovi a Terezce – až dorostou, aby porozuměla a nezapomněla" (Litomyšl: Augusta, 1997).

Memoirs of a Jew born in 1926 in Libáň (Bohemia) to a wealthy family of factory owners. In 1941 they were forced to move to a smaller apartment and to wear the yellow star. Pick became involved with the communist resistance. Pick became involved with the communist resistance. In January 1943 the family was deported to Theresienstadt, and in

September 1944 Pick was deported to Auschwitz, where he found his sister and learnt that his parents had been gassed. He was sent to work in the Meuselwitz labor camp (a satellite of Buchenwald); from there he was taken on a death march and managed to escape. After the war he settled in Prague. His sister also survived. *ID*

0334 *Revue d'Histoire de la Shoah* 171 (Jan-Apr 2001).

This issue is entitled "Des voix sous la cendre: Manuscrits des Sonderkommandos d'Auschwitz-Birkenau." Contents: Saletti, Carlo: A propos des manuscrits des membres du "Sonderkommando" de Birkenau (12-17); Mark, Bernard (Ber), ed.: "Megilot Auschwitz" (Les Rouleaux d'Auschwitz) [Three accounts by members of the "Sonderkommando" – Zalman Gradowski, Leib Langfus, Zalman Levental – translated from Mark's book, in Yiddish, "Megiles Oyshvits" (Tel-Aviv: Yisroel-Bukh, 1977).] (18-126); Gradowski, Zalman: Extraits de "Au coeur de l'enfer," 1944 [Translated from the Yiddish: "Sefer Ushpitzin," eds. Chaim Wolnerman et al. (1977).] (127-148); Mesnard, Philippe: Ecrire au dehors de la mort [On Gradowski's manuscript.] (149-162); Le recueil "Auschwitz" [Translation of a Yiddish account by a member of the "Sonderkommando," written 3 January 1945, and published in "Yivo Bleter" 27 (1946).] (162-167); Les survivants des "Sonderkommandos" au procès de Cracovie en 1946 [Testimonies by Henryk Tauber, Szlama Dragon, and Alter Feinsilber (mistakenly listed as Foicinbler in the testimony), translated from the Polish.] (168-219); Piper, Franciszek: Les conditions de vie et de travail spécifiques du "Sonderkommando" (220-231); Bensoussan, Georges: Les étapes du camp de concentration et du centre de mise à mort de Auschwitz-Birkenau (1940-1945) (232-240); Hazan, Katy: Les lieux d'oubli: Sur les traces de l'extermination (241-247); Greif, Gideon; Gabbay, Yaakov: Témoignage du "Sonderkommando" Yakov Gabbay [An interview conducted with Gabbay in Israel; translated from the Hebrew.] (248-291); Greif, Gideon: La tragédie des hommes du "Sonderkommando" [Translated from his book, in Hebrew, "Bakhinu beli dema`ot" (1999).] (292-303); Saletti, Carlo: A l'épicentre de la catastrophe (304-316); Cohen, Nathan: Manuscrits des "Sonderkommandos" d'Auschwitz: Tenir face au destin et contre la réalité [Appeared in English and Hebrew in "Yad Vashem Studies" 20 (1990).] (317-354); Saletti, Carlo: Tableau synoptique des principaux événements survenus à Auschwitz et au "Sonderkommando" (355-380); Mesnard, Philippe; Saletti, Carlo: Bibliographie sélectionnée (381-388). *HV*

0335 *Revue d'Histoire de la Shoah* 172 (May-Aug 2001).

This issue is entitled "Entre piège et refuge: Sur la traque et le sauvetage des pérsecutés, 1940-1945." Partial contents: Cahen, Fabrice: Le Comité international de la Croix-Rouge (CICR) et les visites de camps: Etude d'une controverse [About the visits of Red Cross delegate Maurice Rossel in Theresienstadt and Auschwitz.] (7-62); Krouck, Bernard: La mission de Victor Martin à Auschwitz (1943) [On the activities of Martin, a Belgian resistant, who visited Auschwitz and brought back information about the fate of the Jews.] (66-96); Pintel, Samuel: Les centres d'accueil du Service social des étrangers sous Vichy (1941-1944) (97-158); Pettinotti, Olivier: Gilbert Lesage, l'âme du Service social des étrangers (SSE) (159-173); Boulet, François: Les Juifs en Isère, Savoie et Haute-Savoie (1940-1944): De la touristophobie aux montagnes-refuges (174-227); Yagil, Limor: Typologie de la résistance sans armes et de l'aide aux Juifs en Limousin (228-265). *HV*

0336 Rodak-Izso, Helen: *The Last Chance to Remember*. Montreal: Concordia University Chair in Canadian Jewish Studies; Montreal Institute for Genocide and Human Rights Studies, 2001. 97 pp. (Memoirs of Holocaust Survivors in Canada, 17).

Memoirs of Rodak-Izso, born Friedmann in the 1920s in Abauj-Kér, Slovakia, and raised in nearby Košice (between 1939-44, Kassa, Hungary). She married in 1942; her husband was sent to a labor camp, then into combat on the Russian front. Describes the German occupation in 1944, after which Rodak-Izso was deported to Auschwitz with her mother, who was killed, and her sister Olly, with whom she survived the war. They were sent to the Kaiserwald concentration camp, then to several labor camps and to Stutthof, and then to the Gloewen labor camp, a satellite of Sachsenhausen. Finally, they were taken on a death march and liberated near Hamburg. One of her brothers also survived the war, while another had immigrated to Palestine. Rodak-Izso includes details about many family members who perished. After the war she and her husband settled in Toronto. *YC*

0337 Rowiński, Aleksander: *Zygielbojma – śmierć i życie* [Zygielbojm – His Death and Life]. Warszawa: Oficyna Literatów "Rój," 2000. 272 pp.

A popular account of the life and death of Szmul Artur Zygielbojm, written by a Polish journalist in the 1980s, based on the Yiddish "Zygielbojm-bukh" (New York, 1947), conversations with Zygielbojm's brother Fayvl in Israel and with other people who knew Zygielbojm, as well as on documents from the Bund archives in New York. Fayvl relates the family history before the war. Pp. 45-103 contain a translation of Szmul's diary (written in Yiddish) describing his travel at the end of 1939 from occupied Poland through Germany to Belgium. The Dutch did not allow him entry to the Netherlands, so returned to Poland. Later, he left Poland for New York, and was then sent to London as a Jewish representative to the Polish National Council. Describes in detail one month (October 1942) of his activities in London. His main goal was to inform and to influence the governments of the free world to stop the Nazi killings of Jews in Poland. Rowinski quotes from conversations he had with people who met Zygielbojm shortly before his suicide in May 1943. Rowinski also quotes from a conversation he had with Jan Karski. *HV*

0338 Saari, Peggy; Saari, Aaron Maurice, eds.: *The Holocaust and World War II Almanac. Vol. 1-3.* Detroit: Gale Group, 2001. lxxxvii, 460; lxxxvii, 470; lxxxvii, 517 pp.

A popular history of World War II and the Holocaust, interspersed with excerpts from historical documents, memoirs, etc., and accompanied by numerous photographs. Vol. 1 discusses, inter alia, the Nazi rise to power, the development of Nazi antisemitism, Nazi anti-Jewish policy in 1933-39, the anti-Jewish measures in occupied Poland, ghettoization, and the murder of Polish Jews. Dwells on the Warsaw ghetto and Jewish resistance. Discusses, also, the Nazi death camp system, focusing on Auschwitz. Reports on the fate of the Jews in the USSR, Denmark, the Netherlands, and France. Vol. 2 discusses rescue of Jews, the Nuremberg trials, and Holocaust remembrance. Vol. 3 contains biographies of 103 personalities, ca. 30 of them related to the history of the Holocaust. *DR*

0339 Saunders, Harold: *Spravodlivosť bola ďaleko* [Justice Was Far Away]. Trans.: Dominik Bartosiewicz. Bratislava: Slovenské Národné Múzeum – Múzeum Židovskej Kultúry, 2001. 120 pp. Originally published as "Zeugnis geben" (Konstanz: Hartung-Gorre, 2001).

0340 Saunders, Harold: *Zeugnis geben: Von Bratislava durch Auschwitz-Birkenau ins Lager Gleiwitz I und zurueck: Juedische Schicksale in der Slowakei, 1938-1945.* Trans.: Avri Salamon. Konstanz: Hartung-Gorre, 2001. 120 pp. Translated from English.

Memoirs of a Jew, born Harold Salamon in 1927 in Bratislava. In September 1944 Saunders was deported, with his parents, two brothers and a sister, to Auschwitz. Another brother (the translator of this book) survived in hiding. His mother and sister were killed on arrival. The male members of the family were sent to the Gleiwitz labor camp. The camp was evacuated on 21 January 1945 and they were taken on a death march; they were liberated by the Red Army, and returned to Bratislava. After the war Saunders' father and brothers immigrated to Palestine, while he settled in England. *ID*

0341 Schindler, Oskar: *Ich, Oskar Schindler: Die persoenlichen Aufzeichnungen, Briefe und Dokumente.* Ed.: Erika Rosenberg. Muenchen: Herbig, 2000. 448 pp.

A collection of letters (from the postwar period) by, to, and about Schindler, with connnective biographical material by the editor. Pp. 23-109 contain Schindler's descriptions of his wartime activities. They include letters written in an attempt to secure compensation for the fortune spent on maintaining his workers and bribing Nazi officials; details requested by writers and filmmakers who wanted to use his story, as well as tables of expenses, a plan of the factory complex, and lists of the workers. Other sections deal with attempts of Schindler and his supporters to gain him recognition and to solve his financial problems, his relationship with his wife, and the failed project (in his lifetime) to produce a film on his rescue of Jews. *RW*

0342 Schneider, Peter: *"Und wenn wir nur eine Stunde gewinnen...": Wie ein juedischer Musiker die Nazi-Jahre ueberlebte*. Berlin: Rowohlt, 2001. 159, [16] pp.

A biography of Konrad Latte, a Jewish musician born in 1922 in Breslau. With the rise of the Nazi regime, Latte was forced to transfer to a Jewish school and his father's business was Aryanized. During the "Kristallnacht" pogrom his father was arrested and deported to Buchenwald. In 1941 the family moved into a "Jewish house" and Latte was assigned to forced labor. In March 1943, warned of impending deportation, the family fled to Berlin. They went into hiding assisted by Harald Poelchau, a priest, and by friends in artistic circles. Latte studied with pianist Edwin Fischer, obtained work at the opera, and served as an organist in a church. In October 1943 his parents were arrested and deported to Auschwitz. Latte managed to avoid deportation; however, he was detained in prison, awaiting trial on suspicion of assisting deserters. He escaped during an air raid and survived with a false identity. After the war he remained in Germany. *ID*

0343 Spiegel, Paul: *Wieder zu Hause? Erinnerungen*. Berlin: Ullstein, 2001. 295 pp.

Memoirs of Spiegel, now president of the Zentralrat der Juden in Deutschland. He was born in 1937 in Warendorf. Pp. 18-83 deal with the Nazi period. His father was arrested during the "Kristallnacht" pogrom, and afterwards left for Belgium. Several months later the rest of the family followed – his wife, seven-year-old daughter, and Paul. A few days after the German occupation, Spiegel's father was arrested and deported to Buchenwald and Auschwitz. The rest of the family found refuge with the Blomme family, the father's former employer. In October 1942 Spiegel's sister Rosa was arrested on the street and disappeared. Spiegel was placed with a Belgian family in Uccle, and later with a peasant family near Namur. In January 1945 the village was liberated.

Spiegel's father survived Auschwitz, and after the war the family returned to Warendorf. Rosa was murdered in Auschwitz in November 1942. *ID*

0344 Srubar, Helena: *Eine schreckliche Zeit: Tschechisch-juedische Ueberlebensgeschichten 1939-1945*. Konstanz: Hartung-Gorre, 2001. 180 pp.

Following an introduction (pp. 11-37), presents the stories of nine Holocaust survivors from Czechoslovakia (based on interviews with them): Hanka Selzerová, Marta Navrátilová, Jiří Kosta, Ruth Šmolková, Eva Roubíčková, and four others who only give their first name. They were deported to concentration camps, mostly to Theresienstadt, and some from there to Auschwitz. All did forced labor and some participated in death marches. During an evacuation from a labor camp, Selzerová jumped from a train and then survived in hiding in Prague. *ID*

0345 Stier, Frank: *Kriegsauftrag 160: Behelfsheimbau im Ghetto Litzmannstadt (Łódź) und im KZ-Aussenlager Koenigs Wusterhausen durch das Deutsche Wohnungshilfswerk*. Berlin: Verlag Wilmuth Arenhoevel, 1999. 128 pp.

While other ghettos served mainly as internment and transit camps, the Łódź ghetto developed under German occupation into a labor camp where Jews did forced labor in more than 130 factories manufacturing goods for the German military. In summer 1944, most of the 70,000 remaining Jews in the ghetto were deported to Auschwitz. Only a few hundred remained as a "clean up" commando; ca. 600 Jews were transferred to Koenigs Wusterhausen, a satellite camp of Sachsenhausen, in order to continue the project Kriegsauftrag 160. This project, undertaken in the Łódź ghetto in 1943, involved construction of pre-fabricated cottages for the Deutsche Wohnungshilfswerk. As a result of the bombing of urban centers by the Allies, the German authorities sought a solution to the problem of homeless persons and opted for the dispersion of the population outside the towns in temporary housing. In April 1945, the camp Koenigs Wusterhausen was liberated by the Red Army. *ID*

0346 Stoecker, Jens: "Wir haben dauernd Hunger...": Die Deportation pfaelzischer Juden in das Camp de Gurs. *Tribuene* 157 (2001) 50-59.

Chronicles the deportation of Jews from the Palatinate to Gurs in October 1940, together with Jews from Baden and Alsace. Describes the lack of food, the miserable living conditions and the high rate of disease and mortality in the camp, and the efforts of French and international aid organizations to ameliorate these conditions. The French considered the upkeep of the camp a burden and agreed to the German proposal to transport the inmates to the East. The deportations, mainly to Auschwitz, began in March 1942. Few survived. The camp has been restored as a memorial site and commemorations are held annually; but in the Palatinate there were few efforts to commemorate its Jews until the 60th anniversary of their deportation in 2000. *RW*

0347 Szegő, György: *Két ima egy istenhez* [Two Prayers to One God]. Budapest: Múlt és Jövő, 1999. 320 pp.

Memoirs of a psychologist now living in Australia. He was born in Mezőberény, Hungary, in 1928. His father converted to Catholicism before his birth, and his mother converted when he was eight years old. Describes internment in the ghetto of Mezőberény in 1944, the deportation to Auschwitz, the transfer to Muehldorf labor camp, the horrors of camp life and his own story of survival. His father also survived. *AV*

0348 Szita, Szabolcs: *Együttélés, üldöztetés, Holokauszt: Segédkönyv a történelem középiskolai tanulásához* [Coexistence, Persecution, Holocaust: A Book to Aid the Study of History in Secondary Schools]. Budapest: Korona, 2001. 288, [40] pp.

The book is divided into two sections: the first describes how Nazi ideology developed into the Holocaust, with a particular focus on the concentration camps, and the second deals with the history of Hungarian Jewry prior to and during the Holocaust. In the second section, discusses the Tiszaeszlár blood libel in 1882, Istóczy's Antisemitic Party, the antisemitism of the Horthy era, pogroms in 1919-20, and the anti-Jewish laws. Focuses on Jews in the Hungarian army after 1938 (pp. 148-170), especially on the military labor service. Stresses the fact that Jews were saved from deportation by being recruited to the labor service units. Relates the consequences of the German occupation in March 1944, the roles of Horthy and Eichmann, and the ghettoization and deportation of the Jews in the provinces. Devotes a chapter to Auschwitz. Describes the wave of persecution after the Szálasi coup in 1944. Deals, also, with acts of rescue and the fate of Budapest's Jews. Provides figures about the estimated loss of human life and about survivors. *AV*

0349 Tomaszewski, Jerzy: *Preludium Zagłady: Wygnanie Żydów polskich z Niemiec w 1938 r.* [Prelude to the Holocaust: Expulsion of the Polish Jews from Germany in 1938]. Warszawa: PWN, 1998. 340 pp.

Presents the situation of Polish Jews living in Germany, and the events which led to the expulsion in October 1938. Ca. 11,000 Jews returned to Poland in 1937, without money or prospects of employment. Poland, desiring to stem this tide, cancelled the citizenship of Poles who had not visited Poland for five years. The Nazis reacted by expelling ca. 17,000 Polish Jews. Describes the expulsion actions in several German towns.

Relates how the Polish consul in Leipzig, Feliks Chiczewski, allowed 1,000 Jews to take asylum in the consulate. Describes conditions in Zbąszyń, where more than 6,000 Jews were interned for months, and the aid offered by Jewish organizations and committees, among them Emanuel Ringelblum and his co-workers. Relates reactions in the press, and negotiations between Poland and Germany concerning the expelled Jews. In January 1939 an agreement allowed some of the Jews to return to Germany temporarily. In spring 1939 there were still ca. 4,000 Jews in Zbąszyń. With the outbreak of war in September, there was no longer any interest in the fate of the Jews there. *HV*

0350 Turski, Marian, ed.: *Losy żydowskie: Świadectwo żywych. Vol. II* [Jewish Fates: Living Witnesses. Vol. II]. Warszawa: Stowarzyszenie Żydów Kombatantów i Poszkodowanych w II Wojnie Światowej, 1999. 408 pp. Vol. I appeared in 1996.

Presents additional war memoirs from the Holocaust period, written by Polish Jewish survivors of the Holocaust. Contains 24 memoirs describing experiences in ghettos, concentration camps, on the "Aryan side," and in the Soviet Union. *HV*

0351 Ueberschaer, Gerd R., ed.: *NS-Verbrechen und der militaerische Widerstand gegen Hitler*. Darmstadt: Primus, 2000. x, 213 pp.

A collection based on papers delivered at a conference at the Fritz Bauer Institut, Frankfurt a.M., May 1998. Partial contents: "Antisemitische Einstellungen im Widerstand gegen Hitler": Dipper, Christof: Der "Aufstand des Gewissens" und die "Judenfrage": Ein Rueckblick (14-18); Wette, Wolfram: Reichswehr, Wehrmacht, Antisemitismus und militaerischer Widerstand (1933-1939) (19-30); Ueberschaer, Gerd R.: Der militaerische Widerstand, die antijuedischen Massnahmen, "Polenmorde" und NS-Kriegsverbrechen in den ersten Kriegsjahren (1939-1941) (31-43). "Die Konfrontation militaerischer Verschwoerer mit den NS-Verbrechen im Vernichtungskrieg an der Ostfront": Gerlach, Christian: Hitlergegner bei der Heeresgruppe Mitte und die "verbrecherischen Befehle" (62-76); Streit, Christian: Angehoerige des militaerischen Widerstands und der Genozid an den Juden im Suedabschnitt der Ostfront (90-103). "Antisemitismus und NS-Verbrechen als Motive fuer den Umsturz gegen Hitler am 20. Juli 1944": Messerschmidt, Manfred: Motive der militaerischen Verschwoerer gegen Hitler (107-118); Mommsen, Hans: Die Stellung der Militaeropposition im Rahmen der deutschen Widerstandsbewegung gegen Hitler (119-134). "Ausgewaehlte Dokumente": Steinkamp, Peter; Ueberschaer, Gerd R.: Die Haltung militaerischer Verschwoerer zum Antisemitismus und zu den NS-Verbrechen im Spiegel von Dokumenten (137-206). *RW*

0352 Uziel, Daniel: Wehrmacht Propaganda Troops and the Jews. *Yad Vashem Studies* 29 (2001) 27-63. Appeared in Hebrew in "Yad Vashem" 29 (2001).

Discusses the role of the Wehrmacht propaganda machine in the dissemination of antisemitic propaganda in the Nazi-occupied countries. The Wehrmacht's main propaganda arm was its propaganda companies (PKs) set up in 1938. As an agency providing news and photographs from the combat area, these companies were involved also in anti-Jewish propaganda aimed at the German population. In 1939-41, the PKs collaborated with the Propaganda Ministry in shooting antisemitic films, notably "Der ewige Jude"; their cameramen and photographers also filmed some of the ghettos in Poland. In 1941-44, the PKs conducted propaganda aimed at the Red Army, at the civilian population in the Balkans, the USSR and France, and at German soldiers as well. The PKs' activities in the rear tried to strengthen the anti-Jewish feelings of the population and to justify the measures taken against the Jews; it is difficult to assess its real impact on the population. The PKs' activities invalidate postwar attempts to whitewash the Wehrmacht in regard to its involvement in the Holocaust. *DR*

0353

עוזיאל, דניאל: גייסות התעמולה של הוורמכט והיהודים. יד ושם 29 (2001) 23-51.

[Uziel, Daniel: Wehrmacht Propaganda Troops and the Jews. *Yad Vashem* 29 (2001) 23-51.] Appeared in English in "Yad Vashem Studies" 29 (2001).

0354 Verolme, Hetty Esther: *The Children's House of Belsen*. Fremantle, W. Australia: Fremantle Arts Centre Press, 2000. 311 pp.

Memoirs of a Jew (née Werkendam) who was born in Belgium in 1930 and raised in Amsterdam. Verolme, her parents, and two brothers were deported to Westerbork in 1943, and sent to Bergen-Belsen in February 1944. Her parents were deported from there in December 1944. Verolme and her brothers, along with other Dutch children, lived in the "children's house" in the camp, where they were cared for by several women, including their provider, Luba Fredrick, a Polish inmate. They were also saved more than once by an SS-man. Verolme served as a beloved big sister to all the children, while taking special care of her brothers. After the liberation, the children returned to the Netherlands, where Verolme discovered that her parents had survived. *YC*

0355 Wajcblum-Heilman, Anna: *Never Far Away: The Auschwitz Chronicles of Anna Heilman*. Ed.: Sheldon Schwartz. Calgary: University of Calgary

Press, 2001. xxv, 159 pp. With an introduction by Dieter K. Buse and Juergen C. Doerr (pp. xi-xxv).

Memoirs of a Jew born in Warsaw in 1928. Describes the German occupation and life in the Warsaw ghetto. Anna and her sister, Esther (Estusia) Wacjblum, joined Hashomer Hatzair and later participated in the ghetto uprising. They escaped from the ghetto and joined their parents on the "Aryan side." The family was deported in May 1943 to Majdanek, where the parents were killed. The two sisters were sent to Auschwitz, where they worked at the Weichsel-Metall-Union munitions factory and were active in the resistance. They participated in preparations for the October 1944 uprising in Auschwitz, smuggling gunpowder from the factory to members of the Sonderkommando. In January 1945 Esther and three other women were executed for their role in the uprising. Anna was sent on a death march in January 1945; she was liberated in Neustadt in May. After the war she immigrated to Israel, and then to Canada. In June 1991 she and her sister Sabina, who had escaped the Holocaust, participated in the dedication at Yad Vashem of a memorial to the four women who were hanged at Auschwitz for their resistance activities. *RE*

0356 Wenkert, Emil: *Czernowitzer Schicksale: Vom Ghetto nach Transnistrien deportiert: Juedische Schicksale 1941-1944*. Konstanz: Hartung-Gorre, 2001. 36 pp.

Memoirs of a Jew born in Czernowitz in 1914. In 1941 he was interned with his parents and his wife in the ghetto. Later they were deported to Transnistria, to the Mogilev ghetto and the nearby Scazinetz labor camp, where Wenkert's stepmother died. In fall 1942 his father was forced to walk 70 km to Tivori, where he perished. Wenkert and his wife managed to escape and returned to the Mogilev ghetto, where they remained until the liberation in March 1944. Wenkert's wife returned to Czernowitz, while he was drafted into the "Arbeitsarmee" ("work army") in Belarus; he returned to Czernowitz in May 1945. In 1990 he immigrated to Israel. *ID*

0357 Westermann, Edward B.: The Royal Air Force and the Bombing of Auschwitz: First Deliberations, January 1941. *Holocaust and Genocide Studies* 15, 1 (Spr 2001) 70-85.

With reference to the controversy over the Allies' failure to bomb Auschwitz-Birkenau in 1944, relates a less well-known episode: in January 1941 the Polish government-in-exile asked the British to bomb the camp, where (at the time) 20,000 prisoners, Jewish and non-Jewish, were confined. The request was rejected on the grounds of technical difficulties, the risk of lives of the prisoners, and the low military priority of such an action. Three years later, similar arguments were put forward by the British when a request to bomb Birkenau came from Jewish leaders. The refusal of 1941 might have been used as a precedent for the refusal of 1944. This fact does not excuse the reluctance of the British military to bomb Birkenau in 1944; the probability of success of such a raid had increased, and it was no longer the rescue of 20,000 slave laborers that was at stake, but the rescue of the Jewish people from annihilation. *DR*

0358 White, Joseph Robert: "Even in Auschwitz... Humanity Could Prevail": British POWs and Jewish Concentration-Camp Inmates at IG Auschwitz, 1943-1945. *Holocaust and Genocide Studies* 15, 2 (Fall 2001) 266-295.

In fall 1943 and winter 1945, ca. 1,400 British POWs, captured in Italy, were sent as a work detachment to Auschwitz-Monowitz to work on construction of the plant of IG Farben. They lived in a separate compound, comprising unit E715 of Stalag VIII B. They were witnesses of the brutal Nazi treatment of the Jewish prisoners. The British POWS, despite mild antisemitism harbored by some of them, extended a helping hand to Jewish inmates in Monowitz. Some of them attempted to report on the mass murder of the Jews through International Red Cross channels; others tried to organize escapes. Dwells on the activities of the POW Charles Coward, who collected information on the Nazi treatment of Jews and helped some of them escape. Concludes that the British challenged the Nazi racial hierarchy at IG Auschwitz by treating the Jews from Monowitz as men, not "Untermenschen." *DR*

0359 Witnesses Speak: *An Anthology*. Montreal: Concordia University Chair in Canadian Jewish Studies; Montreal Institute for Genocide and Human Rights Studies, 2001. Various pagings. (Memoirs of Holocaust Survivors in Canada, 15).

Presents memoirs by nine Holocaust survivors: Minna Aspler, Hyman Gutman, Myra Gutman, Rachel Philipson-Levi (in English and French), Miriam Reich, Andrew Savin, Koineh Schacter Rogel, Veronika Schwartz, and Babey Widutschinsky-Trepman. The accounts discuss experiences of forced labor, and life in ghettos, in hiding, and in concentration camps, mainly in Poland (Warsaw and Radom), France, Lithuania (Kovno and Šiauliai), and Germany. *YC*

0360 Witte, Peter; Tyas, Stephen: A New Document on the Deportation and Murder of Jews during "Einsatz Reinhardt" 1942. *Holocaust and Genocide Studies* 15, 3 (Win 2001) 468-486.

Presents two German secret radio dispatches which were intercepted and decoded by British Intelligence in January 1943. These documents were found among recently declassified material at the Public Record Office.

The messages were sent from Lublin by SS Major Hoefle; one was addressed to Eichmann in Berlin (this one was only partially intercepted) and the second to SS Lieutenant Colonel Heim in Kraków. The second dispatch relates the numbers of Jews murdered in Majdanek, Bełżec, Sobibór, and Treblinka until 31 December 1942; in sum, 1,274,166. This number roughly equals that given in the Korherr Report ordered by Himmler. The number of victims given for Bełżec (434,508), which ceased operation in December 1942, is lower than most historians have calculated. Discusses the historical context of the dispatches. Hoefle's dispatch is only the fourth known report on the murder of Jews in the General Government submitted after Himmler's order to "resettle" its whole Jewish population before 31 December 1942. *DR*

0361 Wodziński, Marcin; Spyra, Janusz, eds.: *Jews in Silesia*. Cracow: Księgarnia Akademicka; [Wrocław]: University of Wrocław, Research Centre for the Culture and Languages of Polish Jews; [Český Těšín]: Muzeum Těšínska, 2001. 468, [24] pp.

A collection of papers delivered at a conference held at the University of Wrocław, May 2000. Partial contents: Wodzinski, Marcin: 'Walking in the Steel Boots of Faith...': Anti-Semitic Journalism in the Voivodship of Silesia 1922-1939 (99-124); Maciejewski, Marek: Silesian Nazi Nationalism and Anti-Semitism 1921-1933 [A Polish version appeared in "Studia nad Faszyzmem i Zbrodniami Hitlerowskimi" 24 (2001).] (141-158); Jonca, Karol: The Final Solution to the 'Jewish Problem' in the Silesian Province 1940-1945 (159-171); Konieczny, Alfred: The 'Schmelt Organisation' in Silesia [A Polish version appeared in "Studia nad Faszyzmem i Zbrodniami Hitlerowskimi" 15 (1992).] (173-179); Borák, Mečislav: The Situation of the Jewish Population in the Territory of Czech Silesia during the Occupation 1938-1945 (181-192); Kozerska, Ewa: Pius XI and German Anti-Semitism (193-200); Hasiewicz, Małgorzata: Cardinal Bertram and the German Episcopate's Attitude towards Nazi Anti-Semitism before 1933 (201-210); Sadowski, Mirosław; Scheffler, Tomasz: From the Studies of the Death Rate of Jewish People in Wrocław in the Period of Nazi Deportations 1941-1944 [Pp. 348-373 present a list of Jews who died in Wrocław between 1941-45, with biographical information. A Polish version appeared in "Studia nad Faszyzmem i Zbrodniami Hitlerowskimi" 24 (2001).] (345-373). *DR*

0362 Wojak, Irmtrud: *Eichmanns Memoiren: Ein kritischer Essay*. Frankfurt a.M.: Campus, 2001. 279 pp.

Analyzes Eichmann's differing accounts of his role in the Holocaust: in his interviews with the former Dutch SS-man Sassen in Argentina shortly before his kidnapping, in his interrogation by the Israeli police and in court, and in the two memoirs written in prison. Especially in the interviews with Sassen, he boasts of his own "creative" proposals for the resettlement of the Jews, and complains about the rivalry within the bureaucracy, which blocked his plans. In his pronouncements in Israel, he tried to represent his organization of migration and resettlement as efforts to help the Jews. Toward the end of 1941, when resettlement proved impracticable, and there was pressure to make room in German-annexed territories for ethnic Germans from the USSR and a demand by governors of the occupied regions to rid them of Jews, Eichmann's duties changed: he was put in charge of organizing the transport of Jews to extermination sites and had to inspect these sites personally. Discusses individual responsibility in the light of Eichmann's assertion that he was only obeying orders, commenting that he had completely assimilated Nazi antisemitic ideology. *RW*

0363 Żbikowski, Andrzej, ed.: *Archiwum Ringelbluma: Konspiracyjne Archiwum Getta Warszawy. Vol. 3: Relacje z Kresów* [Ringelblum Archives: Clandestine Archives of the Warsaw Ghetto. Vol. 3: Testimonies from Kresy]. Warszawa: Żydowski Instytut Historyczny, 2000. xxv, 963 pp. Vol. 1 appeared in 1997; vol. 2 in 2000.

This volume contains 46 testimonies; 26 were written in Polish and 20 in Yiddish (translated here into Polish). The testimonies were written by Ringelblum's co-workers, who transcribed what they were told by Jews who fled from central Poland in 1939 following the Nazi occupation and went to live under Soviet occupation in the eastern parts of Poland called Kresy. Some of them returned to Warsaw in 1941 or 1942. Some of the testimonies also deal with events after June 1941, when the Nazis occupied this region. They describe persecution of Jews in, among other places, Białystok, Grodno, Słonim, Łuck, Kowel. Many testimonies describe the situation in Lvov and Vilna. *HV*

0364 Zilversmit, Guenther Ludwig: *From Holland and Back*. Montreal: Concordia University Chair in Canadian Jewish Studies; Montreal Institute for Genocide and Human Rights Studies, 2001. 56 pp. (Memoirs of Holocaust Survivors in Canada, 20).

Memoirs of a Jew born in Leer, Germany, in 1926. In 1937 his parents sent him and his brother Walter to join their grandparents, who had found refuge in Nunspeet, Netherlands. In 1943 Zilversmit, his brother, and their grandparents were deported to Westerbork and from there to Theresienstadt. In September 1944 the two brothers were deported to Auschwitz, and were sent from there to work at the satellite camp of Gleiwitz. They survived a death march to Blechhammer, where they were liberated. Zilversmit's grandparents were on the last transport from Theresienstadt to Auschwitz, where they were murdered. He returned to Nunspeet, where the neighbors returned his family's property. Relates that during the war the police and mayor of the village refused to arrest Jews. After the war, Zilversmit was reunited with Ilse van Collem, whom he had met in Westerbork; they married and immigrated to Canada. *YC*

0365 Zimmermann, Michael: The National Socialist "Solution of the Gypsy Question": Central Decisions, Local Initiatives, and Their Interrelation. *Holocaust and Genocide Studies* 15, 3 (Win 2001) 412-427.

Describes the Nazi persecution of Gypsies in 1933-45, stressing the interplay between directives from above and initiatives from below. This persecution resulted in the annihilation of a great part of Europe's Gypsies. The same policy inconsistencies were characteristic for the early Nazi handling of both the "Jewish question" and the "Gypsy nuisance"; the process of the Nazi "solution" for both, in the later phase, may be compared in many aspects. However, the Gypsies never constituted the same kind of threat that Nazi ideologists saw in the Jews. Thus, Gypsies were annihilated by the Nazis less systematically than Jews, they became victims of the Einsatzgruppe mass executions in the USSR later than Jews, and no Gypsies were ever deported from occupied Denmark, Greece, Italy, and some other territories. *DR*

Austria

0366 Froehlich, Hannelore: *Spurensuche*. Graz: Steirische Verlagsgesellschaft, 1999. 175 pp.

Describes Froehlich's search for the past of her father, Josef Schleich of Graz, who died in 1949 when she was eight years old. Schleich (a non-Jew) cooperated with the Jewish community of Graz in helping Jews to flee Austria. He provided them with vocational certificates enabling them to immigrate to the U.S., and then organized clandestine border crossings to Zagreb, where the refugees were taken over by Jewish organizations. Between 1938-41, when he was arrested by the Gestapo, he helped 120,000 Jews escape. After the war he was arrested on charges of having enriched himself at the expense of the Jews. The investigation was cut short by his illness and death. Explains that the refugees' property was lost by theft or confiscation, or expended by Schleich to pay bribes or to cover the expenses of poorer refugees. Froehlich also discovered that she was Schleich's daughter by his Jewish secretary; he was not able to save her mother from Auschwitz, but saved his daughter by handing her over, soon after birth, to his gentile secretary, who brought her up as her own. *RW*

0367 Geser, Guntram; Loacker, Armin, eds.: *Die Stadt ohne Juden*. Wien: Filmarchiv Austria, 2000. 487, [16] pp. Accompanied by a videocassette of the film.

A collection of essays which discuss the historical background of Hans Karl Breslauer's film "Die Stadt ohne Juden" (1924). The film was based on Hugo Bettauer's bestselling novel (1922), in which the government decides to expel all the Jewish residents of Vienna in order to forfend a crisis, but this action creates economic and cultural upheaval. Partial contents: Hall, Murray G.: "...alles arische Empfinden verhoehnt, verspottet und in den Kot gezerrt...": Zur Rezeption von "Die Stadt ohne Juden" (105-146); Hoffmann-Holter, Beatrix: "Ostjuden hinaus!" Juedische Kriegsfluechtlinge in Wien 1914-1924 (311-346); Budischowsky, Jens: Politische Stroemungen im Judentum der Ersten Republik (347-383); Caneppele, Paolo; Krenn, Guenter: "Pfiffige Kinojuden": Antisemitismus im Spiegel der kinematografischen Berichterstattung (385-413); Kitzberger, Michael: Bild-Stoerung: Repraesentationen des Juedischen in "Die Stadt ohne Juden" (415-444); Veigl, Hans: Der Weltuntergang als Wiener Gesamtkunstwerk: Apokalyptische Abschiedsprophezeiungen, apriorische Abgrundstimmungen und antisemitische Ausblicksutopien rund um die aesthetische Moderne (445-475). *ID*

0368 Gradwohl-Schlacher, Karin: Gestern wurde Frieden gemacht: August Hermann Zeiz alias Georg Fraser im Dritten Reich. *Jahrbuch fuer Antisemitismusforschung* 10 (2001) 223-238.

Zeiz (1893-1964), a journalist and playwright in Berlin (who at times used the pseudonym Georg Fraser), would normally have been excluded from the Nazi cultural guilds because of his Jewish wife; but he was repeatedly granted exemptions by Hans Hinkel, the Nazi functionary in control of Jewish cultural activities. In 1935, however, conditions in Germany became too difficult, and the family moved to Vienna. There Zeiz and his wife remained, even after the Anschluss, when their son fled to Switzerland. Zeiz was again under the protection of Hinkel, and his plays continued to be performed, while in secret he was supporting the resistance and publishing a resistance newspaper. In 1943 the Gestapo discovered that he had been helping Jews escape. He was interned in Dachau, and his wife was deported and died in Auschwitz. After his release, Zeiz again joined the Austrian resistance. *RW*

0369 Lauritsch, Andrea: *Die Juden in Wolfsberg: Nationalsozialistische Judenverfolgung am Beispiel Wolfsbergs*. Wolfsberg: Kulturamt der Stadtgemeinde Wolfsberg, 2000. 52 pp.

Pp. 3-17 survey the history of Jews in Wolfsberg, Austria, between 1338-1938. Pp. 17-27 discuss the Nazi period, emphasizing the exclusion of Jews from the German economy and the Aryanization of Jewish property. Pp. 27-32 contain short biographies of six former Jewish residents: Adolf and Emma Gross (murdered in Auschwitz in 1942), Lotte (Gross) Roth (immigrated to the USA in 1938), Hermine (Popper) Singer (died in Mauritius in 1944), Hans Singer (immigrated to Palestine in 1938), and Anny (Gross) Junek (deported to Bergen-Belsen with the Kasztner transport from Budapest, and then transported to Switzerland). Includes photographs and facsimiles of documents. *ID*

0370 Lausecker, Werner; Posch, Herbert: "Ich bin nirgendwo zu Hause": Die "Arisierung" der Wiener Universitaet vor und nach 1938. *Das Juedische Echo* 50 (2001) 247-252.

Starting from the case of Stella Rotenberg, whose studies at the Vienna University Medical School were terminated by the Anschluss, never to be resumed, points to the unbroken history of antisemitism among the students through all the changes of regime in Austria since the Habsburg Empire – antisemitism expressed in demands for a numerus clausus as well as in physical attacks, brawls and pogroms. Immediately after the Anschluss, Jewish students were expelled and Jewish faculty and staff dismissed; these (together with those dismissed for their political orientation) had formed almost half of the faculty and student body. Many later perished in the concentration camps; among those who emigrated, many were never able to resume their careers. The authors are seeking to create an archive of the life stories of these lesser-known individuals. Notes that even those who returned to Vienna from exile after 1945 were discriminated against. *RW*

0371 Milchram, Gerhard, ed.: *Judenplatz: Ort der Erinnerung*. Wien: Pichler, 2000. 118 pp. Unseen. See the abstract for the English edition: "Judenplatz: Place of Rememberence [sic]" (Wien: Pichler, 2000).

0372 Milchram, Gerhard, ed.: *Judenplatz: Place of Rememberence [sic]*. Trans.: Nick Somers, David Gogarty. Wien: Pichler, 2000. 118 pp. Opposite title page: Edited by the Jewish Museum of the City of Vienna. Originally published as "Judenplatz: Ort der Erinnerung" (Wien: Pichler, 2000).

In October 2000 a Holocaust memorial, a museum, and an exhibition area were opened at the Judenplatz in Vienna. Pp. 32-89, "The Destruction of Austrian Jews, 1938-1945," describe the Anschluss, the "Kristallnacht" pogrom, deportations and killings of Austrian Jews during the war, and the fate of Austrian Jews in Riga, Minsk, Izbica, Treblinka, Auschwitz, and other ghettos and camps. Discusses, also, the fate of Austrian Jews who fled to other European countries and were caught there by the Nazi occupation. Includes the following essays: Palffy, András: The Design Concept for the Museum on Judenplatz (16-19); Schlieker, Andrea: A Book Must Be the Axe for the Frozen Sea within Us: Rachel Whiteread's

Holocaust Memorial (20-31). Pp. 90-117 contain two essays on the history of the Judenplatz from the Middle Ages to the present. *DR*

0373 Rabinovici, Doron: *Instanzen der Ohnmacht: Wien 1938-1945: Der Weg zum Judenrat*. Frankfurt a.M.: Juedischer Verlag, 2000. 495 pp.

Describes the functioning of the Israelitische Kultusgemeinde (IKG) in Vienna, the first Judenrat and Eichmann's model for the Jewish councils, and attempts to understand its cooperation with the Nazis. Immediately after the Anschluss, the Nazis dispersed the IKG and ousted its officers; a few months later they reconstituted it with Josef Loewenherz at its head. The IKG became a mere channel for the orders of Eichmann's Zentralstelle fuer juedische Auswanderung and the Gestapo. Loewenherz and his staff were under constant pressure; Eichmann and his men systematically humiliated and threatened them in order to break all resistance. The IKG believed that the punctilious observance of discipline could enable it to save Jews as well as continue to carry out its welfare functions. Especially feared and hated for his ruthless enforcement of this policy was Loewenherz's deputy Benjamin Murmelstein, later the last "Judenaelteste" in Theresienstadt. Until 1941 the IKG helped the majority of Vienna's Jews to emigrate. Then came the deportations. The lists of Jews to be deported were drawn up by the Nazis; the IKG had no influence on them, but it had to provide staff for their implementation. The fate of those deported became known to the IKG only in 1943. The Jews of Vienna did not realize the helplessness of the IKG; they blamed it, rather than the Nazis, for refusing their desperate pleas for help and for the selections for deportation. *RW*

0374 Stoffel, Manfred, ed.: *Versoehnung braucht Erinnerung: Juden in Kirchberg/Hunsrueck*. Kirchberg/Hunsrueck: Presbyterium der Evangelischen Kirchengemeinde, 2000. 162 pp. On title page also: Dokumentation aus der Sammlung von Ernst Fuchss, Kirchberg, mit einem Aufsatz von Gustav Schellack "Die juedische Schule in Kirchberg/Hunsrueck."

A memorial booklet published in honor of the Jews who lived in Kirchberg (Tyrol) during the Nazi period. The presence of Jews in Kirchberg is traceable from 1303. Surveys the history of the Jewish community until 1945, focusing on the "Kristallnacht" pogrom. P. 77 contains names of 27 former Jewish residents who were victims of the Holocaust. Pp. 81-115 relate the fate of 18 Jewish families, former citizens of the town. *ID*

0375 Strassl, Harald; Vosko, Wolfgang: Der Arbeitseinsatz und die Lager entlang der Grenze im Bezirk Oberwart (Burgenland). *Holocaust Füzetek* 14 (2000) 62-80.

Describes the events affecting the civilian population of the villages of this district near the Hungarian border during the last months of the war and the subsequent Soviet occupation. The villagers were drafted for labor on border fortifications, along with civilians from other parts of Austria, East European forced laborers, and thousands of Hungarian Jews, most of them transferred from labor camps in Hungary. When the area had to be evacuated, those Jews no longer capable of work were shot. The biggest massacres, of hundreds of Jews, took place in the vicinity of Rechnitz; a massacre of 60-80 Jews took place in Deutsch-Schuetzen. Mentions that some villagers aided Jews with food and some hid Jews to save them from the massacres and death marches. *RW*

0376 Szita, Szabolcs: A mauthauseni koncentrációs táborrendszer altáborai Hirtenberg, Hinterbruehl, Lenzing, Linz, Melk [The Satellite Camps of the Mauthausen Concentration Camp System Hirtenberg, Hinterbruehl, Lenzing, Linz, Melk]. *Holocaust Füzetek* 14 (2000) 34-61.

Describes the respective satellite camps one by one, providing information about the types of work done in each (usually for the war industry), the prisoner population and their treatment, circumstances of evacuation and/or liberation by the American army. Gives the number of victims or number of survivors in each of the camps. Relates stories about saving of lives, getting weapons (in one of the camps), and cases of outwitting the Germans. *AV*

0377 Walzer, Tina; Templ, Stephan: *Unser Wien: "Arisierung" auf oesterreichisch*. Berlin: Aufbau-Verlag, 2001. 292, xxxii pp.

Discusses the expropriation of Jewish property by the Nazi regime in Austria. Contends that the Viennese government and population played a major role in the exploitation of Jews. Following the Anschluss, "wild" Aryanization actions began, leading to legislation which enabled the organized dispossession of Jews. After the war, whilst Germany began to cope with its past, Austria insisted on its status as victim. Antisemitism persisted in politics, the judiciary, the economy, and society, while the suffering, persecution, and robbery of Viennese Jews was erased from the public consciousness. Pp. 109-223, "Topographie des Raubes," contain a list of buildings in Vienna which belonged to Jews, including pharmacies, book stores, cafés, hotels, department stores, cinemas, synagogues, and museums. Presents, also, a list of stolen art collections, and information on places where literary and musical events took place. The last xxxii pp. contain photographs. *ID*

Belarus

0378 Arkadeva, Olga M. et al., eds.:*...na perekrestkakh sudeb: Iz vospominanii byvshikh uznikov getto i pravednikov narodov mira* [...at the Crossroads of Fates: From the Memoirs of Ghetto Inmates and Righteous of the Nations]. Minsk: Chetyre chetverti, 2001. 239 pp. With an English summary (pp. 235-236).

Presents the stories of ten Holocaust survivors and ten rescuers of Jews. Nine of the Jews were interned in the ghetto of Minsk; the tenth was interned in the ghetto of Białystok and various Nazi camps (including Majdanek, the Bliżyn labor camp, Auschwitz, and Bergen-Belsen). One of the rescuers, from the village Vysokii Bereg, hid two Jewish women who fled from Borisov; the others hid Jews from the Minsk ghetto. Some of the Jews joined partisan units after their rescue from the Minsk ghetto. *DR*

0379 Davydova, Galina R.: *Ot Minska do La-Mansha, ili dorogami Kholokosta: Dokumentalnaya povest* [From Minsk to La Manche, or On the Roads of the Holocaust: A Documentary Story]. Minsk: Chetyre Chetverti, 2000. 231 pp.

Memoirs of a Jew (née Nisons) born in 1930 in Minsk. Pp. 14-140 deal with the war period. Davydova and her family were interned in the ghetto. She left the ghetto in summer 1942 and wandered through nearby villages; she could sing well, and this ability helped her provide food for her family. In summer 1943 she left the ghetto again, but was arrested and sent to prison; taken for a Russian, she was sent to an orphanage in Minsk. When she was exposed as a Jew in September 1943, she fled from the orphanage. She was arrested again and, after a long stay in prison and in a camp on Shirokaya Street in Minsk, was deported for forced labor in France and then in Germany. In April 1945 she was liberated in Helmstedt, and returned to Minsk. *DR*

0380 Dmitruk, Yelena: Genezis partii narodnykh demokratov i ikh antisemitskaya deyatelnost v Polesskom voyevodstve (1921-1939 g.) [The Genesis of the National Democratic Party and Its Antisemitic Activities in the Polesian Voivodship (1921-1939)]. *Yevrei Belarusi* 6 (2001) 143-151.

In the late 1920s, the Endeks (National Democrats) in Poland changed their policy from "tactical antisemitism" to inclusion of antisemitic violence in party politics. The Camp of Greater Poland (which became the Endecja after 1933) used aggressive antisemitism in order to mobilize the youth (even schoolchildren) and involve them in nationalist activities. A Polesian branch of Endecja was established in 1936. It distributed antisemitic leaflets, organized the boycott of Jewish trade, and committed acts of violence. The government tried to suppress nationalist activities (e.g. it prevented the establishment of party cells in the district centers of Polesie). In 1938 moderate elements came to power in Endecja, and the party focused on the "economic struggle" against the Jews. *DR*

0381 Fatal-Knaani, Tikva: The Jews of Pinsk, 1939-1943, through the Prism of New Documentation. *Yad Vashem Studies* 29 (2001) 149-182. Appeared in Hebrew in "Yad Vashem" 29 (2001).

Using documents from Belorussian archives, photocopied for the Yad Vashem archives, reconstructs the history of the Jewish community in Pinsk during World War II. These newly available records help to estimate the number of victims of the first mass execution of Jewish men in August 1941 – 7,000-8,000. Describes the forced labor, money and goods extorted from the Jews by the Nazi authorities, murder of Jews, and other aspects of Nazi anti-Jewish measures in Pinsk. The new documentation also shows the struggle for Jewish survival conducted by the Judenrat, which attempted to regulate the forced labor, to provide the Jews with food, and to thwart the ghettoization of Pinsk Jews in April 1942. Some requests of the Judenrat put to the German authorities were answered in the affirmative. But in October 1942 the last 10,000 Jews of Pinsk were shot. *DR*

0382

פתל-כנעני, תקוה: יהודי פינסק לאור תיעוד ארכיוני מהשנים 1939-1943. יד ושם 29 (2001) 117-141.

[Fatal-Knaani, Tikva: The Jews of Pinsk in the Light of Archival Documentation from 1939-1943. *Yad Vashem* 29 (2001) 117-141.] Appeared in English in "Yad Vashem Studies" 29 (2001).

0383 Haberer, Eric E.: The German Police and Genocide in Belorussia, 1941-1944. Pt. I: Police Deployment and Nazi Genocidal Directives. Pt. II: The "Second Sweep": Gendarmerie Killings of Jews and Gypsies on January 29, 1942. Pt. III: Methods of Genocide and the Motives of German Police Compliance. *Journal of Genocide Research* 3, 1 (2001) 13-29; 3, 2 (2001) 207-218; 3, 3 (2001) 391-403.

Discusses the role of the civil administration in helping the Nazi military occupy and maintain order in Belorussia after the first wave of the murder of Jews by Einsatzgruppen in 1941. Deals with the Order Police and its role in supplementing the much smaller military forces. Provides information about the Schutzmannschaften, which in 1942-44 contributed to the massacre of the Jews (1.5 million). Discusses the functions of the local Gendarmerie and Schutzmannshaft, such as rounding up victims and collecting their property, as well as executing them by shooting and catching those that escaped. Focuses on three stages in the genocide in Belorussia: the removal of Jews from the countryside, the reduction of the ghetto population through selections, and the liquidation of the ghettos and Jewish communities. Among the factors that influenced the Gendarmerie to collaborate, notes the authority of the Gebietskommissar, a deeply ingrained sense of duty, and, to a lesser degree, ideological antisemitism. *YC*

0384 Ioffe, Emanuil G.: *Inostrannye yevrei v Trostenetskom lagere smerti* [Foreign Jews in the Death Camp Trostinets]. Minsk: Natsionalnyi Arkhiv Respubliki Belarus, 2000. 19 pp.

The Nazi extermination camp Trostinets, near Minsk, was active from November 1941 to June 1944. More than 200,000 people were killed here. Deportees were brought to Trostinets from Austria, Czechoslovakia, Germany, France, Poland, and other countries. Estimates the number of foreign Jews who perished in Trostinets as 80,000. *DR*

0385 Ioffe, Emanuil G.: *Po dostovernym istochnikam: Yevrei v istorii gorodov Belarusi* [According to Reliable Sources: Jews in the History of the Cities of Belarus]. Minsk: Chetyre chetverti, 2001. 350 pp.

Contains eight essays on the history of Jews in Belarus, particularly the cities of Brest, Vitsebsk, Homel, Minsk, Mahilyow, and Pinsk, and the townlet of Berezino. For each town, describes, also, events which took place during the Holocaust. Many Jews from these towns took part in the anti-Nazi resistance, first and foremost in the partisan movement. *DR*

0386 Kagan, David: *Rasskazhi zhivym. – Belorusskaya madonna* [Relate It to the Living. – A Belorussian Madonna]. Tel Aviv: Biblioteka Matveya Chyornovo, 1999. 301 pp.

Two documentary novels bound together. The second, "Belorusskaya madonna" (pp. 157-301), relates the story of survival of a real Jewish family during the Holocaust. Iosif Sosnovik was born in Germanovichi, Polish Belorussia, in 1919. During the Nazi occupation he, his life partner Lelya (Iokha Pliskina), and their newborn son Yakov were enclosed in the ghetto of Germanovichi. In 1942, after the mass murders of Jews in the surrounding towns, Iosif and Lelya handed over their son to a peasant woman, Marysya (Mariya Kozachonok), and went into hiding in a forest. In 1943 they joined the partisans. All three survived the war. *DR*

0387 Komissarova, Sonya: Gorit na serdtse kostyor [A Bonfire Is Burning on My Heart]. *Mishpokha* 9 (2001) 60-63.

Memoirs of a Jew born in Lyuban (Minsk region). In August 1941 the Nazis shot the Jewish men in the town; then a ghetto was established. A non-Jewish relative warned Komissarova that on 14 December 1941 the ghetto would be liquidated. Komissarova left Lyuban, and was hidden by non-Jews. In order to survive, she passed herself off as a "half-Jew," i.e. a person of mixed origins. *DR*

0388 Minova, Natalya: Problema antisemitizma v mezhnatsionalnykh otnosheniyakh na Gomelshchine v 20-30-ye gody XX veka [The Problem of Antisemitism in Inter-Ethnic Relations in the Homel Region in the 1920s-30s]. *Yevrei Belarusi* 6 (2001) 152-164.

In the early 1920s, Jews in the Homel region fell victim to pogroms. Categorizes the campaign of the Soviet regime against Jewish religious institutions and Zionist organizations in the 1920s as a form of antisemitism. Soviet policies (e.g. the agrarianization of Jews) also incited anti-Jewish sentiments among Belorussians. Popular antisemitism grew stronger in 1929-32, in the wake of collectivization and as a result of industrialization. Pp. 160-164 contain a list of towns in the Homel region where there were pogroms in 1919-22; notes who the perpetrators were (the Polish military, Balakhovich's army, etc.). *DR*

0389 Raban, Yehezkel, ed.: *The Dereczin Memorial Book: A Book of Remembrance Honoring the Communities of Dereczin, Halinka, Kolonia-Sinaiska*. Trans.: Jacob Solomon Berger. Mahwah, NJ: J.S. Berger, 2000. x, 401 pp. Originally published as "Sefer Dereczin," in Hebrew and Yiddish (Tel-Aviv: The Dereczin Organizations in Israel and the United States, 1966).

A collection of brief essays by former residents of Derechin, a town in interwar eastern Poland, now Belarus. Pp. 193-256, "The Destruction," present personal accounts from the Holocaust period. The Jews in the Derechin ghetto, including Jews from the nearby towns of Halynka and Kolonia Sinaiska who had been resettled in Derechin, were murdered in July 1942. Pp. 257-307, "In the Forests," deal with the participation of Derechin Jews in partisan warfare, in particular the Jewish partisan unit headed by Dr. Atlas and Eliyahu Lipshovich. Pp. 309-318 contain a list of Jewish partisans from Derechin who perished. *DR*

0390 Rein, Leonid: Otnosheniye belorusskovo naseleniya k yevreyam vo vremya Katastrofy, 1941-1944 gg. [The Attitude of the Belorussian Population toward Jews during the Holocaust, 1941-1944]. *Yevrei Belarusi* 6 (2001) 181-205.

The spectrum of Belorussians' attitudes toward the Nazi occupiers and the Jewish victims was broad. The predominant attitude toward the Jews was indifference, but denunciations of Jews who tried to hide were widespread. Dwells on the role of the Belorussian collaborators (heads of towns and villages, local police, pro-Nazi periodicals, etc.) in the Holocaust. The Nazis increasingly needed to involve the locals in the genocide because the German forces were limited. Indigenous policemen were sometimes more brutal than the Germans; their excessive brutality toward the Jews can not be explained solely by antisemitism: craving for power or for material gains, immorality and sadism also played a role. Soviet partisans displayed much enmity toward the Jews who were fleeing from the ghettos, and sometimes killed them. Partisans even neglected Stalin's order of 1943 to help the "peaceful population." Nevertheless, many righteous Belorussians rescued Jews during the war. *DR*

0391 Rozenblat, Yevgenii S.: Ekonomicheskiye aspekty Kholokosta v Zapadnoi Belarusi: Sistema kontributsii, nalogooblozheniya i shtrafov [Economic Aspects of the Holocaust in Western Belarus: The System of Enforced Taxation and Fines]. *Vestnik Yevreiskovo Universiteta* 23 (2001) 119-136.

Examines various forms of Nazi looting of Jewish property in Western Belarus in 1941-43. Among them were "nationalization" of Jewish property, levies in cash and kind, various fines, special taxes placed on Jews, and confiscations. The ghettoization of Jews also served as a means of expropriation, because those who were resettled were not permitted to take all their belongings with them. Executions and mass murders were also accompanied by collection of the Jews' property – clothes and valuables. The spoils from the Nazi looting of Jews enriched the Reich and was a substantial factor in the economy of occupied Western Belarus. The looting was also intended to deprive the Jews of their means of subsistence and was, thus, an integral part of the Final Solution. *DR*

0392 Ruess, Hartmut: SD u Baranavichakh (1941-1943) u kantekstse lakalnaha akupatsyinaha rezhymu [The SD in Baranovichi (1941-1943) in the Context of the Local Occupation Regime]. *Belaruski Histarychny Ahlyad* 5, 1 (June 1998) 67-87.

The SD in Baranovichi, a branch of the Security Police and the SD, was in charge of all Jewish matters, including the ghettos, and played a major role in the mass murder of Jews in this region, especially in the liquidation of the Baranovichi ghetto in 1942. The SD's authority in Jewish matters was higher than that of the Gebietskommissar, from whom the SD could demand additional forces during mass "actions." It participated also in the actions of Einsatzgruppe B in this area. The concentration camp Koldychevo was established on its initiative. The Jewish partisans in the vicinity chose the local SD as their first and foremost target. The case of Baranovichi shows how much authority and independence the local chiefs of the SD had. *DR*

Belgium

0393 Delplancq, Thierry: 1940-1942, une cité occupée et ses Juifs: Quelques aspects heuristiques. *Les Cahiers de la Mémoire Contemporaine* 3 (2001) 125-134.

Examines how Belgian national and local government dealt with antisemitic legislation imposed upon the Jews of Belgium, both citizens and refugees, by the German occupying force in 1940-42. Caught between the German anti-Jewish ordinances on the one hand and the Belgian constitution together with the international Hague Convention on the other, the national government opted for a policy of passive collaboration. With the Belgian government relegated to the role of a custodian, this line of policy varied according to the situation. States that the role of the local government is a difficult field of research due in part to its autonomous status and to source material which has been lost, scattered or destroyed. Local reactions varied from zealous execution of German orders to passive collaboration and active resistance. Gives examples of internal conflict among local authorities. Explores the varied motives behind opposing political decision-making on a local level. *EO*

0394 Klein, Maya: Souvenirs de résistants autrichiens en Belgique. *Les Cahiers de la Mémoire Contemporaine* 3 (2001) 73-86.

A study based on interviews conducted in Vienna in 2000 with survivors of a small group of Jewish-Austrian refugees who had come to Belgium as teenagers after the Anschluss in 1938. Cared for at first by the Comité d'Aide et d'Assistance aux Victimes de l'Antisémitisme en Allemagne, the unaccompanied youths soon found a home in the Communist Party, where they received ideological and cultural education. During the German invasion of Belgium the young Austrian communists went into hiding and launched their ideological struggle to incite German troops to rebel. The onset of systematic deportation of Belgian Jewry in August 1942 led to armed resistance. Austrian resistants collaborated with the Belgian Partisan Army. At the end of the war the Austrian resistants participated in the liberation of Yugoslavia and Austria. Though they won the respect of the Belgian authorities, none of those interviewed has received any of the benefits granted to Belgian resistants. *EO*

0395 Schreiber, Jean-Philippe: L'accueil des réfugiés juifs du Reich en Belgique, mars 1933-septembre 1939: Le Comité d'Aide et d'Assistance aux Victimes de l'Antisémitisme en Allemagne. *Les Cahiers de la Mémoire Contemporaine* 3 (2001) 23-71.

Discusses the activities of the Comité, a Jewish organization based in Brussels, whose primary challenge was balancing the interests of the

Belgian government with the needs of the refugees and the Jewish community. Founded in April 1933 and headed by Max Gottshalk, the Comité supported the refugees during their allocated time in Belgium and helped create conditions for their remigration. Sensitive to xenophobic and antisemitic Belgian reactions, the Comité supported the creation of internment camps to accommodate the huge influx of Jewish refugees after the Anschluss and "Kristallnacht." The Comité participated in the running of these camps together with the Belgian authorities. Supported by private Jewish donors and international Jewish organizations, the Comité relocated over 3,000 Jews abroad between 1933-38. With new refugees seeking help at a rate of 500 a day in the summer of 1939, the Comité turned to the Belgian government for funds, which were granted. At the end of 1939 there were 30,000 Jewish refugees in Belgium in addition to the 70,000 local Jews. *EO*

Bulgaria

0396 Baruch, Nir: Territorial Claims as a Basis for Bulgaria's Policy: The Jewish Perspective, 1940-1944. *Shvut* 10 [26] (2001) 124-159.

Examines Bulgaria's wartime policy toward the Jews in the context of the country's history and its national aspirations. Bulgaria's obsessive national ideal was the restitution of the territories of Macedonia, Thrace, and "Pomoravia" (eastern Serbia) to Bulgaria. It was this obsession that predetermined the country's alliance with Germany in the first half of the 20th century and its adherence to Nazi antisemitic policy in 1940-44. The ideological alliance with Nazi Germany concerning Jewish matters began in 1940, when Bulgaria issued three overtly or covertly antisemitic laws. In spring 1943 Bulgaria rounded up and deported the Jews of Macedonia and Thrace, ceded to her by Germany in 1941. At the same time, realizing that Germany would not win the war, Bulgaria did not want to cooperate; thus, the Jews of Old Bulgaria were spared. Concludes that 11,000 Jews of Macedonia and Thrace were the price paid by Bulgaria to Germany in order to secure its share in the spoils of war without actually taking an active part in combat. Pp. 153-159 contain documents. *DR*

0397 Birman, Mikhail: Bulgarian Jewry and the Holocaust (History and Historiography). *Shvut* 10 [26] (2001) 160-181.

Despite Bulgaria's alliance with Germany, the implementation of anti-Jewish measures in the country met with difficulties due to popular opposition to antisemitism. Protests were made by the Kyustendil group, Church leaders, and intellectuals. The opposition caused the government to slow down its anti-Jewish policies. Surveys the historiography of the Holocaust in Bulgaria. There have been various responses to the question of who rescued the Bulgarian Jews – the Bulgarian people, the king, the

Communist Party, and the Church. Post-communist historiography devotes much attention to the king in this respect. Some writers emphasize the role of Dimitar Peshev, neglecting other members of the Kyustendil group. Notes that many aspects of the rescue of Bulgarian Jews have been neglected; e.g. differences among the ruling elites regarding the "Jewish question," the role of the Jewish leadership, and diplomatic negotiations. *DR*

0398 Hakov, Djengiz: The Fate of the Bulgarian Jews during World War II. *Etudes Balkaniques* 34, 1-2 (1998) 122-130.

A paper presented at an international conference held in Skopje in 1993. States that antisemitism did not have broad support in Bulgaria. The first anti-Jewish law, the Defence of the Nation Act passed in January 1941, was met with protests on the part of leading intellectuals, clerics, and some public organizations. In February 1943, the first Bulgarian-German agreement on the deportation of Jews was signed. Part of the quota was filled by the deportation of Jews from Macedonia and Western Thrace in March 1943, but the deportation of 8,500 Bulgarian Jews was postponed. Some groups (e.g. 43 members of the National Assembly, and the Holy Synod of the Church) protested vehemently against this deportation. The deportations were thwarted, but more than 25,000 Jews were expelled from Sofia, Stara Zagora, and Kazanlyk to the countryside. It was protests, as well as the victories of the anti-Nazi coalition, that held back the government from the deportation of Bulgarian Jews. *DR*

0399 Mossek, Moshe: The Bulgarian Church and the Jews during the Second World War. *Shvut* 10 [26] (2001) 280-298.

After the war, the Bulgarian Orthodox Church did not receive the recognition it deserved from the communist government, nor in Bulgarian historiography, for its significant, though indirect, role in the rescue of the Jews because of its influence over the masses of believers, especially the king and circles close to the court. States that there were no differences of opinion among the eleven members of the Holy Synod regarding Bulgarian policy toward the Jews. The Church's resolute and courageous stand against the government's antisemitic policy was unique among the Churches of European countries under Nazi domination. Pp. 287-298 contain documents: the Statement of the Holy Synod to the Prime Minister on 15 November 1940, protesting against the proposed antisemitic law; letters written by hierarchs; and an eyewitness account. *DR*

0400 Todorov, Tzvetan, ed.: *La fragilité du bien: Le sauvetage des Juifs bulgares*. Trans.: Marie Vrinat, Irène Kristeva. Paris: Albin Michel, 1999. 222 pp. Translated from Bulgarian.

The introduction (pp. 9-51) surveys the history of the rescue of Bulgaria's 48,000 Jews during the Holocaust, examines competing interpretations, and discusses the moral question of what allows goodness to triumph. Pp. 55-137 present documents dealing with the exclusion of the Jews in December 1940, the plan to deport them in March 1943, and internment and expulsion of the Jews from Sofia in May 1943. Pp. 139-217 consist of memoirs by Dimo Kazasov, the Metropolitan Stefan, Assen Suichmezov, and Dimitar Peshev (pp. 164-217). Kazasov, Suichmezov, and Peshev have been honored by Yad Vashem as Righteous Gentiles. *EO*

0401 Todorov, Tzvetan, ed.: *The Fragility of Goodness: Why Bulgaria's Jews Survived the Holocaust*. Trans.: Arthur Denner. Princeton, NJ: Princeton University Press, 2001. 197 pp. First published in Bulgarian. See the abstract for the French translation: "La fragilité du bien: Le sauvetage des Juifs bulgares" (Paris: Albin Michel, 1999).

Czechoslovakia
(see also Slovakia)

0402 Adamec, Stanislav: *Židé v Kyšperku, 1700-1945: Příspěvek k historii města Letohradu* [Jews in Kysperk, 1700-1945: A Contribution to the History of the Town of Letohrad]. Letohrad: Josef Marek, 2001. 92 pp.

A compilation from different sources prepared for the commemoration of the 60th anniversary of the deportation of Jews from Letohrad. Includes a brief overview of the history of Jews in Kyšperk and lists its Jewish inhabitants. Describes the deportation of Jews and reproduces several wartime lists, including a list of those who were deported to Terezín. Out of 26 members of the Jewish community listed in Kyšperk in 1941, only four survived. Includes excerpts from testimonies. *MF*

0403 Chládková, Ludmila: Kinder und Jugendliche in Theresienstadt. *Zwischenwelt* 18, 3 (Oct 2001) 48-51.

Describes the organization of the children's homes in Theresienstadt and the many activities of the children and young people, commenting that they made an important contribution to the cultural life of the ghetto. *RW*

0404 Eckert, Rainer: *Emigrationspublizistik und Judenverfolgung: Das Beispiel Tschechoslowakei*. Frankfurt a.M.: Peter Lang, 2000. 346 pp.

Surveys the factions represented among political refugees from Nazi Germany in Czechoslovakia, and analyzes their reactions, as expressed in their press organs, to Nazi persecution of the Jews. The communists saw Nazi antisemitism as a symptom of the weakness of capitalism and a means of diverting attention from the problems of the regime. They accused "Jewish fascists" (nationalists, capitalists, Zionists) of being themselves Nazis. The German people, they argued, was not antisemitic, and Jews should remain in Germany and fight the regime as part of the popular front. The socialist press expressed similar views, but less one-sidedly; it recognized psychological as well as instrumental factors in antisemitism, saw the danger of the extinction of German Jewry, and supported Zionism. *RW*

0405 *Intolerance: Češi, Němci a Židé na Ústecku 1938-1948: Edice dokumentů z fondů Archivu města Ústí nad Labem = Intoleranz: Tschechen, Deutsche und Juden in Aussig und Umgebung 1938-1948: Edition der Dokumente aus den Bestaenden des Archivs der Stadt Aussig*. Ústí nad Labem: Albis International, 1998. 167 pp.

A collection of documents on the coexistence of Czechs, Germans, and Jews in the region of Ústí nad Labem from 1938 to the early 1950s. All the documents are given in Czech and German. They describe the escape and expulsion of Czechs and Jews from the Sudetenland in 1938, the persecution of Czechs and Jews by the Nazis during World War II, and the expulsion of the German settlers after the war. Deals, also, with attempts of Jewish survivors to regain their Aryanized property in the late 1940s. Includes facsimiles of caricatures, some of them antisemitic. *MF*

0406 *Jewish Gold and Other Precious Metals, Precious Stones, and Objects Made of Such Materials: Situation in the Czech Lands in the Years 1939 to 1945*. Prague: Publishing House Sefer, Terezín Initiative Institute, 2001. 138 pp. On title page also: Unlawful Infringement of Property Rights and Its Scope; Subsequent Fate of the Jewish Assets Affected by This Infringement: Report by a Team of Experts Created by the Joint Task Force Based on Decision No. 773 of the Czech Government of November 25, 1998.

Discusses the legal and institutional framework of curtailment of Jewish property rights in regard to the items specified in the title, the role of the Protectorate Bank, sales of the illegally seized items, the scope of the seized assets, the fate of Jewish-owned gold purchased by the Protectorate Bank, and the postwar fate of Jewish movables unsold by the Germans. Notes that Germans, not Czechs, were responsible for the confiscations. Procedures have been established by the Czech government for owners to reclaim these properties or to seek restitution. *YC*

0407 Křesťan, Jiří: Židovské spolky a česká společnost (1918-1940) [Jewish Associations and Czech Society (1918-1940)]. *Židovské spolky v českých zemích v letech 1918-1948* [Jewish Associations in Czech Lands 1918-1948]. Praha: Sefer, 2001. Pp. 15-76.

Relates, also, to the influence of antisemitism in Germany and in Bohemia on the activities of the Czech Jewish associations. Describes reactions to antisemitism during the first years of the Czech republic, including a public protest meeting and a resolution of the Národní rada židovská (National Jewish Committee) in 1922. Discusses the impact of the Nazi accession to power in Germany in 1933 on the associations, citing the resolution of Svaz Čechů-židů (Union of Czechs-Jews) in April 1933 which condemned the situation in Germany and called for support of the Czech national program. Describes the rise of antisemitism in Czechoslovakia and the decline of the associations' activities following the Munich Agreement (1938) and after the establishment of the Protectorate of Bohemia and Moravia (1939), ending with the dissolution of the Jewish associations by the Nazis in the spring of 1940. *JT*

0408 Kryl, Miroslav: *Osud vězňů Terezínského ghetta v letech 1941-1944* [The Fate of the Prisoners in the Theresienstadt Ghetto 1941-1944]. Brno: Doplněk, 1999. 225 pp.

Chs. 1-4 (pp. 7-33) deal with the role of antisemitism in the ideology and policies of Nazi Germany. Discusses conflicting approaches of historians on the question of to what extent the Final Solution was an integral and well-planned part of the strategy of Hitler and the Nazi central command. Ch. 5 (pp. 34-53) outlines the Nazi anti-Jewish policies in the Protectorate and the establishment of the Theresienstadt ghetto. The rest of the book (pp. 54-156) relates the history of the Theresienstadt ghetto from the beginning of 1942 until the last deportations in October 1944, with special attention paid to the visit of the International Red Cross delegation on 23 July 1944 (pp. 92-147). Relates, also, to newly discovered documents, offering new interpretations of various topics. Pp. 161-223 contain endnotes, a bibliography, and an index. *JT*

0409 Potůčková-Taussigová, Jarmila: *Die Taussigs: Juedische Familien- und Leidensgeschichte in Boehmen und Maehren 1909-1989*. Trans.: Liselotte Teltscherová. Konstanz: Hartung-Gorre, 2000. 58 pp. Translated from the Czech manuscript.

The author, a non-Jew, writes about the Jewish family of her first husband František (Franta) Taussig – the Taussigs of Hlinsko. Franta (1909-1941) was a communist party organizer and editor of a communist newspaper in Bruenn (Brno). He was arrested and shot by the Nazis in 1941 for his activities in the resistance. His parents and a younger brother, Josef (Pepek), a writer and artist, were sent to Theresienstadt, where Pepek contributed to the cultural life. The parents died in Auschwitz; Pepek survived Auschwitz and the death march to Flossenbuerg, only to die three days before liberation. Pp. 35-38 contain a biography of Pepek by Elena Makarová, who made a film on theatrical performances in

Theresienstadt. Also contains some of Pepek's verses for the Theresienstadt cabaret and the story of the author, who was interned in Ravensbrueck. *RW*

0410 Roučka, Zdeněk: *Plzeň pod hákovým křížem: Fotografie 1939-1945* [Pilsen under the Swastika: Photographs 1939-1945]. Plzeň: ZR&T, 2001. [not paginated].

Presents a short history of the city of Plzeň (Pilsen) under the Nazi occupation. The introduction and commentary accompanying the photographs are in Czech and English. The introduction and several of the photos relate to the deportation of the Jews from Plzeň. Most of them were deported in January 1942 to Terezín. Out of 2,604 Jews deported, only 204 survived. *MF*

0411 Rupnow, Dirk: *Taeter, Gedaechtnis, Opfer: Das "Juedische Zentralmuseum" in Prag 1942-1945*. Wien: Picus, 2000. 231 pp.

Argues that "musealization" means removing objects from their living context, most often after the destruction of that context. This is especially clear in the case of the "Central Jewish Museum" in Prague, whose collections all came from liquidated Jewish communities whose members had been sent to the concentration and death camps. The Jews and the SS cooperated in this project, each for their own ends. The Jewish specialists who administered the collections hoped to save at least the memory of Jewish life in Czechoslovakia; the Nazis wanted this too, but as testimony to future generations of the adversary that they had vanquished. They did not want to obliterate that memory, but to manipulate it. Today, this musealization persists: the Jewish quarter of Prague and the Jews now living there have been turned into tourist attractions. *RW*

0412 *Theresienstaedter Studien und Dokumente* (2001).

Partial contents: Franková, Anita: Die Vorbereitungen zur Konzentrierung der Juden im Protektorat: Die "Vorgeschichte" des Theresienstaedter Ghettos (49-74); Zemanová, Pavla: Die Theresienstaedter Aussenkommandos (75-105); Tošnerová, Patricia: Die Postverbindung zwischen dem Protektorat und dem Ghetto Theresienstadt (106-147); Přibyl, Lukáš: Die Geschichte des Theresienstaedter Transports "Be" nach Estland (148-229); Hájková, Anna: Die acht Transporte aus dem Reichskommissariat Niederlande nach Theresienstadt (230-251); Woegerbauer, Michael: Das prominente Ehepaar Gutmann und die diplomatischen Bemuehungen zu seiner Befreiung (252-268); Fedorovič, Tomáš: Die Gemeinde Schoenwald und ihre unfreiwilligen Bewohner [On an internment camp in Bohemia.] (269-286); Schmid, Daniel C.: Moeglichkeiten und Grenzen schweizerischer Schutzmachtpolitik im

Protektorat Boehmen und Maehren 1939-1945 [On the activities of the Swiss consul in Prague, Albert Huber, to help Czech Jews who held foreign passports.] (287-304); Jančík, Drahomír et al.: Ein abartiges Monopol: Die "Hadega" Handelsgesellschaft m.b.H. und ihr Geschaeft mit Edelmetallen und Edelsteinen waehrend des Zweiten Weltkriegs [On its role in the confiscation of valuables.] (305-372). *ID*

Denmark

0413 Levysohn, Herbert W.: En skildring om et enkelt tilfälde under Jödeforfölgelsen i Danmark [An Account of a Single Case History during the Persecution of Danish Jews]. *Rambam* 10 (2001) 29-44. With an English summary.

Written in Stockholm in October 1943, this is an account by a young Jew of his experiences in Denmark from the time the country was placed under martial law on 29 August 1943 until his and his family's rescue flight to Sweden. He relates his father's arrest on the same day martial law was declared. Herbert, his mother, and sister had to leave their home. His mother and sister hid at a rest home before they managed to escape to Sweden. Herbert went into hiding in various places. Eventually, he too crossed to Sweden and was reunited with his mother and sister. The manuscript is now in the Danish Royal Library. *SSC*

0414 Ruenitz, Lone: *Danmark og de jødiske flygtninge, 1933-1940: En bog om flygtninge og menneskerettigheder* [Denmark and the Jewish Refugees, 1933-1940: A Book about Refugees and Human Rights]. København: Museum Tusculanums Forlag; Københavns Universitet, 2000. 239 pp. Based on the author's diss. – Københavns Universitet, Åbent Universitet, 1995.

A study of Danish opinions and policies regarding the stream of Jewish refugees arriving from Germany between 1933-40. Sketches the phases of persecution of Jews in Germany and the reactions of the democratic states. Relates the work done for the refugees by the international community and international Jewish organizations. Focuses on the political, social, and economic problems behind Denmark's refugee policy. Immigration restrictions were imposed in order to limit the number of immigrants. Discusses parliamentary debates concerning the refugees. Often, refugees were sent back to Germany. Describes the rescue action of children initiated by Danish women's organizations in 1939. Examines the attitudes of the Jewish leadership in Denmark, and also the rescue work of a committee constructed for this purpose, the "Committee of May 4th." In conclusion, explains Germany's behavior, the options of the refugees, the responses of various authorities to the refugee problem, and the endless obstacles faced by Jews trying to escape. *RSL*

France

0415 Arnold, Edith France: *Enfants sous l'Occupation (1939-1945): Et les amandiers fleurissaient quand même*. Paris: L'Harmattan, 2001. 199, [5] pp.

Memoirs of a Jew (née Weill) born in Mulhouse in 1934. The family (parents and four children) moved around France avoiding deportation, spending a few years in Nice and another few years in a small village in Provence. After the war they returned to Mulhouse. None of the immediate family was killed in the Holocaust. *EO*

0416 Collin, Claude: *Jeune combat: Les jeunes juifs de la MOI dans la Résistance*. Grenoble: Presses Universitaires de Grenoble, 1998. 143 pp.

Examines the role of the youth organizations in the Jewish section of the Main d'Oeuvre Immigrée (MOI), a principal arm of the French resistance. Traces the history of interwar Jewish immigration to France and the creation of the Section de Langue Juive which joined the armed struggle in 1942, centralizing its activities in Lyon. The Union de la Jeunesse Juive was created the following year as a united mass movement of Jewish youth against Nazism. This youth wing of the MOI was affiliated with the Parti Communiste Français, and was subsequently criticized for its allegiance to the Party rather than to the rescue of Jews. Rejects these allegations, citing testimony by former resistance activists and analyses of the press, pamphlets, and reports from that era. *TS*

0417 Deroo, Delphine: *Les enfants de la Martellière*. Paris: Bernard Grasset, 1999. 286 pp.

The story of 16 Orthodox Jewish children who hid in a children's home established by Rabbi Zalman Chneerson in La Martellière near Voiron. The children were arrested by the Gestapo in March 1944 and deported via Drancy to Auschwitz; only one of them survived. The book is based on interviews with 40 other children, who were also sent from occupied France first to Marseille and then to one of Chneerson's four children's homes run by the Association des Israélites Pratiquants. Reconstructs daily life in these homes, where religious practices were strictly observed and the Torah studied. Also describes Chneerson's and the AIP's other efforts to save Jewish children in France. *EO*

0418 Estrade-Szwarckopf, Mouny; Estrade, Paul: *Un camp de juifs oublié: Soudeilles (1941-1942)*. [Treignac]: Editions Les Monédières, 1999. 245 pp.

A study of the camp in Soudeilles (Corrèze) which was reserved exclusively for Jews of foreign extraction residing illegally in France; they were grouped together by the Vichy government in the 665th division of the Travailleurs Etrangers. During the time when the camp was in operation, between June 1941-December 1942, 550 Jewish inmates passed through it. Ca. 50 of them were deported via Drancy to Auschwitz. The poorly guarded camp consisted of a simple shack not far from the center of the small village. Most of the 230-260 Jews interned there at any one time were given temporary assignments in agriculture, communal works, or as lumberjacks. Also gives an overview of the status of Jewish refugees in France and of the Travailleurs Etrangers. Includes a list of Jews interned in Soudeilles, with biographical notes (pp. 101-238), as well as a list of Jews deported from the camp (pp. 239-245). *EO*

0419 Guigueno, Vincent: Les élèves juifs et l'Ecole polytechnique (1940-1943): La reconquête d'une identité. *Vingtième Siècle* 57 (Jan-Mar 1998) 76-88.

Discusses the impact of the discriminatory laws under Vichy on the Jewish students at the Ecole Polytechnique. Following the transfer of the Ecole to the Lyon region in 1940, Jews were required to register as such; their requests for financial aid were rejected, and the obligatory practical experience in their fields of study was cancelled. An official numerus clausus was installed in 1942, and by 1943 study was further restricted to correspondence. The decision by the administration to return to Paris in 1942 inspired many to join the resistance or seek options for emigration. Relates the case of André Bollier, a Jewish student who became a journalist for the resistance and died in 1944. Contends that the extent of hostility towards Jews at the Ecole was difficult to measure. On one hand, the Ecole was affiliated with the military, which implied strong ties with the Vichy government and severe constraints regarding protest against antisemitic legislation. Nevertheless, testimonies from Jewish alumnae attest to solidarity of members of the faculty with their Jewish students. *TS*

0420 Henry, Patrick: Daniel's Choice: Daniel Trocmé (1912-1944). *The French Review* 74, 4 (Mar 2001) 728-739.

Daniel Trocmé, cousin of Pastor André Trocmé, was head of one of the houses of refuge established on the plateau Vivarais-Lignon in southern France. In March 1943 he agreed to become head of another house, La Maison des Roches, which housed thirty young male students, most of them Jews. In June 1943 the Gestapo arrested 18 students who were in the house and Daniel Trocmé. All were deported; seven of the students survived. Trocmé died in the gas chamber in Majdanek on 4 April 1944. Relates how his service to the young people became an avocation that combined the values of his parents with his own search for meaning in life. He was honored posthumously by the French resistance, and as one of the "Righteous among the Nations" by Yad Vashem in 1976. *YC*

0421 Klarsfeld, Serge, ed.: *Lettres au Premier ministre des orphelins des déportés juifs de France.* Paris: Association "Les Fils et Filles des Déportés Juifs de France," 1999. 252 pp.

Presents 280 testimonies by orphans whose parents were deported from France and killed in the Shoah, published to back up claims for restitution addressed to the French government. Whereas Holocaust survivors and widows of victims have received decent French pensions since 1998, most orphans have so far received only a small pension averaging 700 francs per month. In addition to this, children of French parents (one-third of those deported had French citizenship) received a modest lump sum as stipulated in a French-German agreement signed in 1960. French-Jewish children of foreign parents, the overwhelming majority of the orphans, were not included in this agreement. Between 5,000-10,000 orphans are still alive today. *EO*

0422 Klarsfeld, Serge: *La Shoah en France. Vol. 1-4.* Paris: Fayard, 2001. 391; 1000; 1001-2029; 1256 pp. 4 vols. in a case.

A reprint of books published previously by Klarsfeld. Contents: Vol. 1: Vichy-Auschwitz: La "solution finale" de la question juive en France [Appeared in two vols. in 1983-1985]. Vols. 2-3: Le calendrier de la persécution des Juifs de France, 1940-1944 [Appeared in 1993]. Vol. 4: Le mémorial des enfants juifs déportés de France [Appeared in 1995]. *SSC*

0423 Kluger, Daniel; Sullaper, Victor: *Vigtor le rebelle: La résistance d'un Juif en France: Récit biographique.* Paris: L'Harmattan, 1999. 229 pp.
Memoirs of Sullaper, a Jew born in Sosnowiec in 1921, who was active in the French resistance and instrumental in helping hundreds of Jewish children escape across the Swiss border. The Szulklapers settled in Paris in 1930 with four children; one daughter remained in Poland, where she and her family were killed by the Germans in 1939. An ardent Zionist, Vigtor spent three years in Palestine, returning to France in 1939. He joined the French Polish army, escaped from an internment camp in Switzerland, returned to France, was interned in Pithiviers, and escaped from there as well. He helped his family hide in Lyon and joined the resistance in 1943. He forged identity papers and worked for the UGIF to save Jewish children. His family was deported to Auschwitz in August

1944 with the last train from Lyon. The parents, a brother and a sister survived. After the war Vigtor (Victor) returned to Paris. *EO*

0424 Poznanski, Renée: The Geopolitics of Jewish Resistance in France. *Holocaust and Genocide Studies* 15, 2 (Fall 2001) 245-265.

Focuses on the role of the cities in the development of Jewish resistance in France. The anti-Jewish laws and the implementation of the Final Solution in 1942 overturned the fundamentally urban nature of Jewish life in France. Two types of migration of Jews took place: from Paris to the unoccupied zone, and from the cities to the countryside. This movement was determined by changes in the anti-Jewish policy and by the Jewish organizations' transition from mainly economic support to rescue of lives. However, the cities preserved their central role in the organization of rescue-oriented resistance. When resistance efforts began to focus on armed struggle, the role of the cities increased. Jews constituted a great majority of the men in communist-led armed underground groups in the Paris area. In cities of the Vichy zone, Jewish armed resistance was on a lower scale. *DR*

0425 Poznanski, Renée: *Jews in France during World War II*. Trans.: Nathan Bracher. Hanover, NH: Brandeis University Press, in association with the United States Holocaust Memorial Museum; University Press of New England, 2001. xxv, 601 pp. Originally published as "Les Juifs en France pendant la Seconde Guerre mondiale" (Paris: Hachette, 1997). First published as "Etre Juif en France pendant la Seconde Guerre mondiale" (1994).

0426 Schwarzfuchs, Simon: *Aux prises avec Vichy: Histoire politique des Juifs de France 1940-1944*. Paris: Calmann-Lévy, 1998. 444 pp.

A study of the activities of the Jewish institutions in France during the Vichy regime, based on archival material, especially protocols of meetings of the Consistoire Central des Israélites de France. Argues that activities of the Consistoire and of the Grand Rabbinat helped UGIF (Union Générale des Israélites de France), created in November 1941, not to become a Judenrat. The Consistoire, through its protests, declarations, and negotiations with Vichy authorities was able to slow the actions of the Commissariat général aux Questions juives, to curb the application of discriminatory measures, and to help some Church representatives to take a stance against anti-Jewish measures. Concludes that Jewish leaders in France prevented the UGIF from becoming a corrupt Judenrat, and in this way contributed to the fact that three-quarters of French Jewry was saved. *HV*

0427 Tessler, Gloria: *Amélie: The Story of Lady Jakobovits*. London: Vallentine Mitchell, 1999. xviii, 318 pp.

A biography of Amélie Jakobovits, born in 1928 in Ansbach, Bavaria. Chs. 1-7 (pp. 1-64) relate to antisemitism in the pre-Nazi and early Nazi periods in Ansbach. Jakobovits's grandfather, president of the Nuremberg Jewish community, died of a heart attack after an encounter with Nazis. Jakobovits's family moved to Paris in 1936; her father was Rabbi Elie Munk. With the Nazi invasion, they fled to southern France, finally settling in Nice. When the Nazis occupied the south, they were threatened with deportation and had some narrow escapes. With false papers the family escaped to Switzerland in September 1943. In 1945 they returned to Paris. As a member of the Orthodox youth movement Yeshurun, Jakobovits helped care for survivors. She later married Immanuel Jakobovits, Chief Rabbi of Ireland and, subsequently, of Great Britain. *YC*

Germany

0428 Bankier, David, ed.: *Probing the Depths of German Antisemitism: German Society and the Persecution of the Jews, 1933-1941*. New York: Berghahn Books; Jerusalem: Yad Vashem; Leo Baeck Institute, 2000. 585 pp.

Contents: Bauer, Yehuda: Overall Explanations, German Society and the Jews or: Some Thoughts about Context (3-16). "Party and State Antisemitic Policy": Gerstenberger, Heide: Acquiescence? (19-35); Yahil, Leni: The Double Consciousness of the Nazi Mind and Practice (36-53); Meyer, Beate: The Mixed Marriage: A Guarantee of Survival or a Reflection of German Society during the Nazi Regime? (54-77); Gruner, Wolf: Public Welfare and the German Jews under National Socialism: On Anti-Jewish Policies of the Municipal Administrations, the German Council of Municipalities and the Reich Ministry of the Interior (1933-1942) [Appeared in German in the "Zeitschrift fuer Geschichtswissenschaft" 45 (1997).] (78-105). "Nazi Antisemitic Policy on the Regional Level": Lohalm, Uwe: Local Administration and Nazi Anti-Jewish Policy (109-146); Obenhaus, Herbert: The Germans: "An Antisemitic People": The Press Campaign after 9 November 1938 (147-180); Wildt, Michael: Violence against Jews in Germany, 1933-1939 [Appeared in German in "WerkstattGeschichte" 18 (1997).] (181-209). "The Policy of Expropriation": Fischer, Albert: The Minister of Economics and the Expulsion of the Jews from the German Economy (213-225); Bajohr, Frank: The "Aryanization" of Jewish Companies and German Society: The Example of Hamburg (226-245); Ladwig-Winters, Simone: The Attack on Berlin Department Stores ("Warenhaeuser") after 1933 [Focuses on the Wertheim department store chain.] (246-267).

"Popular Attitudes to Nazi Antisemitism in Wartime": Kulka, Otto Dov: The German Population and the Jews: State of Research and New Perspectives (271-281); Wickert, Christl: Popular Attitudes to National Socialist Antisemitism: Denunciations for "Insidious Offenses" and "Racial Ignominy" (282-295); Luedtke, Alf: German Work and German Workers: The Impact of Symbols on the Exclusion of Jews in Nazi-Germany (296-311); Heim, Susanne: The German-Jewish Relationship in the Diaries of Victor Klemperer (312-325). "Jewish Society under the Nazi System": Walk, Joseph: Reactions of the Jewish Press in Germany to the Nuremberg Laws (329-338); Fraenkel, Daniel: Jewish Self-Defense under the Constraints of National Socialism: The Final Years of the Centralverein (339-359); Weiss, Yfaat: The "Emigration Effort" or "Repatriation": The Issue of the Relationship between Jewish-Polish Emigrants and the Jewish-German Establishment (360-370); Paucker, Arnold; Kwiet, Konrad: Jewish Leadership and Jewish Resistance (371-394); Kaplan, Marion A.: Jewish Daily Life in Wartime Germany (395-412). "Responses of the Churches": Bacharach, Walter Zwi: The Catholic Anti-Jewish Prejudice, Hitler and the Jews (415-430); Buettner, Ursula: "The Jewish Problem Becomes a Christian Problem": German Protestants and the Persecution of the Jews in the Third Reich [Appeared in German in the "Zeitschrift fuer Geschichtswissenschaft " 45 (1997).] (431-459). "Responses of German Resistance": Hoffmann, Peter: The German Resistance to Hitler and the Jews (463-477); Dipper, Christof: 20 July and the "Jewish Question" (478-497). "Responses of the Bystanders": Breitman, Richard David: American Diplomatic Records regarding German Public Opinion during the Nazi Regime (501-510); Bankier, David: German Social Democrats and the Jewish Question (511-532); Aronson, Shlomo: British Policy, Allied Intelligence and Zionist Dilemmas in the Face of Hitler's Jewish Policies [Appeared in Hebrew in "Mediniyut milhamah, mediniyut shalom" (2000).] (533-569). *DR*

0429 Behne, Ulrich: *Die Viehhaendlerfamilien Gerson und das Schicksal der juedischen Gemeinde zu Vechta*. Diepholz: Schroederscher Buchverlag; Verlag fuer Regionalkultur, 2001. 204 pp.

A book published in conjunction with an exhibition held at the municipal museum on this topic between November-December 2001. Surveys the history of Jews in Vechta (Oldenburger Muensterland) between 1920-42, focusing on the fate of the Gerson family. Emanuel Gerson and his brother Adolf moved to Vechta in 1911 from Ostfriesland, and became cattle-dealers. In 1932 the Jewish community consisted of four families. Gradually the Gerson brothers were excluded from the cattle-dealing trade; during the "Kristallnacht" pogrom they were arrested and deported to Sachsenhausen. After their release, Adolf and his family left for Palestine, via Switzerland. Emanuel and his three daugthers (their mother died in 1934) left for Oldenburg in June 1939 and were forced to move into a crowded "Jewish house"; in April 1940 they were transferred to Hamburg, where Emanuel committed suicide. Two of his daughters were deported in November 1941 to Minsk, and the third was deported in December 1941 to Riga. They did not survive. *ID*

0430 Behr, Hartwig; Rupp, Horst F.: *Vom Leben und Sterben: Juden in Creglingen*. Wuerzburg: Koenigshausen & Neumann, 1999. 278 pp.

Surveys the history of Jews in Creglingen (Wuerttemberg) from the beginning of their settlement there in 1532 until 1938. Pp. 135-151, by Behr, discuss a pogrom which took place on 25 March 1933 in Creglingen, during which Hermann Stern was murdered and several other Jews were injured. Pp. 153-181, also by Behr, describes events between 1933-38. Despite the pogrom, few of the 70 Jewish residents left the town in 1933; but in 1937-38, due to the antisemitic legislation, several families emigrated. After November 1938 the remaining 32 Jews left Creglingen. Among those who stayed in Germany, 16 were deported. Pp. 183-211 report on two local Nazi war criminals: Fritz Klein, the main perpetrator of the pogrom, and Karl Stahl. *ID*

0431 Bertram, Ellen: *Menschen ohne Grabstein: Die aus Leipzig deportierten und ermordeten Juden*. Leipzig: Passage-Verlag, 2001. 255 pp. On title page also: Herausgegeben von Rolf und Brigitte Kralovitz in Verbindung mit der Ephraim Carlebach Stiftung und der Israelitischen Religionsgemeinde zu Leipzig.

The introduction (pp. 11-51) discusses the Nazi period in Leipzig: the discrimination, forced labor, and deportations to Riga, Bełżec, Auschwitz, and Theresienstadt, which took place between January 1942-July 1943. In 1933, there were 11,564 Jews in Leipzig; in 1939 their number decreased to 4,284 (mainly due to emigration); in August 1941, 1,912 remained. Pp. 55-252 contain a list of 1,998 victims, with information on birth date, last known address in Leipzig, profession, nationality, place of deportation, date and cause of death. The fate of many others remains unknown. *ID*

0432 Bucholtz, Erika: *Henri Hinrichsen und der Musikverlag C.F. Peters: Deutsch-juedisches Buergertum in Leipzig von 1891 bis 1938*. Tuebingen: Mohr Siebeck, 2001. viii, 367 pp. (Schriftenreihe wissenschaftlicher Abhandlungen des Leo-Baeck-Instituts, 65). Based on the author's diss. – Technische Universitaet Berlin, 2000.

Describes the history of the music publishing house and of its Jewish owners, the Abraham-Hinrichsen family. Ch. 6 (pp. 284-313), "Die Zeit ab 1933," chronicles the family's slow awakening to the developments in Germany. Boycotted by the central organizations of the branch, Henri Hinrichsen solicited the aid of influential Nazi musicians, pleading that he could not imagine being excluded from the current rebirth of Germany and that he and his firm should not be put in the same category as "rootless international Jewry." Hinrichsen was also excluded from cultural and civic institutions which he had founded and nurtured. He found no customers for his editions of Jewish composers, whose works were no longer performed. Intimidated "Aryan" composers withdrew their works. In the "Kristallnacht" pogrom, the offices were demolished and the warehouse plundered; the firm was Aryanized. Several sons had already emigrated; in January 1940, Henri with his wife and another son fled, penniless, to Belgium. All three, as well as another son, died in the Holocaust. A daughter survived Auschwitz. *RW*

0433 Davis, Hadassah: What Was the Role of the Wannsee Conference in the Final Solution? *European Judaism* 32, 2 (Fall 1999) 26-36.

Argues that the "Final Solution of the Jewish question" in the form of mass murder was centrally planned well before the Wannsee Conference took place. All the machinery of destruction was under preparation, and the mass murder was already under way in 1941. The Wannsee Conference merely updated the participants in regard to the plan to remove Western European Jews to Eastern Europe for hard labor and death. *DR*

0434 Doerr, Thomas: *"Muehsam und so weiter, was waren das fuer Namen...": Zeitgeist und Zynismus im nationalistisch-antisemitischen Werk des Graphikers A. Paul Weber.* Luebeck: Erich-Muehsam-Gesellschaft, 2000. 55 pp.

Questions Weber's reputation as an opponent of Nazism. From 1918 through the 1920s he illustrated antisemitic books such as Dinter's "Die Suende wider das Blut" or Manfred von Killinger's "Ernstes und Heiteres aus dem Putschleben," which contained a diatribe against the leaders of the Munich republic, noting their Jewish names. Weber depicted Jews as monstrous, repulsive reptiles or "Stuermer"-like figures sucking the blood of heroic Germans. In the 1930s he published, together with Ernst Niekisch, the nationalist-antisemitic "Widerstand," whose only objection to Nazism was that it was not radical enough; throughout the war, he contributed to Nazi publications. After the war he republished his drawings, with alterations that enabled liberals to interpret them as anti-Nazi and humanistic. *RW*

0435 Duewell, Kurt et al., eds.: *Vertreibung juedischer Kuenstler und Wissenschaftler aus Duesseldorf 1933-1945.* Duesseldorf: Droste, 1998. 244 pp.

A collection based on papers delivered at a conference in Duesseldorf, February 1998. Partial contents: Benz, Wolfgang: Emigration aus

Deutschland: Vertreibung juedischer Kuenstler und Wissenschaftler 1933-1945 (15-27); Baumeister, Annette; Kleinbongartz, Sigrid: Juedische Maler und Bildhauer in der Duesseldorfer Kunstszene: Wege in die Emigration (29-46); Meiszies, Winrich: "...bluten wir nicht? ...lachen wir nicht?" Juedische Theaterkuenstler in Duesseldorf und im Exil (47-66); Suchy, Barbara: Juedische Musiker, Komponisten und Musikwissenschaftler in Duesseldorf und in der Emigration: Fuenf Portraitskizzen (97-140); Griese, Kerstin; Woelk, Wolfgang: Juedische Aerztinnen und Aerzte in Duesseldorf und in der Emigration (177-205). *RW*

0436 Fassl, Peter, ed.: *Geschichte und Kultur der Juden in Schwaben II: Neuere Forschungen und Zeitzeugenberichte*. Stuttgart: Jan Thorbecke, 2000. 444 pp. Vol. I was published in 1994.

A collection of articles on the history of the Jews in Swabia. The following deal with the Nazi period: Rosenau, Paul; Landman, Henry: Erinnerungen an die 20er und 30er Jahre in Augsburg (319-337); Immenkoetter, Herbert: Die Ermordung juedischer Geisteskranker aus Bayern in der NS-Zeit (339-351); Roemer, Gernot: Die letzten schwaebischen Rabbiner in der NS-Zeit (353-364); Richter, Martin: Juedische Refugees aus Bayerisch-Schwaben in Grossbritanien (365-377); Pemper, Mieczyslaw: Zwangsarbeiter bei Goeth, Mitarbeiter bei Schindler, Zeuge und Uebersetzer in den NS-Prozessen der Nachkriegszeit: Ein Zeitzeugenbericht [Testimony of a former Polish prisoner in the Płaszów labor camp and in Kraków, on Amon Goeth, Oskar Schindler, and war crimes trials in Kraków.] (379-399); Sommer, Karin: Vom Vernichtungslager ins Fluechtlingslager: Juedische Ueberlebende des Holocaust als "Displaced Persons" (401-412). *ID*

0437 Gellately, Robert: *Backing Hitler: Consent and Coercion in Nazi Germany*. Oxford: Oxford University Press, 2001. xvi, 359, [24] pp.

Examines the extent of knowledge amongst the German populace about the concentration camps and other aspects of Nazi terror and how they knew about them. Germans were in fact meant to know that their country had a secret police and a concentration camp system. Hitler was largely successful in getting the backing, one way or another, of the great majority of citizens. The press was used to foster a consensus of support for Nazi policies. Highlights widespread active collaboration via denunciations, including against the Jews. The invasion of the USSR accelerated the murderousness of the Nazis, which was accompanied by increased stress on the Jews (along with the Bolsheviks) as arch-enemies of the German people. While other "enemies" (e.g. foreign workers and POWs) were treated harshly, the Jews suffered disproportionately. Ch. 6 (pp. 121-150), "Injustice and the Jews," focuses on antisemitism. *YC*

0438 Gellately, Robert: *Hingeschaut und Weggesehen: Hitler und sein Volk.* Trans.: Holger Fliessbach. Stuttgart: Deutsche Verlags-Anstalt, 2001. 456 pp. Originally published as "Backing Hitler: Consent and Coercion in Nazi Germany" (Oxford: Oxford University Press, 2001).

0439 Glass, James M.: Normal Politics as Murder: The Holocaust and Group Psychosis. *JPCS* 5, 1 (Spr 2000) 115-120.

Interprets Nazi antisemitism and the Nazis' genocide of the Jews as a case of group psychosis. The psychotic group develops its own insane epistemology, redefines normalcy, and is able to harness quite normal people in order to achieve its insane goals. In Nazi Germany, the Jews were reinterpreted as a biological menace to the German nation and were dehumanized. Their murder became a normal function of normal people, as defined by Nazi society. The Nazi German group psychosis effectively impeded resistance to the regime; a person resisting Nazism had to defy its group insanity and could be regarded as insane himself. *DR*

0440 Gomel, Ilana: From Dr. Moreau to Dr. Mengele: The Biological Sublime. *Poetics Today* 21, 2 (Sum 2000) 393-421.

Traces the roots of the "bio-ideologies" of fascism and German Nazism to late 19th century ideas of Social Darwinism and eugenics. Josef Mengele symbolized Dr. Death. His medical experiments and the Nazi genocide of the Jews reflected a view of science that was based on the assumption of the superiority of the Aryan race. This idea of a superman who experiments on and tortures "lower" creatures was first embodied in H.G. Wells' "The Island of Doctor Moreau" (1896). Discusses, also, the short story "Mengele" (1989) by Lucius Shepard. *YC*

0441 Gross, Raphael: *Carl Schmitt und die Juden: Eine deutsche Rechtslehre.* Frankfurt a.M.: Suhrkamp, 2000. 442 pp. Based on the author's diss. – Universitaet Essen.

Asserts that antisemitism was a fundamental element in Schmitt's political doctrine. It appeared in his writings even before the Nazi takeover, and made it easy and natural for him to join the party in March 1933. He identified the Jews with "normative," abstract, universally valid law, against which he set German "Nomos," decisions made by the leader. The state is based on the political principle, whose essence is the distinction between friend and foe: the Jews are the foe, they are "artfremd," foreign and dangerous to German nature ("Art"). Traces the sources of Schmitt's antisemitism in Catholicism, the Action Française, the Young Hegelians, and Weimar Protestant political theology. Schmitt profited from the dismissal of Jewish colleagues. In a conference on Jews in jurisprudence convoked by him in 1936, he called for the removal of works by Jewish authors from law libraries. In his "Leviathan," he uses the Jewish mythical monster to denote the state, which the Jews dismember and devour. After the war, he complained that he was being persecuted by returning Jewish émigrés, and that the Jews were the real victors of the war, posing as victims in order to pocket reparations, on their way to rule the world. *RW*

0442 Haffner, Sebastian: *Geschichte eines Deutschen: Die Erinnerungen 1914-1933.* Stuttgart: Deutsche Verlags-Anstalt, 2000. 239 pp. A paperback edition appeared in Muenchen: Deutscher Taschenbuch Verlag, 2002 (304 pp.), with a new preface and an epilogue by Oliver Pretzel, and additional chapters which were discovered after the first edition.

The author (1907-1999), a non-Jew from Berlin, began after his immigration to England to write his memories of the First World War, the struggle between good sense and fanatic nationalism in the Weimar Republic, and the hyper-inflation – and the psychological influence these had on his generation; along with certain peculiarly German character traits, they prepared young Germans to become convinced Nazis. Pp. 137-177 describe the blithe antisemitic slogans of the Hitler Youth; the brutal ousting by the SA of Jewish judges and lawyers from the court building where the author was a law intern, and his shame at having answered "yes" when asked whether he was an Aryan; the fear among his Jewish friends (and his own fear for them) on the eve of the boycott of 1 April 1933, and the hasty departure abroad, that same day, of his best friend, a Jew. *RW*

0443 Hansert, Andreas: Zum Schicksal der Sammlung Alfred Oppenheims waehrend und nach der NS-Zeit = The Fate of the Alfred Oppenheim Collection during and after the Nazi Period. *Moritz Daniel Oppenheim: Die Entdeckung des juedischen Selbstbewusstseins in der Kunst: Jewish Identity in 19th Century Art,* eds. Georg Heuberger, Anton Merk. Frankfurt a.M.: Wienand Verlag, 1999. Pp. 304-325. German and English in parallel columns.

Alfred Oppenheim, grandson of Moritz Oppenheim and a resident of Frankfurt, was himself a painter and also an art collector. His collection included his grandfather's paintings and his own, as well as French and Oriental prints. When he emigrated in February 1939, he was not permitted to take his collection with him. On the initiative of Ernst Holzinger, director of the Staedel Art Institute and the local expert for the evaluation of confiscated art objects (appointed by the Nazis, although not a Nazi himself), important parts of the collection were taken over by Frankfurt museums and restored to Oppenheim after the war. The rest were auctioned off; the location of many of these remains unknown. *RW*

0444 Harpprecht, Klaus: Un jour très particulier: L'histoire du pasteur Julius von Jan et de sa prédication contre les pogroms nazis: Un récit instructif. *Cahiers de l'IRP* 37 (Oct 2000) 25-33. First appeared in German in "Die Zeit" (4 November 1999).

Discusses the case of Pastor Julius von Jan who was born in and served in the village Oberlenningen in Germany. A few days after the "Kristallnacht" pogrom, Jan gave a sermon in his church against the anti-Jewish violence. Following this sermon, Jan was imprisoned and tortured by the SA. He remained in prison for 16 months. In 1943 he was called up by the Wehrmacht. After the war he returned to Oberlenningen. He died in 1964. Pp. 30-33 present the text of the sermon in French translation. *HV*

0445 Heinemann, Stephan: *Juedisches Leben in den nordostniedersaechsischen Kleinstaedten Walsrode und Uelzen.* Walsrode: Bund der Freunde des Heidemuseums Walsrode, 2001. 422 pp. Based on the author's diss. – Universitaet Hannover, 2001.

Discusses Jewish life in two small towns of Lower Saxony between 1705-1943. Pp. 275-336 deal with the Nazi period in Uelzen and Walsrode: the boycott of Jewish enterprises, Aryanization, exclusion of Jews from social and economic life, the "Kristallnacht" pogrom, emigration, and deportation. Between 1933-38, four out of nine Jews living in Walsrode and eight out of 32 in Uelzen emigrated. Those remaining were moved into "Jewish houses"; between 1941-43 they were deported and murdered. *ID*

0446 Hilgenfeld, Martin, ed.: *Lebensspuren: Juedische Woehlerschueler, Opfer des Terrors, 1938-1945.* Frankfurt a.M.: Woehlerschule, 2001. 131 pp. On title page also: "Begleitheft zur Ausstellung der Spurensuche-AG der Woehlerschule Frankfurt a.M." On back of title page: Zusammengestellt von Waltraud Giesen, Felicitas Guersching, Stefan Reinhardt.

A brochure accompanying an exhibition on the fate of 27 Jews, former students of the Woehlerschule in Frankfurt, who were victims of the Holocaust. The introduction (pp. 9-16) traces the history of the Woelerschule. In January 1933, 14% of the students were Jewish. After the Nazi takeover, the school authorities were ordered to reduce the number of Jewish students to 1.4%. Some of those who were expelled transferred to a Jewish school. Many of the students emigrated. Pp. 23-108 present 27 short biographies of the students who were murdered in the Holocaust, accompanied by facsimiles of documents. *ID*

0447 Hyams, Helge-Ulrike et al., eds.: *Juedisches Kinderleben im Spiegel juedischer Kinderbuecher: Eine Ausstellung der Universitaetsbibliothek Oldenburg mit dem Kindheitsmuseum Marburg. Vol. 1.* Oldenburg:

Bibliotheks- und Informationssystem der Universitaet Oldenburg, 2001. 352 pp.

A collection of essays published as a catalogue accompanying the 17th exhibition of children's literature, organized by the university library in Oldenburg in 1998. The following essays discuss the Nazi period: Foelling, Werner: Das Leben juedischer Kinder im "Dritten Reich" bis 1939 (101-108); Meseberg-Haubold, Ilse: Das juedische Kinder- und Landschulheim Caputh: Wirkungen und Nachwirkungen (109-114); Fritsche, Michael; Peper, Ulrike: Juedische Kindheit im Schatten des Holocaust (115-124); Klattenhoff, Klaus; Wissmann, Friedrich: Juedische Kindheit im Nationalsozialismus (137-160); Nussbaum, Laureen: Anne Frank, zur Symbolfigur erhoben, als Schriftstellerin verunglimpft (305-314); Hyams, Helge-Ulrike: Kindliche Reaktionen auf Antisemitismus in Deutschland vor 1933 (315-335). *ID*

0448 Johnson, Eric Arthur: *Der national-sozialistische Terror: Gestapo, Juden und gewoehnliche Deutsche*. Trans.: Udo Rennert. [Berlin]: Siedler, 2001. 638 pp. Originally published as "Nazi Terror: The Gestapo, Jews, and Ordinary Germans" (New York: Basic Books, 1999).

0449 Karay, Felicja: *Wir lebten zwischen Granaten und Gedichten: Das Frauenlager der Ruestungsfabrik HASAG im Dritten Reich*. Trans.: Susanne Plietzsch. Koeln: Boehlau, 2001. 223 pp. Translated from the Hebrew. Originally published as "Pagazim ve-haruzim: Mahane nashim Hasag-Leipzig" (Tel-Aviv: Moreshet, 1997).

0450 Kwiet, Konrad: Leo Baeck und der deutsch-juedische Widerstand. *Neuer Anbruch: Zur deutsch-juedischen Geschichte und Kultur,* eds. Michael Brocke, Aubrey Pomerance, Andrea Schatz. Berlin: Metropol, 2001. Pp. 77-94.

Based on a lecture delivered in Jerusalem in February 1997. A version revised by the author, together with Arnold Paucker, appeared in "Probing the Depths of German Antisemitism" (2000). The Reichsvereinigung, under Leo Baeck, opposed Nazi persecution by self-assertiveness and protests, but rejected militant resistance or illegal actions which might endanger the community. Jewish resistance had no chance, since it had practically no support from the German population or the resistance movements. Contends that because of its insistence on legality, the Reichsvereinigung missed real possibilities of rescue. This policy changed only when the full desperation of the situation became clear. The Reichsvereinigung was informed about the extermination camps, but felt it would serve no good purpose to inform the Jewish public; besides, it hoped that the reports were unfounded or exaggerated. (Baeck took a similar view in Theresienstadt.) Also discusses the accusation that Baeck wrote his treatise on the history of Jewish rights in Germany at the behest of the SS; considers it more likely that he wrote it on his own initiative, or, as he himself claimed, at the invitation of the conservative resistance. *RW*

0451 LeBor, Adam; Boyes, Roger: *Seduced by Hitler: The Choices of a Nation and the Ethics of Survival.* Naperville, IL: Sourcebooks, 2001. xi, 356, [16] pp. First published in the UK as "Surviving Hitler" (London: Simon & Schuster, 2000).

Argues that the Germans were "seduced by Hitler," i.e. that they basically surrendered psychically, going along with the mass deportations and murders of the Jews. Stresses the conformity of the German people, noting that some degree of civil disobedience or low-level resistance was possible. Cites the case of the Rosenstrasse protest and Germans who helped Jews in hiding. A major motive for ordinary Germans to ignore information about the Holocaust was consumerism – i.e. they were bought off by perks, goods, money, etc.; among those affected were women, workers, high-ranking officers, and industries (like IG-Farben). Refers briefly to different responses toward the Jewish problem in the occupied countries and in the neutral countries. Relates, also, to the impossible choices faced by the Judenraete. Concludes that the dilemmas of individuals, groups, and nations are not unique to Germany. *YC*

0452 LeBor, Adam; Boyes, Roger: *Surviving Hitler: Choices, Corruption and Compromise in the Third Reich.* London: Simon & Schuster, 2000. ix, 338, [8] pp. Unseen. See the abstract for the American edition: "Seduced by Hitler: The Choices of a Nation and the Ethics of Survival" (Naperville, IL: Sourcebooks, 2001).

0453 Licht, Baruch: "Kraft durch Freude" und Auschwitz: Die NS-Sozialpolitik und die "Endloesung." *Tribuene* 159 (2001) 50-60.

Surveys the social benefits bestowed on Germans by the Nazi regime, often at the expense of Jews who were gradually deprived of all benefits and rights, and whose homes and property were expropriated for the sake of Germans. *RW*

0454 Listmann, Anja: *Beinahe vergessen: Juedisches Leben in Bad Salzschlirf.* Huenfeld: Rhoen Verlag, [2000?]. 180 pp.

Discusses Jewish life in Bad Salzschlirf (Hesse) between 1890-1941. Pp. 48-115 deal with the Holocaust period. In 1933 there were 23 Jews in the town. By 1938, 21 of them had moved or emigrated. Describes the persecutions suffered by the Jews and their reactions. 12 former Jewish residents of Bad Salzschlirf died in concentration camps. *ID*

0455 Longerich, Peter: *Der ungeschriebene Befehl: Hitler und der Weg zur "Endloesung."* Muenchen: Piper, 2001. 240 pp.

An expanded version of two expert reports submitted by Longerich for the defense in the Irving v. Lipstadt trial. States that Hitler obviously did not leave a written order authorizing the mass murder of the Jews in Europe, and this fact gives revisionist historians a pretext to deny the Holocaust. Other historians also deny Hitler's responsibility for the Holocaust. Argues that the Final Solution was classified as a secret operation, and therefore the Nazis preferred oral instructions to explicit written directives. Antisemitism lay at the core of Nazism from its early days; the motif of elimination of the Jews was present in Hitler's public pronouncements of the 1920s. Once he achieved power, Hitler initiated all the anti-Jewish actions. Moreover, he used the anti-Jewish policy to gain advantages for his regime both in Germany and abroad. During the war, it was he who steadily radicalized Germany's anti-Jewish policy leading to the desire for the total annihilation of the Jews. He had full knowledge of the practical measures against the Jews and bears full responsibility for the genocide. *DR*

0456 Longerich, Peter: *The Unwritten Order: Hitler's Role in the Final Solution.* Trans.: Catherine Stodolsky, Robert Vilain. Stroud, Gloucestershire: Tempus, 2001. 160 pp. Originally published as "Der ungeschriebene Befehl: Hitler und der Weg zur 'Endloesung'" (Muenchen: Piper, 2001).

0457 Matthaeus, Juergen: Abwehr, Ausharren, Flucht: Der Centralverein deutscher Staatsbuerger juedischen Glaubens und die Emigration bis zur "Reichskristallnacht." *Exilforschung* 19 (2001) 18-40.

Since their founding in the 1890s, the Centralverein Deutscher Staatsbuerger Juedischen Glaubens (CV) and the German Zionist organization opposed each other. The CV's aim was to foster Jewish life in Germany and to combat antisemitism by emphasizing the Jews' German patriotism; the Zionists fostered Jewish nationalism and emigration to Palestine. The Nazis exploited the differences in order to speed up Jewish emigration: they encouraged the Zionists and obstructed the activities of the CV, which at first clung to the hope that Jewish life in Germany was still possible. As Nazi persecution escalated, however, the CV also turned its efforts to emigration; it began to cooperate with the Zionists in a division of labor: while the Zionists prepared Jews, especially youth, for aliyah, the CV prepared them for emigration to other countries. *RW*

0458 Meinl, Susanne: "'Schalom' – meine Heimat": Stationen der Flucht aus Deutschland. *Exilforschung* 19 (2001) 41-64.

Describes the means by which the Nazi authorities extracted money from Jews planning emigration: freezing of their bank accounts and a series of taxes that together amounted to almost total expropriation. The Jews, already empoverished by Aryanization, reached their countries of refuge without minimal means of existence. Traces the process of empoverishment in Germany and the struggle to make a living in the new country of four Jewish families from Hesse: Margarete Sallis-Freudenthal of Frankfurt, in Palestine; Meta Frank, of the Koenigsthal family of Karlshafen, in Palestine; Ruth L. David, of the Oppenheimer family of Fraenkisch-Crumbach, who was sent on a Kindertransport to England and was the sole survivor of her family; and the Goldmann family of Reinheim and Frankfurt, with their son Robert, who were able to establish themselves in the U.S. with the aid of smuggled currency. Afterward, when the survivors applied for restitution of their property, they often found that they had to deal with the same officials who had carried out the expropriation, and who were unrepentant, obstructive, and still antisemitic, as were the new owners of the property. *RW*

0459 Metternich, Wolfgang: Begegnung. *Tribuene* 159 (2001) 164-176.

Based on an interview with Otto Schiff (b. 1921), traces the history of Jews in Hoechst, a town annexed to Frankfurt am Main, and of the Schiff family, founders and owners of its leading department store. The Nazis enjoyed widespread support in Hoechst, and although the Schiffs were highly respected both for their business methods and their philanthropy, in 1933 they, together with other Jews, immediately became victims: of the boycott, of imprisonment and mistreatment by the SA in a local "wild concentration camp," and of attacks by Nazi rowdies. Many Jews left Hoechst, at first only to bigger cities within Germany. The Schiffs moved to Frankfurt but retained their store until its Aryanization in 1938. This, and the exorbitant taxes levied on Jews after the "Kristallnacht" pogrom and in connection with their emigration, left the family penniless. Describes the successful struggle to build up a new existence in the United States. *RW*

0460 Mierzejewski, Alfred C.: A Public Enterprise in the Service of Mass Murder: The Deutsche Reichsbahn and the Holocaust. *Holocaust and Genocide Studies* 15, 1 (Spr 2001) 33-46.

During World War II, the Deutsche Reichsbahn used ca. 2,000 trains to move ca. 3,000,000 people to ghettos and death camps in the East. Despite this fact, a myth was created after the war that the DRB leadership did not know that it was instrumental in the mass murder. Examining previously unused documents, contends that both the upper-level management of the railway and its operating personnel had to know that the DRB was involved in transporting Jews to their deaths and their property to the Reich. Some of the DRB staff even visited Auschwitz. The traffic involving Jews was small, however, in comparison to other business conducted by the Reichsbahn; transports to camps usually had a low priority. The result was an indifference of the majority of German railway workers toward the genocide. *DR*

0461 Petuchowski, Elizabeth: Gertrud Luckner: Widerstand und Hilfe. *Freiburger Rundbrief* 7, 4 (2000) 242-259.

Translation of a lecture held at Seton Hall University, October 1998. Describes the aid given by Luckner to Jews in Nazi Germany, quoting from interviews with her. Luckner traveled from city to city, with only limited backing from the Catholic Caritas organization and Archbishop Groeber of Freiburg, to find out the immediate needs of Jews and possible ways to alleviate their plight: moral and material support, warning of impending danger, using connections abroad to obtain visas, sending parcels to Jews, etc. In March 1943 she was arrested by the Gestapo and sent to Ravensbrueck. Notes her social consciousness, her keen understanding of the situation and grasp of practical details, and her ability to react swiftly and effectively. *RW*

0462 Phayer, Michael: Questions about Catholic Resistance. *Church History* 70, 2 (June 2001) 328-344.

Although Catholic resistance to the Nazi dictatorship was feeble, it existed. This resistance was limited, in particular, due to the ardent German patriotism of many clergymen. Dwells on protests made by some Church leaders against Nazi racism and antisemitism, as well as against the deportations of Jews. However, no bishop clearly crossed the line from bland protest to resistance, and this lessened the chance that Catholics at large would do so. Nevertheless, there were cases of Catholic aid to Jews. Describes the rescue activities of Gertrud Luckner and Margarete Sommer, who tried to rescue Jews and Jewish converts. *DR*

0463 Pine, Lisa: *German Jews and Nazi Family Policy*. London: Holocaust Educational Trust, 2000. ii, 22 pp. (Holocaust Educational Trust Research Papers, 1, 6).

Analyzes the personal experiences of Jews who lived in Nazi Germany between 1933-39. Reflects on the impact of the Nazi anti-Jewish measures (e.g. the April 1933 boycott, the Nuremberg Laws, and the "Kristallnacht" pogrom) on Jewish family life and Jewish-German relations. Despite the Nazi measures and rising social ostracism, many Jewish families clung to their old patterns of life. Impoverishment, loss of social status, and general powerlessness led to an increasing dependence on women in Jewish families and a feeling of insecurity among children.

While Jewish-German ties were weakened or cut, Jewish intra-familial ties were strengthened. The "Kristallnacht" pogrom was a heavy blow to Jewish family life – many families suffered when husbands, fathers, etc. were detained in concentration camps, and houses and businesses were destroyed. *DR*

0464 Radvan, Florian: "...Mit der Verjudung des deutschen Theaters ist es nicht so schlimm!" Ein kritischer Rueckblick auf die Karriere der Literaturwissenschaftlerin Elisabeth Frenzel. *German Life and Letters* 54, 1 (Jan 2001) 25-44.

Frenzel (b. 1915) studied theater history in Berlin, and in 1938 submitted her dissertation, "Judengestalten auf der deutschen Buehne" (published by the Deutsche Volksverlag, 1940). She contended that German theater had always been antisemitic, thus legitimizing Nazi antisemitism. Philosemitic theater she denounced as dangerous propaganda, due to machinations of Jews who had excessive influence on the theater. The philosemitic play of an admired author like Lessing she explained away as grounded in the mistaken mentality of the period. Subsequently, she published many articles in Nazi journals on this and related themes. Rosenberg's office appointed her director of its branch for theater. In this capacity she compiled preliminary files for a dictionary of theater people of Jewish descent, based on genealogical research. After the war, she and her husband published popular reference works on German literature. *RW*

0465 Reemtsma, Jan Philipp: *"Wie haette ich mich verhalten?" Gedanken ueber eine populaere Frage*. Muenchen: Lehrstuhl fuer Juedische Geschichte und Kultur, Ludwig-Maximilians-Universitaet, 2000. 29 pp.

A lecture delivered at the University of Munich in January 2000. Discusses possible norms of behavior of German citizens during the Nazi period. States that one cannot really know what a person (or he himself) would have done in a given situation, but only what he should have done. Criticizes the statement that "people who did not approve of Nazi policy had to resist," judging it too simplistic. Notes, though, that only a few acts of resistance by Germans incurred capital punishment (e.g. desertion), but not refusal to shoot civilians or to serve as a guard in a concentration camp. Concludes that this line of questioning points only to a potential: the fact that a person feels able to commit a crime does not mean that he probably will carry it out. As an illustration of the impossibility of resolving this question, discusses, also, the case of Elisabeth Langgaesser, who, married to a Jew, found herself in the situation of "agreeing" to the deportation of her daughter (Cordelia Edvardson) to Theresienstadt in order to avoid her own prosecution for violation of the racial laws. *ID*

0466 Ritter, Wolfgang: Ein Haus in der Burgstrasse: Esens wird nationalsozialistisch. *Das August-Gottschalk-Haus in Esens: Gedenkstaette und Ausstellung zur neueren Geschichte der ostfriesischen Juden: Zwei Aufsaetze zur Erinnerung.* Esens: Oekumenischer Arbeitskreis Juden und Christen, 2000. Pp. 25-47.

Describes the persecution of the Jews of Esens (East Frisia; 82 Jews in 1933) by the town council as well as the townspeople. Jews expelled from their homes lived in crowded conditions in the former Jewish school and community house. In the "Kristallnacht" pogrom, the synagogue was destroyed; the Jews (men, women, children, and infants) were driven through the streets, beaten and humiliated, while their homes were vandalized; the men were sent to Sachsenhausen. Recounts the fate of the different families: some managed to escape abroad, many perished in the East. In 1990 the school and community house was turned into a museum of Jewish life and of the Holocaust in East Frisia. *RW*

0467 Roseman, Mark: *The Past in Hiding.* London: Allen Lane, 2000. xiii, 576 pp.

A biography of Marianne Ellenbogen (née Strauss), born in Essen in 1923, based on interviews with Ellenbogen and others, her diary, and archival materials. Her family was protected after almost all other Jews were deported from their city in 1941. This unusual fate was due to protection they received from a top member of the Nazi bureaucracy. In 1942 Ellenbogen's fiancé, Ernst Krombach, and his parents were deported to Izbica in Poland (as described in Krombach's letters, printed here); they did not survive. In 1943 Ellenbogen's parents and brother were deported to Theresienstadt, and later perished in Auschwitz. From October 1943 until the liberation, Ellenbogen survived on the run, saved mostly by the efforts of a leftist resistance group called "Bund: Gemeinschaft fuer sozialistisches Leben." *YC*

0468 Sax, Boria: *Animals in the Third Reich: Pets, Scapegoats, and the Holocaust.* New York: Continuum, 2000. 206 pp.

Nazi ideologues regarded Nazism as "applied biology"; their attitude toward animals, complex and paradoxical as it was, is crucial for understanding their ideology. Pretending to follow and implement the implacable laws of nature, the Nazis drew the main distinction not between humans and animals, but between victor and vanquished in the struggle for existence, i.e. predator and prey. Thus, Nazi culture exalted animals such as the wolf or horse as symbolizing Aryan man and likened the Jews to apes, swine, lambs, etc. Compares Nazi laws concerning animal protection (which, inter alia, introduced "humane" methods of slaughter and led to the prohibition of Jewish ritual slaughter) and

Auschwitz, where mass murder was carried out in a "technological" and dispassionate way, ostensibly without cruelty and regret. *DR*

0469 Scales-Trent, Judy: Racial Purity Laws in the United States and Nazi Germany: The Targeting Process. *Human Rights Quarterly* 23, 2 (May 2001) 259-307.

Compares U.S. racial legislation and practices from the colonial era until the post-World War II period to Nazi anti-Jewish legislation and practices of 1933-43. The two legal systems reveal many similarities in the ways in which they defined the targeted groups, tried to effect the racial purity of the society (e.g. to control miscegenation between the racial groups), denied the human rights and citizenship of the targeted groups, and even practiced sterilization. Both legal systems were confronted with the problem of those who fell between the statutory racial groups ("Mischlinge" in Germany, "mulattoes" in the USA) and both coped with it in comparable ways. Both systems could change the boundaries of the targeted groups out of pragmatic considerations. Pp. 292-296 deal with the segregation and persecution of Afro-Germans by the Nazis, who saw links between Jews and "Negroes"; Hitler often commented on the relationship between the two groups. Jewish refugees and survivors of the Holocaust were compensated in some way by the West, but Afro-German victims of Nazism were not. *DR*

0470 Schrader, Baerbel: Spielerlaubnis mit Sondergenehmigung: Juedische Schauspieler im "Dritten Reich." *Exil* 21, 1 (2001) 5-13.

Based on a lecture delivered in Hamburg, April 2001. States that non-Aryan actors – including half-and quarter-Jews as well as Aryans married to non-Aryans – were not admitted to Goebbels' "Kulturkammer" and thus were not allowed to appear on the stage. However, since Goebbels was ambitious for German theater and there was a shortage of good actors, he issued long-term or short-term special permits to some of these non-Aryans. Aryans in mixed marriages were pressured to divorce or send their partners to live abroad, but in the meantime these partners were protected. In every other respect, the actors and their families were still subject to the racial laws. Tells the stories of a number of actors in Germany and Austria who worked under special permits. *RW*

0471 Schramm, Reinhard: *Ich will leben...: Die Juden von Weissenfels.* Koeln: Boehlau, 2001. 194 pp.

Pp. 11-16 present a brief history of the Jews in Weissenfels (Sachsen-Anhalt). In 1933, 165 Jews lived in the town; most of them emigrated after the "Kristallnacht" pogrom. Deportations began in 1940; altogether, 38 Jews from Weissenfels were murdered in the Holocaust. The rest of the book deals with the fate of Schramm's extended family. Friedrich Murr (Schramm's maternal grandfather), who died in 1932, was a Catholic, but his wife, Emma, was Jewish and their children were "Mischlinge." Their son Rudolf was deported in 1940 to Sachsenhausen and murdered in April 1941 in Neuengamme. Their daughter Rosel was married to an Aryan. In 1945 she and her baby son Reinhard were hidden by friends until the war's end. Emma and her sister were gassed in the euthanasia killing center Bernburg in March 1942, and another sister disappeared in October 1944. Pp. 69-102 relate the fate of friends of the Murr family; pp. 103-138 report on other victims and survivors from Weissenfels. Pp. 146-182 present facsimiles of documents. Pp. 183-189 contain a list of Jews who lived in Weissenfels during the Nazi period, relating their fate. *ID*

0472 Specht, Franziska: *Zwischen Ghetto und Selbstbehauptung: Musikalisches Leben der Juden in Sachsen 1933-1941*. Altenburg: Kamprad, 2000. 235 pp. On title page also: Hrsg. Ephraim Carlebach Stiftung. Based on the author's Promotionsschrift – Universitaet Leipzig, 1999.

Organized Jewish musical activity in Nazi Germany developed because of the prohibition of Jewish performers from appearing before the general public. In Saxony, new Jewish ensembles formed, at first, under the auspices of the religious communities, the only Jewish organizations permitted to function by the Saxon authorities; only with the formation in 1935 of the Reichsverband der juedischen Kulturbuende, under the supervision of Hans Hinkel, did Saxony also consent to autonomous Jewish Kulturbuende as they existed elsewhere in the Reich. Hinkel protected the Kulturbuende as his fiefdom, but imposed restrictions: Aryans were not allowed to attend or participate in performances; music by composers like Beethoven, who expressed the German soul, could not be performed by Jews. Nevertheless, Jewish musical life flourished, not only in the big cities (Leipzig, Dresden, Chemnitz), but also in the smaller towns – an affirmation of the German Jews' identification with German and European culture. Includes many facsimiles of documents, concert programs, and calendars of performances. *RW*

0473 Steinecke, Hartmut, ed.: *"Warum immer Vergangenheit?" Leben und Werk Jenny Alonis*. Muenster: Ardey-Verlag, 1999. 99 pp.

A collection of essays accompanying an exhibition held in Paderborn in 1997 on the life and work of the German-Israeli writer Jenny Aloni (née Rosenbaum), who was born in Paderborn in 1917 and immigrated to Palestine in 1939. Some of the essays discuss the Nazi period and remembrance of the Holocaust. Also contains excerpts from Aloni's works, diaries, and letters. Partial contents: Lettermann, Petra; Rochell, Gabriele: Ausgrenzung – Verfolgung (25-30); Olbrich, Andreas:

Auswanderung (31-36); Rochell, Gabriele; Lettermann, Petra: Erinnerung – Schuld (49-54). *ID*

0474 Sternberg-Siebert, Elisabeth: *Juedisches Leben im Huenfelder Land: Juden in Burghaun*. Petersberg (Fulda): Michael Imhof Verlag, 2001. 288 pp.

Surveys the history of Jews in the Huenfelder Land district, near Fulda, between the 16th century and World War II. Reports on antisemitic occurrences in 1819-49, focusing on Burghaun and Steinbach in 1847. Pp. 58-71 discuss the Nazi period. In 1933, 90 Jews lived in the area. In 1939, after some of them emigrated, 64 remained; 26 were deported in December 1941 to Riga (four survived). In September 1942, 19 Jews were deported to Theresienstadt. Pp. 72-153 discuss the history of the Jewish community in Burghaun; pp. 114-153 focus on the Nazi period and include excerpts of testimonies. Pp. 154-236 include brief essays on the fate of 30 Jewish families. *ID*

0475 Stimpfig, Karl Ernst: *Die Juden in Sugenheim und Ullstadt: Eine Dokumentation*. Scheinfeld: Hans Meyer Verlag, 2001. ix, 262 pp.

Presents documents on the history of Jews in Sugenheim and Ullstadt (Franconia) between 1611-1945. With increasing antisemitism after World War I and the rise of the Nazi regime, most of the Jews in Sugenheim (ca. 80) emigrated; only eight Jews remained after the "Kristallnacht" pogrom. Pp. 197-198 comprise a list of 60 former residents of Sugenheim who were victims of the Holocaust. Pp. 201-212 contain memoirs of Herbert Kolb (b. 1920). He was deported in 1942 to Theresienstadt and from there in 1944 to a labor camp in Wulkow. He survived, as did his parents. His sister died in Bergen-Belsen. In Ullstadt there were five Jews in 1930; all of them emigrated. *ID*

0476 Teichert, Carsten: *Chasak! Zionismus im nationalsozialistischen Deutschland, 1933-1938*. Koeln: Elen-Verlag, 2000. 560 pp.

Discusses the activities of the Zionist movement in Nazi Germany between 1933-38, focusing on the ability of German Zionists to exploit the non-uniformity of German racial policy for the purpose of the Palestine project. Tenacious negotiations of the Zionistische Vereinigung fuer Deutschland with various German authorities led to the Ha`avarah agreement, and enabled ca. 50,000 Jews to emigrate to Palestine, along with some of their property, before the outbreak of World War II. Notes that the effort to "zionize" German Jewry failed. Most German Jews believed that the political crisis was only temporary. *ID*

0477 Victor, George: The Phantom Jew: From Spoiled Identity to Genocide. *The Journal of Psychohistory* 28, 4 (Spr 2001) 469-479.

Examining one of the earlier Nazi racial prohibitions, that Jews may not employ German women under the age of 45 as maids, infers that Hitler was obsessed with the rumor of his father's and his own Jewish origin. Stories of the seduction of his grandmother by a Jewish employer (which led to the birth of his father) and the seduction of his mother by his father (they later married), as well as his own early sexual obsessions, made Hitler believe in the sexual depravity of Jews and people of Jewish descent, and drove him to plan the racial purification of Germany. The Nazi euthanasia program was in fact aimed at elimination of all people of mixed parentage. Many other leading Nazis (e.g. Streicher, Heydrich, Rosenberg) had similar fears concerning their alleged Jewish origins and negative inherited Jewish traits. By killing Jews, the Nazis were trying to obliterate the people who reminded them of what they hated in themselves. *DR*

0478 Voglmayer, Anne: *Mein Name ist Meta Herz: Erinnerungen an die juedische Gemeinde Nuembrecht*. Nuembrecht: Martina Galunder-Verlag, 2000. 152 pp.

Presents a survey of the history of the Jewish community in Nuembrecht, Germany, where four Jewish families lived from the 18th century. Pp. 67-95 chronicle events of the Nazi period, between 1932-42. Pp. 95-104 trace the life story of Meta Herz, who was born in Nuembrecht in 1905. During the "Kristallnacht" pogrom her family home was plundered and demolished, her mother was beaten, and her brother was arrested and interned briefly in Dachau; in July 1942 they were deported to Theresienstadt. Herz was murdered in Auschwitz in 1944. *ID*

0479 Weiner, John: Todesmarsch. *Dachauer Hefte* 17 (Nov 2001) 162-170.

Weiner, born in 1926 in Szombathely, Hungary, relates how his father was separated from him in Buchenwald and sent to Auschwitz to be gassed, and then chronicles the death march, in April 1945, of 1,650 Jewish prisoners in the direction of Theresienstadt. Almost all the prisoners broke down and were shot on the way. Weiner and twelve others, who had been arrested after trying to flee near Oberlindhart in Bavaria, were saved by a policeman, Max Maurer, who was supposed to take them to a jail to be shot and brought them instead to a friendly farmer near Ergoldsbach, who sheltered and fed them until they were freed by the Americans. In the same neighborhood there is a cemetery of 67 prisoners from the same transport who died a day before or after liberation. *RW*

0480 Wiesen, S. Jonathan: Morality and Memory: Reflections on Business Ethics and National Socialism. *The Journal of Holocaust Education* 10, 3 (Win 2001) 60-82.

After World War II, German industrialists were accused of complicity in Nazi crimes. The most damning aspects of their behavior were their involvement in the policy of Aryanization and the use of slave labor during the war. States that West German industrialists addressed the implications of their accommodations to Hitler; however, they described their behavior as highly moral, claiming that they were apolitical during those years and acted in accordance with business ethics. Contends that these claims are largely correct, and that their business behavior was defined by a fundamental pragmatism. However, business ethics did not have to exclude humane treatment of slave workers or efforts to withstand Nazism. Points out the horrifying divergence between Nazi business ethics and Western morality. Contends that the recent German business initiative to compensate survivors of forced labor is also motivated by business ethics and pragmatism. *DR*

0481 Wurm, Siegfried: *Die finanzielle Vernichtung der Juden im Dritten Reich: Wie vollzog sich der Griff der Nationalsozialisten nach dem juedischen Vermoegen? (Eine dokumentarische Skizze)*. Berlin: Kronen, 1999. 151, [74], vi pp.

Following an introduction (pp. 1-16), presents 31 documents which show the process of destruction of Jewish economic life in the Third Reich. In 1932, ca. 100,000 Jewish enterprises functioned in Germany; ca. 50,000 of them dealt in retail trade. Discusses the boycott of Jewish enterprises, the dismissal of Jewish functionaries, the Nuremberg Laws, and the Aryanization of Jewish property. By the time World War II began, the exclusion of Jews from the German economy was largely completed. On 25 November 1941 the property of the Jews who had emigrated or were deported was appropriated by the state. Includes four appendixes (74 pp.). *ID*

Great Britain

0482 Bergen, Amanda: *Leeds Jewry, 1930-1939: The Challenge of Anti-Semitism*. Leeds: Thoresby Society, 2000. 40 pp.

Presents reactions of the Jewish community in Leeds first to lingering local antisemitism, then to the rise of international fascism. For most of this period, well-to-do community members tended to turn a blind eye to social antisemitism. Active opposition, including street fights against Mosley's Blackshirts, was undertaken by members of the working class and youths. Broader communal responses emerged from the popular communist and Zionist movements, especially the latter. Strong local community leadership included factory owner Sir Montague Burton, Professor Selig Brodetsky, and Alderman Hyman Morris. The Jewish community united to oppose antisemitism in England, but the most significant unifying factor was the crisis European Jewry was undergoing as Hitler's regime increased its persecution. The Jews of Leeds mobilized in an impressive way to aid Jewish refugees from Germany. *YC*

0483 Fraser, David: *The Jews of the Channel Islands and the Rule of Law, 1940-1945: 'Quite Contrary to the Principles of British Justice.'* Brighton: Sussex Academic Press, 2000. x, 262 pp.

Shows how local British authorities on the Channel Islands (Jersey, Guernsey, and Sark) collaborated with German occupation forces during World War II. Due to antisemitism and legalism, they subjected local Jews to discrimination, dispossession, and deportation. Although the total number of Jewish victims may have been less than 20 (with only a few killed), the case exemplifies how the Nazis extended their antisemitic policy to all the territories they conquered. Channel Islands residents have been slow to deal with this aspect of their past and the issues involved, including blind acceptance of the law regardless of its immorality. *YC*

Greece

0484

מואיסי, עזרא: יהודי יוון בהתנגדות הלאומית. לא נשכח! 16 (2001) 32-37.

[Moissi, Ezra: The Jews of Greece in the National Resistance. *Lo Nishkach!* 16 (2001) 32-37.]

Discusses the participation of Greek Jews in underground organizations and partisan groups, which were first established there at the end of 1942. Since the Jews of northern Greece (including Salonika) were deported soon after, in spring 1943, they did not have time to escape. However, the Germans invaded the other regions of the country only in September 1943; ca. 80% of the Jews in these areas survived. Many of them fled to the mountains and some joined partisan groups. Pp. 35-36 contain names of Jews who fought with the partisans; many of them were killed in battle. States that some of the partisan groups made special efforts to save Jews. *RE*

0485

נקב, מימי ניצחון: על יהודי טריקאלה שניספו בשואה. לא נשכח! 16 (2001) 8-13.

[Nakav, Mimi Nitzahon: On the Jews of Trikala Who Perished in the Holocaust. *Lo Nishkach!* 16 (2001) 8-13.]

Pp. 9-13 contain a list of names of Jews from Trikala, Greece, who perished in concentration camps during World War II. Nakav's father, Ya`akov Nitzahon, was born in Trikala and returned there in 1943. In April 1944 he, his wife, and his son were deported; his wife and son were killed. After the war he immigrated to Israel. *RE*

0486 Wiehn, Erhard Roy: *Juden in Thessaloniki: Die alte sephardische Metropole im kurzen geschichtlichen Ueberblick unter besonderer Beruecksichtigung der Schoáh 1941-1944*. Konstanz: Hartung-Gorre, 2001. 50 pp.

Pp. 11-22 survey the history of Jews in Salonika from the ancient period to the 20th century. Pp. 23-39 discuss the Nazi period. Between April 1941-October 1944 Salonika was occupied by the German army. Anti-Jewish measures against the 49,000 local Jews began with arrests of Jewish notables, confiscation of apartments and the Jewish hospital, and the closing of the three Jewish newspapers. In April-May 1941, the Einsatzstab Rosenberg confiscated art and literary treasures from private and public libraries and synagogues. From July 1942, 9,000 Jewish men were taken for forced labor. The arrival of Dieter Wisliceny and Alois Brunner in February 1943 marked the start of the systematic annihilation; 45,620 Jews were deported in 18 transports to Auschwitz. Includes excerpts from the memoirs of Erika Kounio-Amariglio, whose family survived Auschwitz. *ID*

Hungary

0487 Ács, Irén; Levendel, Júlia: *"Őrizd meg..."* ["Cherish It..."]. Szécsény: Liget, 2000. 57 pp.

An account by Ács, as related to Levendel. Recalls Jewish life in the town of Szécsény (northern Hungary) before the war and how it changed with the advent of the Holocaust. Describes the implementation of the anti-Jewish laws and how they affected daily life. In 1941 there were 280 Jews in the town. Describes ghettoization, torture and humiliation of local Jews by the gendarmes. Ács was deported in 1944; she is the only survivor of her family. Includes many photos, facsimiles of newspaper items, and letters relating individual stories of survival. A list of Szécsény victims is provided, with the name and profession of each. *AV*

0488 Bar-Shaked, Gavriel, ed.: *Nevek. [Vol. V]: A Zala megyéből kiüzött zsidók nevei* [Names. Vol. V: Names of the Deported Jews from Zala County].

New York: Beate Klarsfeld Foundation; [Jerusalem]: Yad Vashem, 2001. [12], 282 pp. With an added introduction in English.

The fifth in a series of publications listing the names of Jews deported from Hungary. Zala county (western Hungary) consisted of 14 districts. In 1941, there were ca. 7,000 Jews in Zala. This volume contains 6,200 names. The list includes (when known) the spouse's name, mother's name, date and place of birth, and the name of the village or town where he/she lived. *SSC*

0489 ברהם, רנדולף ל.: הכנסיות הנוצריות בהונגריה והשואה. יד ושם 29 (2001) 187-218.

[Braham, Randolph L.: The Christian Churches of Hungary and the Holocaust. *Yad Vashem* 29 (2001) 187-218.] Appeared in English in "Yad Vashem Studies" 29 (2001).

0490 Braham, Randolph L.: The Christian Churches of Hungary and the Holocaust. *Yad Vashem Studies* 29 (2001) 241-280. Appeared in Hebrew in "Yad Vashem" 29 (2001).

The climate of opinion in Hungary had been conditioned by centuries of Christian hostility to the Jews and Judaism. The leadership of the three major Christian denominations (Catholic, Reformed, and Evangelical) barely protested against the government's anti-Jewish actions. As the main bulwark of Horthy's regime, they approved the first and second anti-Jewish laws (1938-39), and opposed the third law (1941) only inasmuch as it affected Christian converts. In 1944, the leadership of the Churches tried to protect the converts and protested mainly against the methods of the Final Solution. The Churches undoubtedly contributed to Horthy's decision to halt the deportations in July 1944; nevertheless, their reliance on covert negotiations with the government and their failure to offer clear-cut guidance to the clergy and masses also contributed to the climate that made the Holocaust possible. Many individual clergymen rescued Jews and converts; others supported Horthy's and Szálasi's policies. After the war, the Churches denied their co-responsibility for the Holocaust; only in the post-communist period did Church leaders express repentance. *DR*

0491 Braham, Randolph L.: *Studies on the Holocaust: Selected Writings. Vol. 1.* New York: Rosenthal Institute for Holocaust Studies, Graduate Center, City University of New York; Boulder, CO: Social Science Monographs; distributed by Columbia University Press, 2000. ix, 158 pp.

A collection of seven essays, some of them published previously. Contents: The Holocaust in Hungary: A Retrospective Analysis [An earlier version appeared in "The Holocaust in Hungary: Forty Years

Later" (1985).] (1-20); What Did They Know and When? [Appeared in "The Holocaust as Historical Experience" (1981).] (21-49); The Hungarian Press, 1938-1945 (51-67); The Holocaust in Hungary: An Historical Interpretation of the Role of the Hungarian Radical Right [Appeared in "Societas" 2 (1972).] (69-97); The Official Jewish Leadership of Wartime Hungary [Appeared in "Patterns of Jewish Leadership in Nazi Europe, 1933-1945" (1979).] (99-116); The Rescue of the Jews of Hungary in Historical Perspective [Appeared in "The Historiography of the Holocaust Period" (1988).] (117-134); The National Trials Relating to the Holocaust in Hungary: An Overview (135-158). *DR*

0492 Egressy, Gergely: A Statistical Overview of the Hungarian Numerus Clausus Law of 1920 – a Historical Necessity or the First Step toward the Holocaust? *East European Quarterly* 34, 4 (Jan 2001) 447-464.

The law of 1920 curtailed the number of students admitted to schools of higher education in Hungary. It established a quota of 6% for Jews. Using statistical data, examines how well the law was implemented and the extent to which it put Hungarian Jews in a disadvantaged position. Relating to the historical context of the law, notes Hungary's problems, especially economic ones, resulting from the treaty of Trianon. Hungary could no longer provide the same degree of higher education as previously nor give employment to the same number of educated professionals. Thus, reform was necessary. What was antisemitic about the 1920 law was the racial categorization. Concludes that since antisemitism was rampant after the abortive revolutions of 1918-19, for which many blamed the Jews, the law "was not the cause of antisemitism but rather the result of it." However, statistics indicate that after 1920 the admission of Jews to universities was not as affected as might have been expected. In 1928 the law was amended to remove its antisemitic bias. Concludes that there was no clear continuity between the law of 1920 and the anti-Jewish laws of the late 1930s. *YC*

0493 Fenyvesi, Charles: *Mikor az angyalok túljártak a világ eszén: Akik zsidókat mentettek a háborús Magyarországon*. Trans.: Ágnes Walkóné Békés. Budapest: Európa, 2001. 348 pp. A translation of "When Angels Fooled the World: Rescuers of Jews in Wartime Hungary" (Madison: University of Wisconsin Press, 2003).

Focuses on five Rightous Gentiles, three of whom were prominent – Raoul Wallenberg, Gábor Sztehlo, and Endre Bajcsy-Zsilinszky. Wallenberg used his authority as a Swedish diplomat to save thousands of Jews from deportation; Sztehlo was a Lutheran priest who set up homes for endangered Jewish children; and Bajcsy-Zsilinszky was a politician who intervened on behalf of maltreated Jewish labor-servicemen, demanded investigation of the massacre in the southern provinces that were returned to Hungary, and organized anti-Nazi resistance in the country. Describes their activities and discusses their possible motives. Includes two unknown persons who saved Jews – the janitor of Fenyvesi's house in Budapest (Mr. Kanalas) and a Debrecen local council official (Érzsebet Dávid). The janitor saved Fenyvesi and his mother from the Arrow Cross who came to take away Jews. David provided Jews with false (Christian) birth certificates and was imprisoned for her activities. *AV*

0494 Gregus, Emery: *Occupation and Liberation 1944-1945, Aftermath: The Postwar Years, Remembrances*. Montreal: Concordia University Chair in Canadian Jewish Studies; Montreal Institute for Genocide and Human Rights Studies, 2001. 92, [5] pp. (Memoirs of Holocaust Survivors in Canada, 18).

An account based on memoirs originally taped and transcribed in Hungarian. Gregus was born as Imre Guttmann in 1922 in Košice, Slovakia. Pp. 1-72 deal with the Holocaust, from the German occupation of Hungary in March 1944 to its liberation by the Red Army. Gregus fled Košice (then Kassa, Hungary) to Budapest, where he survived on "Aryan" papers, helped by some Christians and other Jews in hiding. His parents and one brother perished in Auschwitz, while another brother died in a Hungarian military labor service unit in Russia. His married sister, Valeria, and her family were gassed. Another sister survived with help from non-Jews. In 1951 Gregus and his wife immigrated to Canada. *YC*

0495 Halasz, Dorottya: *The War Refugee Board and the Destruction of Hungarian Jewry*. Diss. – Texas Christian University, 2000. 263 pp. Unseen.

In January 1944, President Roosevelt established the War Refugee Board as a rescue and relief agency to aid the oppressed people of Europe, including the Jews. At the time, Hungarian Jewry was the largest surviving Jewish community in Europe. In March 1944, German troops occupied Hungary and Adolf Eichmann set in motion his machinery of destruction. Because of its inexperience and the sudden turn of events, the WRB could not adequately deal with the problem during the first phase of the crisis. More than 400,000 Jews were deported between 15 May-8 July, most of them to Auschwitz. Yet, during the same period, the WRB devised and started to use the procedures, both legal and illegal, that ultimately saved the remnant of Hungarian Jewry. These included psychological warfare, cooperation with neutral governments and international and American private agencies, the use of foreign protection documents, as well as participation in rescue and in ransom talks. These efforts brought success between July 1944 and January 1945.

0496 A holocaust győri gyermekmártírjai [Child-Victims of the Holocaust from Győr]. *Holocaust Füzetek* 13 (1999) 118-131.

An alphabetical list of 299 names provides data on place and date of birth, mother's maiden name, and the exact age of the children on 15 June 1944, the day of their selection in Auschwitz. The introduction to the list (pp. 118-121) mentions the protest by the Győr Bishop Vilmos Apor against the ghettoization and the sources used by the researcher Sándor Lónyai to make the list. *AV*

0497 Komoróczy, Géza: *Holocaust: A pernye beleég a bőrünkbe* [Holocaust: The Soot Is Burning into Our Skin]. Budapest: Osiris, 2000. 182 pp.

A collection of speeches and articles (some of them published previously) about the Holocaust, calling on the Hungarian public, intellectuals and institutions, thinkers and politicians to face the past and accept responsibility for the Holocaust of Hungarian Jews. Describes antisemitic phenomena in Hungary during the early 1990s. Includes a summary of the Holocaust (pp. 9-59) written for a lexicon but never published, and an essay on traditional Christian antisemitism. *AV*

0498 Kramer, Thomas D.: *From Emancipation to Catastrophe: The Rise and Holocaust of Hungarian Jewry*. Lanham, MD: University Press of America, 2000. xi, 404 pp.

The first chapter discusses the "Jewish question" in Hungary and the rise of antisemitism in the 19th-early 20th centuries; the rest of the book deals with the period 1919-45. Hungary was the first European country after World War I to introduce antisemitic laws (in 1920). However, the Jews maintained their patriotism. Although he was an antisemite, Horthy was favorably inclined toward the assimilated and "useful" Budapest Jews. Discusses the anti-Jewish legislation in 1938-41, military labor service, and the deportations in 1944. Dwells on the behavior of Jewish leaders, particularly the Zsidó Tanács (Jewish Council) instituted in 1944. The leaders' failure to warn the Jews of the impending danger may be attributed to their inability to comprehend it, their belief that Hungarian Jews had a special status, or to other factors. Their strategy was based on maintaining good relations with the Horthy regime. Discusses activities of Zionist youth movements, which rescued thousands of Jews, and of György Gergely, a member of the Council, who tried to improve the lot of the military labor servicemen. Examining the aborted "blood for trucks" agreement, contends that it failed because of the Allies' reluctance to receive a million Hungarian Jews. *DR*

0499 Molnár, Judit: A Zsidó Tanács megalakulása – cionista szemmel (Dr. Kahan Nison visszaemlékezése) [The Establishment of the Jewish

Council – in the Eyes of a Zionist (as Dr. Nison Kahan Remembered It)]. *Holocaust Füzetek* 13 (1999) 93-117.

The correspondence of Kahan, a Hungarian Zionist leader and member of the Jewish Council in 1944, is found in the Yad Vashem archives and the Central Zionist Archives. Two of the letters, written in May-June 1945, are published here, along with an introduction (pp. 93-96) describing the circumstances in which the Jewish Council was established in Hungary, how it worked, its sphere of action, and the motivation of its members. The order for a Jewish census revealed the murderous intention of the Germans, so the Council decided to sabotage the order. Relates how Kasztner contacted Wisliceny in order to save Jews, attempts of the Council to send food to Jews in internment camps and ghettos, how the Council sent information abroad about the deportation of Hungarian Jews and appealed for help to the Swiss and Swedish embassies, passing on the text of the "Auschwitz protocols." Kahan contends that international protests and the personal intervention of the King of Sweden saved Budapest Jews from deportation. *AV*

0500 A munkaszolgálatban és a deportálásban 1941 és 1945 között elpusztult csornai zsidók névsora [A List of Names of Those Jews from Csorna Who Died between 1941-1945 in Labor Service or Deportation]. *Holocaust Füzetek* 13 (1999) 176-184.

634 Jewish victims and 9 Christian Jews who perished in the Holocaust are listed separately, in alphabetical order, along with their year of birth. *AV*

0501 Orbánné Szegő, Ágnes: *A Heves megyei zsidóság története a XVIII. századtól a Holocaustig* [The History of the Jews in Heves County from the 17th Century to the Holocaust]. Tiszafüred: Tiszafüredi Menóra Alapítvány, 2001. 259, [10] pp. Based on the author's diss. – Debreceni Kossuth Lajos Tudományegyetem. With an English summary (pp. 234-238).

Focuses on the role of Jews in the economy of the county. Gives a summary of local reactions to the Tiszaeszlár blood libel in the 1880s and to the antisemitism of the Horthy regime in 1920 (incitement in the first, pogrom in the second case). Deals with the consequences of the Numerus Clausus Law on the Jews, and their sacrifices to send their sons abroad to study. Describes the antisemitism of the 1930s, the anti-Jewish laws and their effects. In Gyöngyös alone, Szálasi's Arrow Cross numbered about 400 members. Pp. 206-227 deal with the Holocaust period. In 1941, there were more than 7,000 Jews in Heves – in 1949, only 868. *AV*

0502 Pelle, János: *A gyűlölet vetése: A zsidótörvények és a magyar közvélemény 1938-1944* [The Harvest of Hatred: The Jewish Statutes and Hungarian Public Opinion 1938-1944]. Budapest: Európa, 2001. 345 pp.

Based on unpublished archival material and articles which appeared in the antisemitic press in Hungary during 1938-44, shows the influence of Hungary's antisemitic policies and legislation on public opinion. Quotes widely from various periodicals, from the diaries and letters of Miklós Horthy, Mihály Károlyi, Ferenc Szálasi, Ferenc Rajniss (a minister in Szálasi's government), and from the protocols of the conferences of Hungarian Catholic bishops from 1919-44. *MA*

0503 Pelle, János: A zsidótörvények és a Hajdú vármegyei közvélemény [The Jewish Statutes and Public Opinion in Hajdú County]. *Debreceni Szemle* 9, 2 (June 2001) 219-233.

The aim of the antisemitic laws of 1938-39 in Hungary was to exclude the Jewish population from the economic and cultural life of the country. Examines the reactions of authorities in Hajdú County, based on the press and archival documents. The monthly "Hajduvármegye és Debrecen th. j. és sz. kir. Város vitézi székéhez" reflects the antisemitic mood of the local police and the university staff. A letter written by the police-chief to the county's high-sheriff on 9 January 1940 mentions the distribution of the "Protocols of the Elders of Zion" among police officers. *MA*

0504 Salamon, Andrew: *Childhood in Times of War*. Montreal: Concordia University Chair in Canadian Jewish Studies; Montreal Institute for Genocide and Human Rights Studies, 2001. 189, [8] pp. (Memoirs of Holocaust Survivors in Canada, 19).

Memoirs of a Jew born in 1932 in Budapest. In 1941 Salamon's father was drafted into a military labor service unit. After the Nazi takeover in 1944, Salamon and his mother were forced to live in a "Jewish house" in a Budapest suburb. Salamon managed to obtain the papers of a young Nazi and hid among his "comrades" until discovered; he was saved from death by Zionists who were also masquerading as Nazis. His father was saved by Raoul Wallenberg. The family, including his brother Mordi, who had been in hiding, hid in a lumberyard together during the last weeks before the liberation. *YC*

0505 Szita, Szabolcs, ed.: *A humánum példái: Dokumentumok, emlékezések a magyarországi embermentő akciók 1944-1945. évi történetéhez* [Examples of Humanity: Documents and Memoirs about Human Rescue in Hungary in 1944-45]. Budapest: Magyar Auschwitz Alapítvány, Holocaust Dokumentációs Központ, 1998. 149 pp.

A collection of 58 accounts and testimonies relating to rescue actions in Hungary. The introduction (pp. 5-35) stresses the role of Hungary as a haven for Jewish refugees before the Nazi occupation. Quotes from the Veesenmayer report to Berlin in June 1944, blaming the Hungarian army for the fact that only 289,357 Jews were deported in 1944 instead of the planned 310,000. Notes that army officers saved Jews by conscripting them. Stresses that Jews who escaped were able to survive mostly due to help from non-Jews. Army officers, foreign diplomats, the Hungarian representative of the International Red Cross, priests, doctors, and the Zionist underground all took part in rescue activities. The emigration office of the Swiss Embassy provided protection for thousands. Schools, hospitals, and orphanages under the auspices of the International Red Cross saved lives after the Szálasi coup. *AV*

0506 Vertes, Leslie: *Can You Stop the Wind? An Autobiography*. Montreal: Concordia University Chair in Canadian Jewish Studies; Montreal Institute for Genocide and Human Rights Studies, 2001. 134 pp. (Memoirs of Holocaust Survivors in Canada, 21).

Memoirs of a Jew, born László Winkler in 1924 in the Hungarian village of Ajak. In 1935 his family moved to Budapest, where he experienced antisemitism that forced him to leave school. Pp. 19-38 relate his experiences under antisemitic Hungarian and Nazi rule. Vertes escaped service in a military labor service unit temporarily, as a member of a fire brigade. His father was inducted and sent to the eastern front. Vertes was eventually taken for forced labor, but managed to escape and went into hiding under a false identity. His fiancée was killed in a hospital by Hungarian soldiers in 1945. Vertes was arrested by the Red Army and spent several years in Russian labor camps. His mother and sister survived and his father, whom he believed dead, also survived. Vertes and his wife left Hungary in 1956 and settled in Canada. *YC*

Italy

0507 Debenedetti, Giacomo: *October 16, 1943. – Eight Jews*. Trans.: Estelle Gilson. Notre Dame, IN: University of Notre Dame Press, 2001. viii, 100 pp. Originally published as "16 ottobre 1943". – "Otto ebrei" (Palermo: Sellerio, 1993).

0508 Debenedetti, Giacomo: *16 octobre 1943, suivi de Huit Juifs*. Trans.: Monique Baccelli. Paris: Editions Allia, 2001. 110 pp. Originally published as "16 ottobre 1943". – "Otto ebrei" (Palermo: Sellerio, 1993).

0509 Kufeke, Kay: Rassenhygiene und Rassenpolitik in Italien: Der Anthropologe Guido Landra als Leiter des "Amtes zum Studium des Rassenproblems." *Jahrbuch fuer Antisemitismusforschung* 10 (2001) 265-286.

Landra (b. 1913) was commissioned by Mussolini in June 1938 to draw up a manifesto as the basis for Italy's racial policy. In August he was named director of the newly established Ufficio Studi del Problema della Razza in the Propaganda Ministry. He held that the Italians were a Nordic Aryan race and that Jews were not of the same race. He was in close contact with German race theorists such as Eugen Fischer and with the officials of the Nazi Rassenpolitisches Amt, and promoted institutional cooperation with them. However, his overweeningness made him many enemies, and in February 1939 he was removed from office. *RW*

0510 Luzzatto Voghera, Gadi: Dario Disegni: Un rabbino "ottocentesco" nell'Italia fascista. *Cattolici, ebrei ed evangelici nella guerra: Vita religiosa e società, 1939-1945,* eds. Bartolo Gariglio, Riccardo Marchis. Milano: FrancoAngeli, 1999. Pp. 92-107.

Dario Disegni (1878-1967) became the Chief Rabbi of Torino in 1935. He was a strictly observant Jew, and also an Italian patriot and an adherent of fascism. States that, like many other Italian Jews of his generation (who inherited assimilationist tendencies from the time of the "Risorgimento"), he did not realize the terrible danger the Jews were facing. After the enactment of the Racial Laws in 1938, he chose a policy of silence in public, carrying out clandestine educational activities and assisting Jews whenever he could. At the time of the German occupation of Torino (1943), he escaped deportation, hiding in the homes of non-Jews. In 1945 he returned to his post as Chief Rabbi. *MZ*

0511 *La persecuzione degli ebrei durante il fascismo: Le leggi del 1938.* Roma: Camera dei Deputati, 1998. xi, 191, [16] pp.

A publication of the Italian parliament on the occasion of the 60th anniversary of the enactment of the Racial Laws. Pp. 3-29 present reflections about racism by Jacques Chirac, Bill Clinton, Vaclav Havel, Roman Herzog, Oscar Luigi Scalfaro, and Ezer Weizman. Pp. 111-181 contain documents related to the Racial Laws, decrees and parliamentary proceedings in 1938. Presents, also, three essays: Vivanti, Corrado: Gli ebrei nella storia d'Italia (33-49); Luzzatto Voghera, Gadi: L'antisemitismo in Europa e in Italia tra le due guerre (51-77); Sarfatti, Michele: La persecuzione degli ebrei in Italia dalle leggi razziali alla deportazione (81-107). *MZ*

0512 Vincenti, Lucia: *Storia degli ebrei a Palermo durante il fascismo: Documenti e testimonianze*. Introd.: Gabriela Portalone. Palermo: Offset Studio, 1998. xxii, 221 pp.

Relates the fate of the Jews in Palermo from 1938 (enactment of the Racial Laws) until 1943, based on 11 interviews carried out between 1996-98 with Jews and non-Jews in the city. The introduction (pp. vii-xxii) states that the inconsistencies in Mussolini's policy toward the Jews were a result of his opportunism rather than a consequence of his ideological convictions. Pp. 1-21 include the texts of the Racial Laws. Pp. 23-121 deal with their application in Palermo and with the reactions of the press. Jewish ownership of property was restricted, some properties were confiscated, and restrictions were imposed on daily life (e.g. prohibition to own a radio or a car, or to appear in the telephone book). Some of the Jews were interned in concentration camps. The press (since 1926 under state control) acted as an instrument of fascist propaganda. Pp. 93-154 contain the interviews; pp. 155-219 present documents and photos. *MZ*

0513 Zuccotti, Susan: *Il Vaticano e l'Olocausto in Italia*. Trans.: Vittoria Lo Faro. Milano: Bruno Mondadori, 2001. x, 374 pp. Originally published as "Under His Very Windows: The Vatican and the Holocaust in Italy" (New Haven: Yale University Press, 2000).

Lithuania

0514

סמאליאקאוו, צבי הירש, עורך: ניט ווי שאף צו דער שחיטה: זכרונות פון אנטייל-נעמערס אין אקטיוון ווידערשטאנד קעגן די נאציס אין קאוונער געטא (1944-1941). תל-אביב: פארלאג י.ל. פרץ, 2001. 304 עמ'. בשער גם: 60 אקא [=אנטיפאשיסטישע קאמף ארגאניזאציע].

[Smoliakovas, Grigorijus, ed.: *Not as Sheep to the Slaughter: Memoirs of Participants in Active Resistance against the Nazis in the Kovno Ghetto (1941-1944)*. Tel-Aviv: I.L. Peretz Pub. House, 2001. 304 pp.] On title page also: 60 AKO [=Antifashistishe Kamf Organizatsie].

Presents 38 essays by former members of the Kovno ghetto resistance group, published in honor of the 60th anniversary of its founding. The writers describe their experiences in the Holocaust, particularly in the resistance. Led by Chaim Yellin, the members focused on acquiring arms and escaping from the ghetto to join partisan groups in the Rudninkai forest. They worked with the Judenrat and its head, Elkhanan Elkes, the Jewish ghetto police, who trained Jews in the use of arms and helped them escape from the ghetto, and underground Zionist organizations. *RE*

0515 Čiplytė, Joana Viga: *Amžina knygos tauta: Lietuvos žydų gyvenimo puslapiai* [The Eternal People of the Book: Pages of Jewish Life in Lithuania]. Vilnius: Homo Liber, 2001. 95 pp.

A brief history of the Jews in Lithuania in the 20th century, focusing on the Jewish community of Panevėžys and of the nearby town of Ramygala. A ghetto was established in Panevėžys on 7 July 1941; on 23 August it was liquidated and its inmates were murdered. Notes that there were Lithuanians who rescued Jews during the Holocaust, and that Holocaust memory is now preserved in Lithuania. *DR*

0516 Cohen, Rose Lerer; Issroff, Saul: Remembering Lithuanian Shoah Victims: A Research Project. *Avotaynu* 15, 1 (Spr 1999) 21-25.

Discusses a project aimed at creating a comprehensive list of names of Jews killed in the Holocaust in Lithuania and Lithuanian Jews killed elsewhere. Describes the sources used in the implementation of the project and the problems with which the researchers have been confronted. Pp. 24-25 contain a list of publications on the Holocaust in Lithuania which can be used in carrying out the project. *DR*

0517 Czapski-Holzman, Helene: *"Dies Kind soll leben": Die Aufzeichnungen der Helene Holzman, 1941-1944.* Eds.: Reinhard Kaiser, Margarete Holzman. Frankfurt a.M.: Schoeffling, 2000. 384, 32 pp.

Memoirs written immediately after the liberation of Kovno, describing life in the town and in the ghetto under Nazi occupation. Holzman, a German woman, moved to Kovno with her Jewish husband in 1922. Immediately after the German invasion, her husband was seized on the street; he was shot soon after. Later, her older daughter (aged 17) was arrested and executed as a "communist." Holzman and a younger daughter survived. She joined an informal network of women who tried to help Jewish acquaintances in the ghetto, to smuggle them out and hide them, or to intervene for them with the authorities. One of these was Edwin Geist, a German-Jewish composer who was married to a Lithuanian. The Nazis' condition for allowing him to live with his wife outside the ghetto was that she be sterilized. In the end he was shot and she committed suicide. Also tells the stories of Jews from other ghettos, especially Vilna. Pp. 319-342 relate the story of Tolja Szabszewicz of Łask in Poland; she was 13-years-old when the Germans invaded. Describes the persecution of Jews in the ghetto; the massacre of most of the Jews of Łask, while Tolja was among those selected for work in Łódź; her deportation to Auschwitz, and then to Stutthof and to the labor camp Schippenbeil; and her escape from a death march. *RW*

0518 Eidintas, Alfonsas, ed.: *Lietuvos žydų zudynių byla: Dokumentų ir straipsnių rinkinys* [The Case of the Massacre of the Lithuanian Jews: Selected Documents and Articles]. Vilnius: Vaga, 2001. 823 pp.

The book consists of three parts: a historical introduction by Eidintas (pp. 21-279), documents (pp. 281-351), and a collection of essays by Lithuanian authors, published between 1941-2000 (pp. 353-794). The first section of the introduction discusses Lithuanian-Jewish relations from the late 19th century until 1914, dwelling on popular anti-Jewish images and stereotypes. The second section deals with the situation of the Jews in independent Lithuania in the 1920s-30s. Economic competition led to a rise in tensions between Lithuanians and Jews. The third and fourth sections cover the World War II period and examine how it is reflected in Jewish and Lithuanian émigré writings. Both Jewish and Lithuanian writers present their respective nationalities as the victims, and ascribe the negative activities of its own members (Jewish participation in the Sovietization of Lithuania in 1940-41, Lithuanian participation in the massacres after 22 June 1941) to an unrepresentative minority. Admits, however, that the Jews had good reason to rejoice in the Soviet occupation, as it was preferable to Nazi rule, and that the Jewish role in Sovietization was not excessive. Suggests that Lithuanians participated in anti-Jewish campaigns under German control and direction. The fifth section deals with Lithuanian-Jewish relations after 1990, dwelling on some scandalous affairs, like that of Lileikis. *DR*

0519 Faitelson, Aleks (Alter): *Nepokorivshiyesya: Letopis yevreiskovo soprotivleniya* [Those Who Didn't Submit: A Chronicle of Jewish Resistance]. Tel-Aviv: The Author, 2001. 608 pp.

Memoirs interspersed with the history of the Holocaust and Jewish resistance in Lithuania. Faitelson (b. 1923), a Jew with prewar experience in Zionist youth and communist organizations, was interned in the Kovno ghetto. Describes pogroms perpetrated by the Lithuanians in Kovno in June-July 1941; the ghetto and forced labor; and mass murders perpetrated in the VIIth and IXth Forts. In December 1941 the pro-Soviet resistance organization Antifashistishe Kamf Organizatsie was established in the ghetto; describes its activities and relations with other Jewish resistance organizations such as the ABC (Irgun Brit Zion). In October 1943, Faitelson and other AKO members were sent to the forest as partisans; the group was intercepted by Lithuanians, and Faitelson was imprisoned in the IXth Fort, where he was assigned to a detail that burned the corpses of those killed by the Nazis. In January 1944, Faitelson and 63 other members of this detail escaped. Most of them joined the Soviet partisan movement in Lithuania and Belarus. After the war he settled in Israel. *DR*

0520 Gasparaitis, Siegfried: "Verraetern wird nur dann vergeben, wenn sie wirklich beweisen koennen, dass sie mindestens einen Juden liquidiert haben": Die "Front Litauerischer Aktivisten" (LAF) und die antisowjetischen Aufstaende 1941. *Zeitschrift fuer Geschichtswissenschaft* 49, 10 (2001) 886-904.

Describes the formation of the LAF by exiled Lithuanian politicians in Berlin during the Soviet occupation of Lithuania. Its ideology called for the liberation of Lithuania from the "Jewish-Bolshevik" yoke, with the aid of Nazi Germany. LAF declarations and propaganda smuggled into Lithuania equated Jews with communists and warned that not only would they be excluded from the new Lithuania but that their lives would be in danger. With the German invasion of the USSR, LAF partisan groups emerged from underground to fight for freedom and, as an integral part of this fight, to massacre communists and Jews. Notes that these massacres were not ordered by a central command but were carried out at the initiative of local groups, whose motives were plunder as well as ideology. *RW*

0521 Ginaitė-Rubinsonienė, Sara: *Atminimo knyga: Kauno žydų bendruomenė 1941-1944 metais* [Book of Remembrance: The Jewish Community of Kaunas 1941-1944]. Vilnius: Margi raštai, 1999. 227 pp. With an English summary (pp. 206-209).

Memoirs of a Jew who was born in Kovno, interspersed with a history of the Holocaust in Lithuania. Ginaitė and her family were interned in the Kovno ghetto. In 1943, as a member of the Jewish underground. Ginaitė escaped to the forests, where she joined a partisan unit. Describes the massacre of Jews in Kovno perpetrated by the Nazis and Lithuanian collaborators after the Nazi occupation; dwells on the massacre of 9,200 Jews at the Ninth Fort on 28-29 October 1941. Describes the underground organizations, the leftist AKO and the Zionist Matsok, as well as the actions of the partisan unit which she joined. Places responsibility for the genocide of Lithuanian Jews not only on the Nazis but also on the Lithuanian Activist Front and the Provisional Government, which had actual power in June-July 1941 and could have tried to stop the massacre of Jews. After the war Ginaitė lived in Lithuania; in 1983 she immigrated to Canada. *DR*

0522 Goldstein, Jonathan: Lithuania Honours a Holocaust Rescuer. *Polin* 14 (2001) 249-255.

By May 1940 at least 10,000 Jews had fled from occupied Poland into Lithuania; in summer 1940, when Lithuania was annexed by the USSR, many of these Jews felt vulnerable. The acting consul of the Netherlands in Kaunas, Jan Zwartendijk, helped many Jews by writing in their passports or other documents that a visa was not required to enter Curaçao, Surinam, or other Dutch possessions. At least 2,345 of these permits were issued; although they did not actually allow their bearers entry to Curaçao, they did help them obtain transit visas through the USSR and Japan. In 1997, Zwartendijk was recognized as Righteous among the Nations by Yad Vashem. In 1999 three stone monuments were dedicated in his memory in Vilnius. *DR*

0523 Gurevičius, Antanas: *A. Gurevičiaus sąrašai: Tūkstančiai lietuvių, kurie gelbejo tukstančius Lietuvos žydų Antrojo pasaulinio karo metais* [The Lists of A. Gurevičius: Thousands of Lithuanians Who Rescued Thousands of Lithuanian Jews in the Second World War]. Introd.: Gintutis Procuta. Vilnius: Protėvių Kardas, 1999. 189 pp.

Pp. 5-70 contain a foreword by Procuta, in Lithuanian and in English, in which he contends that Lithuanians saved many Jews. In 1939-early 1940 the Lithuanian government accepted thousands of Jewish refugees from Poland, thus allowing avenues of escape to Japan and other countries. Claims that thousands of Lithuanians saved thousands of Jews during the war, and deplores the distortion of the history of the Holocaust in Lithuania by Western and Israeli historiographers. Pp. 71-152 contain cumulative lists of 6,271 Lithuanians who rescued Jews; pp. 153-189 contain cumulative lists of 10,137 Jews who were saved by Lithuanians. *DR*

0524 Kuodytė, Dalia; Stankevičius, Rimantas, eds.: *Išgėlbeję pasaulį...: Žydų gelbėjimas Lietuvoje (1941-1944)* [Those Who Rescued a World...: The Rescue of Jews in Lithuania (1941-1944)]. Vilnius: Lietuvos gyventojų genocido ir rezistencijos tyrimo centras, 2001. 371 pp.

Following an introduction by the editors (pp. 5-33), presents articles published in Lithuanian émigré periodicals, as well as texts of letters sent by Lithuanian Jews and non-Jews living abroad (mainly in the USA) to two active members of Vyriausias Lietuvos Islaisvinimo Komitetas (The Main Committee for the Liberation of Lithuania), Domas Jasaitis and Juozas Prunskis, in the postwar period. These documents deal with Lithuanians who rescued Jews during the Nazi occupation. Pp. 334-340 contain a list of 474 Lithuanian rescuers of Jews. *DR*

0525 Kuodytė, Dalia; Stankevičius, Rimantas, eds.: *Whoever Saves One Life...: The Efforts to Save Jews in Lithuania between 1941 and 1944*. Trans.: Laima Junevičienė, Aldona Matulytė. English edition edited by Alexander Fortescue. Vilnius: Genocide and Resistance Research Centre of Lithuania; Garnelis, 2001. 231 pp. An abridged translation of the Lithuanian edition: "Išgėlbeję pasaulį" (Vilnius: Lietuvos gyventojų genocido ir rezistencijos tyrimo centras, 2001).

0526 Melamed, Joseph A.: The Malicious Face and Criminal Acts of the "Provisional Government" of Lithuania, June 23-August 5, 1941, or "The Insurrection of the Lithuanian Nation." *Lithuania – Crime and Punishment* 6 (Jan 1999) 36-41.

Describes the ideological and political platform of the Lithuanian Activist Front, the antisemitic nationalist organization which collaborated with the Nazis in the preparations for the invasion of Lithuania and the annihilation of Lithuanian Jewry. Also relates to the Provisional Government founded by the LAF, which was supposed to take over the control of an independent Lithuania after its liberation from the Soviets. Includes a description of the Provisional Government's divisions, its legislative acts, main figures, police and military formations and their part in the murder of Lithuania's Jewry during the first weeks of the Nazi occupation. *LFo*

0527 Melamed, Joseph A.: Soviet Lithuania and the Jews. *Lithuania – Crime and Punishment* 6 (Jan 1999) 7-15.

Refutes allegations disseminated by Lithuanians after World War II regarding the decisive part played by Lithuanian and Soviet Jews in the persecution of Lithuanians after August 1940. States that Soviet Lithuania was actually controlled by Lithuanians who promoted, as always, nationalistic policies. The accusations against the Jews are still disseminated today, to cover up and minimize the leading role played by Lithuanians in the extermination of the Jews during the Nazi occupation. *LFo*

0528

פרייס, לאה: סוגיות של רפואה, אתיקה ומוסר בגטאות וילנה, קובנה ושאוולי. ילקוט מורשת 71 (אפר' 2001) 23-41.

[Preiss, Leah: Questions of Medicine, Ethics and Morals in the Vilna, Kovno and Shavli Ghettos. *Yalkut Moreshet* 71 (Apr 2001) 23-41.]

Based on diaries and memoirs of Jewish doctors who worked in the ghettos of Vilna, Kovno, and Shavli, and testimonies of survivors, discusses medical dilemmas faced during the Holocaust. In 1942 Jewish women were forbidden to give birth; while some gave birth in secret (e.g. in the ghetto hospital in Vilna), many others had abortions, even in late pregnancy. Judenrat members, doctors, and mothers deliberated over whether or not to kill babies that were born, since the Jews would suffer collective punishment if the babies were discovered; some were given away and some were killed. Discusses other cases in which doctors were forced to make life-or-death decisions (e.g. in distributing the small amounts of medicine available, granting exemptions from work, and hiding children in the hospital in order to prevent their deportation), and arguments between doctors as to whether they had the right to make such decisions. *RE*

0529 Sakowicz, Kazimierz: *Dziennik pisany w Ponarach od 11 lipca 1941 r. do 6 listopada 1943 r.* [A Diary Written in Ponary from 11 July 1941 to 6 November 1943]. Ed.: Rachilė Margolis. Bydgoszcz: Tow. Miłośników Wilna i Ziemi Wileńskiej, 1999. 144 pp.

Sakowicz was a Polish journalist who, during the war, lived in the village of Paneriai, close to the forest where the massacres of the Jews took place. In the 1990s the diary was deposited in the Jewish Museum in Vilnius and deciphered (it was in poor physical condition) in 1998 by Margolis. In the preface (pp. 5-40) Margolis describes events in Vilna after the Nazi occupation as described by Sakowicz and other sources. Sakowicz recorded what he saw and heard from his house: the transports of the Jews, the beatings and shootings, and the trading of the goods after the Jews were killed. He noted the number of people brought in trucks to be killed and some names of Lithuanian collaborators. Sakowicz was killed by Lithuanians in July 1944 because he was an "inconvenient" witness of the massacres. *HV*

0530 Shner-Nishmit, Sara: Jewish-Lithuanian Relations during World War II: History and Rhetoric. *Lithuania* 6 (Jan 1999) 15-23. An English version appeared in "The Historiography of the Holocaust Period: Proceedings of the Fifth Yad Vashem International Historical Conference" (1988).

Gives examples of murderous acts against the Jews perpetrated by Lithuanians during the first weeks of the Nazi occupation, as well as the involvement of Lithuanian police and military formations controlled by the Nazis in the annihilation of Jews in Eastern Europe. Surveys historical writings published in postwar Lithuania and by Lithuanian émigrés in the West which relate to the extermination of the Jews. Summarizes the public discourse between Lithuanians and Jews around the world regarding the part played by Lithuanians in the annihilation of Jews. *LFo*

0531 Sirovich, Livio Isaak: *Ihr Lieben, schreibt mir nicht alles: Ein juedische Familie in Litauen 1935-1941.* Trans.: Friederike Hausmann. Muenchen: Verlag Antje Kunstmann, 2001. 301 pp. Originally published as "Cari, non scrivetemi tutto: Gli Isaak – una famiglia in trappola fra Hitler e Stalin" (Milano: Arnoldo Mondadori, 1995).

0532 The Torture and Massacre of the Jewish Population of the District of: Biržai, Kėdainiai, Lazdijai, Marijampolė, Mažeikiai, Panevėžys, Raseiniai, Šakiai, Telšiai, Ukmergė, Žagarė and Other Districts. *Lithuania – Crime and Punishment* 6 (Jan 1999) 75-122.

For each of the places mentioned above, presents information on the murder of the Jews during World War II as well as a partial list of the names of Lithuanian perpetrators who were involved in the massacre. Includes partial lists of Lithuanian murderers of Jews in Alytus, Kaunas, Kretinga, Rokiškis, Šiauliai, Tauragė, Trakai, Utena, Vilkaviškis, Vilnius, Zarasai and their districts. *LFo*

0533 Zak, Joseph: The Murder of Lithuanian Rabbis. *Lithuania – Crime and Punishment* 6 (1999) 124-128.

Quotes from unidentified sources which describe the circumstances in which the rabbis of the communities of Alytus, Anykščiai, Biržai, Butrimonys, Girkalnis, Josvainiai, Kaišiadorys, Kelmė, Lazdijai, Luokė, Marijampolė, Papilė, Pumpenai, Radviliškis, Ramygala, Rietavas, Seda, Širvintos, Tauragė, Telšiai, Tytuvėnai, Ukmergė, Utena, Užpaliai, and Žagarė were murdered by Lithuanians during the Nazi occupation. *LFo*

Netherlands

0534 Brinks, Jan Herman: The Dutch, the Germans, & the Jews. *History Today* 49, 6 (June 1999) 17-23.

After World War II a myth arose that the Netherlands was a country of resistance, whose citizens helped their Jewish compatriots en masse. Contends that collaboration with the Nazis was widespread amongst the Dutch population. The collaboration began in the 1930s, when Dutch authorities impeded the entry of refugees from Germany and handed over illegal refugees to the Gestapo. During the Nazi occupation, Dutch officials, from the highest ranks to minor railway clerks and policemen, collaborated with the Nazis in everything, including the deportation of Jews. Immediately after the war there was no expression of guilt on the part of Dutch officials or most of the Dutch citizens; hardly any of the main collaborators were prosecuted. *DR*

0535 Jakobs, Rose: *De roos die nooit bloeide: Dagboek van een onderduikster, 1942-1944* [The Rose Which Never Blossomed: A Diary of a Girl in Hiding, 1942-1944]. Ed.: Edith Samuel-Jakobs. Amsterdam: De Arbeiderspers, 1999. 137 pp.

Pp. 23-126 contain the diary of Jakobs (1925-1944), who emigrated from Oldenburg, Germany to the Netherlands in 1938 with her parents and three siblings. The diary was found by her sister Edith after the war; it is now being published for the first time. The diary covers the period 30 August 1942-1 October 1944. The members of the family all went into hiding in different towns. Rose was with an aunt, uncle, and cousin in

Nijmegen. She went out into the street on 2 October 1944 and was instantly killed by bomb fragments. All the other members of the family survived. *MF*

0536 *Kaddisj: Ter nagedachtenis van de joodse Rotterdamse burgers 1940-1945* [Kaddish: In Memory of the Jewish Citizens of Rotterdam 1940-1945]. Introd.: A.W. Oosthoek. Rotterdam: Ad. Donker, 2000. 192 pp.

The introduction (pp. 11-32) describes the fate of the Jews of Rotterdam during World War II, focusing on the process of registration, roundup, and deportation of the Jews in 1942-43. The first transport went to Westerbork on 30 July 1942. The rest of the book (pp. 37-192) contains a list of 6,302 names of Jews from Rotterdam who were deported and murdered, giving their dates and places of birth and death. *MF*

0537 Lebeau, Paul: Un itinéraire spirituel européen: Etty Hillesum (Amsterdam 1941 – Auschwitz 1943). *Nouvelle Revue Théologique* 121, 3 (July-Sept 1999) 397-416.

Reflects on the central themes contained in the diary of Etty Hillesum (1914-1943); considers her one of the foremost young literary voices of Western European Jewry during the Holocaust period. Underscores the relevance of the theological, philosophical, and political issues discussed in the diary. Recounts her emotional and intellectual relationship with her mentor, Julius Spier, as well as her reactions to the deteriorating conditions of Dutch Jewry, which culminated in her deportation to Auschwitz, where she perished. Emphasizes Hillesum's spiritual awakening, despite her assimilated background, which can be perceived as her belief in a divine presence in her life. *TS*

0538 Makkonen, Anna: Holocaust Chronicle, Spiritual Autobiography, Portrait of an Artist, Novel in the Making: Reading the Abridged Diary of Etty Hillesum. *Biography* 22, 2 (Spr 1999) 237-261.

Focuses on the diary of Etty Hillesum (a Dutch Jew born in 1914, who died in Auschwitz in 1943) as literature rather than as a chronicle of the Holocaust. The diarist wrote little about the transit camp Westerbork, where she volunteered to work as a nurse. Nor did she present an account of everyday life under the Nazi occupation. Despite her activism in helping others, Hillesum emphasized her inner life. She felt the inadequacy of language to express the apocalyptic reality of the Holocaust. *YC*

0539 Sagiv, Shaul: *In ungleichem Kampf: Christlich-juedische Rettungsaktion der Westerweel-Gruppe: Von Koeln nach Holland durch Westerbork ueber Frankreich und Spanien nach Israel 1924-1947*. Trans.: Avri

Salamon. Konstanz: Hartung-Gorre, 2001. 262 pp. Unseen. Originally published as "Katarim moshkhim keronot" (Tel-Aviv: Moreshet, 1997).

0540 Wessels, Benjamin Leo: *Ben's Story: Holocaust Letters with Selections from the Dutch Underground Press.* Ed.: Kees W. Bolle. Carbondale: Southern Illinois University Press, 2001. xxiv, 150, [15] pp.

Wessels, born in 1926 in Oostvoorne, was a boyhood friend of Kees Bolle. In October 1942 his family was forced to move to Amsterdam. Bolle presents letters written by Wessels in 1942-43 to Johan Schipper, another friend of theirs, sent from Amsterdam to Oostvoorne. Interspersed among the letters are excerpts from the Dutch underground press, dealing with the Nazi anti-Jewish policy, the deportations, and the genocide. In August 1942 Wessels' older brother was deported to Westerbork; his parents were deported in August 1943. All three died in Auschwitz. In September or October 1943, Wessels was deported to Westerbork; he died in Bergen-Belsen in March 1945. His sister, Carla, was rescued by a non-Jew. *DR*

Poland

0541

אלתר, אברהם מרדכי: בדרכי ווארשא האבלות: סיפורה של ווארשא החרבה. תל-אביב: פאר, תשסא. 597 עמ'.

[Alter, Avraham Mordechai: *On the Mourning Roads of Warsaw: The Story of the Warsaw That Was Destroyed.* Tel-Aviv: Pe'er, 5751 [=2001]. 597 pp.]

Relates the history of Jewish life in Warsaw from an ultra-Orthodox point of view. Pp. 48-111 and 467-597 describe religious life in the ghetto during World War II, e.g. activities in synagogues, religious schools, and other institutions, as well as deportations. Includes short biographies of rabbis who were interned in the ghetto. *RE*

0542

בשביל הזיכרון 43 (ספט'-אוקט' 2001)

[*Bishvil ha-Zikaron* 43 (Sept-Oct 2001).]

This issue, entitled "The Pogrom in Jedwabne – History and Memory," discusses the massacre of the Jews in the town in 1941 by their Christian neighbors, Jan Tomasz Gross' book on the subject, "Sąsiedzi" (2000), and reactions to the book in Poland. *SSC*

0543 Celemenski, Jacob: *Elegy for My People: Memoirs of an Underground Courier of the Jewish Labor Bund in Nazi-Occupied Poland, 1939-45*. Trans.: Gershon Freidlin. Melbourne: Jacob Celemenski Memorial Trust, 2000. xiii, 240 pp. Originally published in Yiddish as "Mitn farshnitenem folk" (New York: Unzer Tsait, 1963).

Celemenski was born in 1904 in Warsaw. As a courier of the Bund during the war, he traveled to Kraków, Piotrków, Zagłębie district, Międzyrzec, and many other places in Poland. Between 1942-44 he lived on the "Aryan side" of Warsaw with false identity papers. He had good contacts with the underground Polish Workers Party and the AK (Polish Home Army). He also took part in the preparations for the Warsaw ghetto uprising. In fall 1944 Celemenski was arrested by the Nazis as a Polish participant in the Warsaw uprising, and sent to Auschwitz. From there he was transported to Mauthausen and liberated in May 1945. *DR*

0544

אלי, קרולינה: עקבות. תל-אביב: עקד, 1999. 152, [15] עמ'.

[Ely, Karolina: *Footsteps*. Tel-Aviv: Eked, 1999. 152, 15 pp.]

Memoirs of a Jew from Łódź, born in 1935 as Karolina Matecka. Pp. 5-39 describe her experiences in the Holocaust. After the Nazi occupation the family fled to Lvov, where her parents were killed (her mother was killed by Ukrainians in July 1941). Ely was cared for by an uncle and aunt, with whom she moved to Warsaw, all of them posing as non-Jews. Ely attended a Polish underground school and experienced the Warsaw uprising in 1944. The family escaped from a forced march out of Warsaw and lived in various places until the liberation, when they returned to Warsaw. Ely's uncle and aunt died shortly after the war, and she entered an orphanage in Otwock, where she first found out that she was Jewish. She immigrated to Israel in 1960. *RE*

0545 Engel, David: On the Bowdlerization of a Holocaust Testimony: The Wartime Journal of Calek Perechodnik. *Polin* 12 (1999) 316-329.

Presents the story of the original manuscript of Perechodnik's journal, kept by his brother Pesach who lives in Israel, and of three copies of the journal which differ one from another and can be found in the Yad Vashem archives and in the Jewish Historical Institute in Warsaw. Analyzes two Polish editions of the journal (entitled "Czy ja jestem mordercą?") from 1993 and 1995, edited by Paweł Szapiro, and states that in many places Szapiro distorted the original meanings of the text through interventions which are significantly greater than he suggested in his preface to the book. Analyzes omissions and changes in the text which concern themes such as Jewish attitudes toward Polish independence and

Polish-Jewish relations during the war. Szapiro softened many of Perechodnik's negative statements about Poles. Concludes that the Hebrew translation of Perechodnik's journal (1993), which is based on the original typescript, is the only reliable version of the journal. *HV*

0546 Engelking-Boni, Barbara: *Holocaust and Memory: The Experience of the Holocaust and Its Consequences: An Investigation Based on Personal Narratives*. Ed.: Gunnar S. Paulsson. Trans.: Emma Harris. London: Leicester University Press, 2001. xx, 348 pp. Originally published as "Zagłada i pamięć: Doświadczenie Holocaustu i jego konsekwencje opisane na podstawie relacji autobiograficznych" (Warszawa: IFiS PAN, 1994).

0547 Engelking-Boni, Barbara: Sześć listów z warszawskiego getta, 9 VIII 1941 – 25 VI 1942 [Six Letters from the Warsaw Ghetto, 9 VIII 1941 – 25 VI 1942]. *Kwartalnik Historii Żydów* 198 (2001) 229-240.

Presents six letters written in the Warsaw ghetto and sent to the "Aryan side" by a Jewish woman named Hala (ca. 22 years old). The letters were addressed to Stefania, her teacher in the gymnasium she attended before the war. The letters contain information about Hala's life in the ghetto; her fate is not known. Copies of the letters are found in the archives of the Jewish Historical Institute. *HV*

0548

שם עולם: רשימת תושבי גטו זמושץ' – 1940. תל-אביב: ארגון יוצאי זמושץ', הסביבה ודורות המשך, 2001. 225, 15 עמ'.

[*Everlasting Name: Zamość Ghetto Population List – 1940*. Tel Aviv: Israeli Organization of Zamość Jewry, Vicinity & Their Descendents[!], 2001. 225, 15 pp.]

In April 1940, the Judenrat of Zamość, apparently at its own initiative, compiled a list of all the Jews living in the city, including refugees from other places who had settled there. Pp. 11-225 contain the list of 10,086 names (including occupation, date and place of birth, and address), now held in the Lublin archives. In 1999-2001 the list was copied by hand; in 2001 it was exhibited in Tel Aviv, supplemented by photographs and documents. The introduction by Adam Kopciowski, in Hebrew (pp. 3-7) and English (pp. 4-10), discusses the demographic structure of the ghetto population. The ghetto was annihilated in five deportations between 11 April-19 October 1942. The 15 pp. at the end contain introductory material in English. *DR*

0549 Fuks, Marian: *Z dziejów wielkiej katastrofy narodu żydowskiego* [From the History of the Great Catastrophe of the Jewish People]. Poznań: Wyd.

Sorus, 1999. 341, [68] pp. At head of title: Żydowski Instytut Historyczny INB.

A historical survey of the main events of the Holocaust in Poland, including the resettlements of Jews, the founding of the ghettos, and the organization of the Judenraete. Describes Adam Czerniaków's diary and Emanuel Ringelblum's writings as important sources of information. Discusses the "Gazeta Żydowska," which appeared in Kraków between July 1940-mid-1942; it was founded by the Nazis and edited by Jews under Nazi censorship. States that, despite Nazi control, the paper managed to convey some information about the fate of the Jews. Deals, as well, with the deportations and extermination, and with Jewish resistance. Discusses Marek Edelman's opinions on the Warsaw ghetto uprising. Describes participation of the Jews in the Warsaw uprising in August 1944. Mentions the fate of the Jews under Soviet occupation (1939-41) and their participation in battles on various fronts. A list of the Jews killed in Katyn is annexed. The book is richly illustrated. *HV*

0550 Gerson, Daniel: Antisemitische Erfahrungen in Lodz zwischen den beiden Weltkriegen. *Polen, Deutsche und Juden in Lodz 1820-1939: Eine schwierige Nachbarschaft,* ed. Juergen Hensel. Osnabrueck: Fibre, 1999. Pp. 257-268.

Jews originally from Łódź who were interviewed by the author recall attacks by street gangs on their way home from school; gangs driving Jews out of the public parks on Sundays; mutually accepted segregation on promenades; men standing in front of Jewish shops to keep away customers; discrimination in the civil service and in access to secondary education; and the persistence of belief in the blood libel. Jews defended themselves and took pride in physical prowess. However, increasing antisemitism after 1935 shook their sense of security. *RW*

0551 Gross, Jan Tomasz: *Neighbors: The Destruction of the Jewish Community in Jedwabne, Poland.* Princeton: Princeton University Press, 2001. x, 261 pp. Originally published as "Sąsiedzi: Historia zagłady żydowskiego miasteczka" (Sejny: Fundacja Pogranicze, 2000). The English edition contains an added introduction (pp. 3-13).

0552

גרוס, יאן טומש: שכנים: השמדתה של הקהילה היהודית בידווובנה שבפולין. תרגום: יז'י מיכאלוביץ'. ירושלים: יד ושם; תל-אביב: משכל, 2001. 175 עמ'.

[Gross, Jan Tomasz: *Neighbors: The Destruction of the Jewish Community in Jedwabne, Poland.* Trans.: Jerzy Michałowicz. Jerusalem: Yad Vashem; Tel-Aviv: Miskal, 2001. 175 pp.] Originally published as

"Sąsiedzi: Historia zagłady żydowskiego miasteczka" (Sejny: Fundacja Pogranicze, 2000).

0553 Gross, Natan: *Who Are You, Mister Grymek?* Trans.: William R. Brand. London: Vallentine Mitchell, 2001. [15], 330 pp. Originally published as "Mi atah adon Grymek?" (Tel-Aviv: Moreshet – Beit Edut al shem Mordechaj Anielewicz, 1986).

0554 Grupińska, Anka: *Ciągle po kole: Rozmowy z żołnierzami getta warszawskiego* [Still on the Circle: Conversations with Warsaw Ghetto Fighters]. Warszawa: Twój Styl, 2000. 396 pp.

A revised and expanded edition of her book "Po kole: Rozmowy z żydowskimi żołnierzami" (1991). The present edition includes conversations with Kazik Ratajzer (Simcha Rotem), Halina Bełchatowska, and Bronek Szpigel. A second talk with Marek Edelman, compiled from interviews conducted in 1999-2000, is included. *HV*

0555

גוטרמן, בלה; מורגנשטרן, נעמי, עורכות: מחזור וולפסברג, תש"ה: מחנה לעבודת כפייה, גרמניה, 1944. ירושלים: יד ושם, 2000. 80 עמ'.

[Gutterman, Bella; Morgenstern, Naomi, eds.: *Mahzor Wolfsberg, 5705: A Forced Labor Camp, Germany, 1944*. Jerusalem: Yad Vashem, 2000. 80 pp.]

התוכן: כהנא, נחמן: רבי נפתלי שטרן ז"ל, חזן ראשי של קהילת יראים בסאטמר, והתפילות שכתב במחנה וולפסברג (24-21); הלבני, דוד וייס: על השואה ועל תפילתה (43-25); גוטרמן, בלה: מחנה וולפסברג (61-45); שן-דר, יהודית: תמימי אמונה ותמימי אמנות (78-65).

Pp. 13-20 contain the facsimile of a handwritten mahzor for Rosh Hashanah, written by Cantor Naftali Stern (1910-1989) in the Wolfsberg labor camp in Upper Silesia in 1944; it is now held at Yad Vashem. Born in Hungary, Stern was deported in May 1944 to Auschwitz, where he lost his wife and four children. He was liberated at Wolfsberg and later immigrated to Israel. The facsimile reprint is accompanied by essays, some of which include illustrations. Contents: [Kahana, Nahman: Rabbi Naftali Stern z"l, Chief Cantor of Kehillat Yere'im in Satmar, and the Prayers He Wrote in the Wolfsberg Camp (21-24); Halivni, David Weiss: On the Shoah and on Its Prayer (25-43); Gutterman, Bella: The Wolfsberg Camp (45-61); Shen-Dar, Judith: The Resolute in Faith and the Resolute in Art [On contemporary art works portraying prayer in ghettos and Nazi camps.] (65-78).] *RE*

0556 Gutterman, Bella; Morgenstern, Naomi, eds.: *Plegarias de Rosh Hashana, Wolfsberg, 5705: Campo de trabajo, Alemania, 1944*. Trans.: Ethel Katz. Jerusalén: Yad Vashem, 2001. 102 pp. Originally published as "Mahzor Wolfsberg, 5705: Mahaneh le-avodat kfiyah, Germania, 1944" (Jerusalem: Yad Vashem, 2000).

0557 Isakiewicz, Elżbieta: *Harmonica: Jews Relate How Poles Saved Them from the Holocaust*. Trans.: Piotr Szymor. Warsaw: PAI, 2001. 280 pp. Originally published as "Ustna harmonijka: Relacje Żydów których uratowali od Zagłady Polacy" (Warszawa: Niezależne Wydawnictwo Polskie, 2000).

0558 Isakiewicz, Elżbieta: *Ustna harmonijka: Relacje Żydów których uratowali od Zagłady Polacy* [Harmonica: Accounts of Jews Who Were Saved from the Holocaust by Poles]. Warszawa: Niezależne Wydawnictwo Polskie, 2000. 262 pp.

Presents memoirs of 19 Holocaust survivors who were rescued by Poles and who now live in Israel. The survivors were interviewed by Isakiewicz in May 2000. Their stories deal with events in Lvov, Kraków, and Warsaw, as well as Otwock, Wieliczka, Włodzimierz Wołyński, and other places (some of them now in Ukraine). *DR*

0559 Jakubowicz, Josef: Juedische Buerger der Stadt Auschwitz: Erinnerungen von Josef Jakubowicz. *Dachauer Hefte* 17 (Nov 2001) 141-156.

Memoirs based on interviews conducted by Sybille Steinbacher in 1994 and 1995. Jakubowicz, born in Oświęcim in 1925, describes the prosperous Jewish community that made up half the population of the town before the Holocaust. When the Germans entered, they immediately destroyed the synagogues, Aryanized Jewish businesses, concentrated the Jews in another section of the town, and drafted them for forced labor in the construction of the concentration camp. In April 1941, after the completion of the camp, the Jews from Oświęcim were evacuated to Sosnowiec and Będzin. From there Jakubowicz was sent from one labor camp to another of the Organisation Schmelt. He was liberated in Bergen-Belsen; most of his family perished. Mentions, also, the fate of other Jews from Oświęcim. *RW*

0560

קראי, פליציה: הנשים בגטו קרקוב. ילקוט מורשת 71 (אפר' 2001) 109-129.

[Karay, Felicja: Women in the Kraków Ghetto. *Yalkut Moreshet* 71 (Apr 2001) 109-129.]

Discusses the views and experiences of Jewish women in the Kraków ghetto. Most of the women opposed resistance; they realized that the Nazi war was not like other wars, and that they would have to concentrate on the struggle for survival. Women were able to cope better than men with the feeling of helplessness and they adjusted more easily to the worsening conditions. They smuggled food, bribed officials, dealt in the black market, found hiding places, and organized cultural and welfare activities. Maintains that general Israeli condemnation of the weak Jewish "diaspora mentality" is unwarranted, especially since millions of non-Jewish victims of the Nazis were never criticized for failing to offer resistance. While there were some women who fought in the ghetto resistance, the passive resistance of many women who struggled to save their families and friends is just as praiseworthy. *RE*

0561 Kosmala, Beate: Lodzer Juden und Deutsche im Jahr 1933: Die Rezeption der nationalsozialistischen Machtuebernahme in Deutschland und ihre Wirkung auf das Verhaeltnis von juedischer und deutscher Minderheit. *Polen, Deutsche und Juden in Lodz 1820-1939: Eine schwierige Nachbarschaft,* ed. Juergen Hensel. Osnabrueck: Fibre, 1999. Pp. 237-245.

The Nazi accession to power in Germany brought an end to the cooperation between the two main national minorities in Łódź, the Jews and the Germans. The Jews, with the support of many Poles who were appalled by Hitler's threats against Poland, called for a boycott of German imports; most of the German press in Łódź accused them of spreading "Greuelpropaganda" and justified Nazi anti-Jewish measures as self-defense. When Polish patriots, together with Jews, demonstrated against Nazism on 9 April 1933, and some of the crowd went on to wreck German institutions in Łódź, the German press blamed the Jews. The event was commemorated every year thereafter as "Black Palm Sunday." Among the Łódź Germans, only socialists sympathized with the Jews, but even they were split on the issue. *RW*

0562 Kotula, Franciszek: *Losy Żydów rzeszowskich 1939-1944: Kronika tamtych dni* [The Fate of the Jews of Rzeszów, 1939-1944: A Chronicle of Those Days]. Rzeszów: Społeczny Komitet Wydania Dzieł Franciszka Kotuli, 1999. 196, [64] pp.

The diary of a Pole, a teacher, who during the Nazi occupation of Rzeszów was the director of a museum in which many Jews who lived in the ghetto worked. Kotula, who noted every event in the town in his "Dziennik okupacyjny" ("Diary of the Occupation"), realized very quickly that the Jews were being treated differently by the Nazis and began to write a separate chronicle about the fate of the Jews. He noted everything that was told to him about life in the ghetto by the Jews working in the museum, and what he and others witnessed. He mentions measures against the Jews and Jewish reactions to them, including the founding of the ghetto, its liquidation in December 1941, and terrorization from the beginning of 1942. Traces the figure of Heinz Ehaus, Kreishauptmann in Rzeszów up to June 1942, and describes his role in the violence against the Jews. After the war Kotula revised the text and added various comments. *HV*

0563

קרקובסקי, שמואל: כפר נידח באירופה: חלמנו (קולמהוף) – מחנה ההשמדה הנאצי הראשון. ירושלים: יד ושם, 2001. 208 עמ'.

[Krakowski, Shmuel: *A Small Village in Europe: Chełmno (Kulmhof) – the First Mass Extermination Camp*. Jerusalem: Yad Vashem, 2001. 208 pp.]

The first Nazi extermination camp was established in the village of Chełmno. It began to operate on 8 December 1941; Jews were gassed in trucks and buried in mass graves in the nearby forest. The Nazis intended to use the site to murder the ca. 420,000 Jews who lived in the Warthegau region. Based on testimonies of Jews and of Nazi war criminals, describes the extermination process at Chełmno, deportations from Łódź, and the liquidation of other ghettos in the area (listed on pp. 75-88). In fall 1942, after most of the Jews in the area had been killed, the Nazis began to remove the traces of the mass murders. In March 1943 the camp was closed, but it was reopened in April 1944, with some changes: the gassing was done in the forest and the bodies were burned in two crematoria. Based on the Łódź ghetto chronicles and other sources, describes deportations to Chełmno in June-July 1944 and the liquidation of the ghetto in August, as well as Rumkowski's cooperation with the Nazis. The few Jews who remained in Chełmno were killed in January 1945. *RE*

0564 Kunert, Andrzej Krzysztof, ed.: *Polacy – Żydzi, 1939-1945: Wybór źródeł = Polen – Juden, 1939-1945: Quellenauswahl = Poles – Jews, 1939-1945: Selection of Documents*. Introd.: Władysław Bartoszewski. Warszawa: Rada Ochrony Pamięci Walk i Męczeństwa; Instytut Dziedzictwa Narodowego; Oficyna Wydawnicza Rytm, 2001. 528 pp. In English, German, and Polish.

A collection of documents, declarations, speeches, protests, decrees, resolutions of different political bodies, notes to the Allies, articles from the underground press, etc. concerning the situation of the Jews in Poland during the Nazi occupation. The book is divided into five sections: documents of the authorities of the Polish government-in-exile and Poles abroad; documents produced in Poland by the Polish underground; documents of the Council for Aid to Jews – "Żegota"; documents and articles written by Jews in Poland and abroad; German documents concerning the Jews in Poland. *HV*

0565 Laskowska-Gielo, Justyna: Pogrom w Mińsku Mazowieckim (czerwiec 1936 r.) [The Pogrom in Mińsk Mazowiecki (June 1936)]. *Biuletyn Żydowskiego Instytutu Historycznego* 196 (Dec 2000) 485-495.

Describes the pogrom in Mińsk Mazowiecki (a small town near Warsaw) mainly on the basis of information in the Warsaw press. There is very little archival material on the pogrom. The immediate cause of the disturbances was the killing of a Polish army sergeant by a mentally disturbed Jew. The pogrom lasted from 1-4 June, when groups of hooligans beat up Jews and vandalized shops and apartments. About 40 Jews were injured. Many fled from Mińsk to Warsaw. After a few days the situation calmed down. The police arrested dozens of participants in the pogrom. The victims received assistance from the Jewish community in Warsaw. *HV*

0566 Lewandowski, Józef: A Fish Breaks through the Net: Sven Norrman and the Holocaust. *Polin* 14 (2001) 295-305.

The first report on the Nazi mass murder of Jews in Poland to reach the West was the second letter of the Bund to the Polish government-in-exile, which arrived in London in May 1942. The letter, which related the murder of 700,000 Jews, mainly in Eastern Poland, prompted General Sikorski to condemn the Nazi genocide in a BBC broadcast. Neither the Bund report nor Sikorski's declaration succeeded in breaking the wall of indifference to the Holocaust. The courier of the Polish underground who delivered the Bund report, as well as numerous other dispatches, was a Swede named Sven Norrman (1892-1979), a representative of the Swedish electronic corporation ASEA in Poland. Among other aspects of Nazi-occupied Poland, Norrman was shocked by the Nazi treatment of Jews. He was first contacted by the Jewish underground in 1942; he also managed to enter the Warsaw ghetto and take some photographs there. *DR*

0567 Loew, Andrea: "Wozu noch Welt?" Tagebuecher aus dem Ghetto Lodz. *Sachor* 11 (2001) 111-128.

Criticizes the great majority of historians for dealing only with the perpetrators of the Holocaust and not with the subjective experience of their victims. This perspective can be gained from reading victims' diaries. Diaries from the Łódź ghetto describe the effects of unbearable hunger, which could break up families because of the competition for food; the apathy in the face of the death of others, but also the pain over the deportation of loved ones; and the despair that hastened death from hunger and disease. *RW*

0568 Nirensztain, Albert: Ruch oporu Żydów w Zagłębiu Dąbrowskim pod okupacją hitlerowską [Jewish Resistance in Zagłębie Dąbrowskie under Nazi Occupation]. *Kwartalnik Historii Żydów* 200 (2001) 587-606.

This article was written in the 1950s for the "Biuletyn ŻIH" but was never published. Describes the founding of the resistance movement Żydowska Organizacja Bojowa (ŻOB) by Jewish youth in the ghettos of the Zagłębie region, spurred by information about the extermination camps and a visit of Mordechaj Anielewicz to Zagłębie in June 1942. Discusses the youth movements and their members, noting that the strongest organizations were in Będzin and Sosnowiec. Describes various actions undertaken against the Nazis and against the Judenrat. During the liquidation of the Będzin ghetto there were four bunkers in which fighters tried to resist the Nazis. Concludes that the Jewish underground in Zagłębie did not have a clear plan to fight the Nazis and that they could save only some individuals. This article is a fragment of the text written in the 1950s and is strongly marked by the Marxist ideology of that period. The Jewish youth movements' struggle was perceived as a struggle of poor Jews with rich Jews (the Judenrat) who collaborated with the Nazis. *HV*

0569 Nowak, Felicja: *Mein Stern: Erinnerungen einer Holocaust-Ueberlebenden*. Trans.: Anna Kaiser. Gerlingen: Bleicher, 2001. 349 pp. Originally published as "Moja gwiazda" (Białystok: Versus, 1991). Translated from the English: "My Star" (Toronto: Polish Canadian Pub. Fund, 1996).

0570 Perz, Bertrand: Das Ghetto in Lodz: Die Grenzen des Ghettos als Grenzen der Erklaerung. *Wiener Jahrbuch fuer juedische Geschichte, Kultur & Museumswesen* 5 (2000-2001) 107-114.

Attempts to explain why mortality among the "Western" (German, Austrian, and Czech) Jews deported to the Łódź ghetto in fall 1941 was almost twice as high as among Polish Jews; and why, in May 1942, twice as many of them were transported to Chełmno than were Polish Jews during the whole existence of the ghetto up to then: by the end of May only 6,300 of the original 20,000 were left. Among the causes are the wholly inadequate physical conditions of the housing allotted to them, and their significantly higher age, making them unfit for the hard labor demanded in the ghetto. But Oskar Singer, a Prague Jew employed in the ghetto archives, author of a 15-chapter treatise "On the Problem of East and West," stated that the main reason for the failure of the "Western" Jews to integrate in ghetto society was mutual stereotypes and prejudices. Argues that Singer confines his view to the boundaries of the ghetto, ignoring the German officials who had created it and who were the real arbiters of life and death. *RW*

0571 Piotrowski, Jacek: The Policies of the Sanacja on the Jewish Minority in Silesia, 1926-1939. *Polin* 14 (2001) 150-155.

While the 1926 coup d'état of General Pilsudski in Poland removed the antisemitic National Democrats from power, the attitude toward Jews of the Silesian branch of his Sanacja party was complex. The latter was concerned with the Polishness of the Silesian border area and viewed it as threatened by the German minority there. The Jews were, with some justification, seen as identifying with the Germans; thus, they were viewed with suspicion. However, antisemitism emanating from Germany in the 1930s led Sanacja to condemn it, as well as the antisemitism among Germans and some Poles in Silesia. The Zionist idea of emigration was popular among Poles seeking a non-violent solution to the "Jewish problem." The place of the Jews between Poles and Germans in Silesia remained problematic until it was overshadowed by the threat of war. *YC*

0572 Rosenthal, David: The Ghetto Press. *Jewish Frontier* 66, 2 (Mar-Apr 1999) 22-25.

Despite Nazi suppression of the press in Poland, including the Jewish press, the latter continued to exist under the German occupation. The first Jewish newspaper that appeared in occupied Poland was "Gazeta Żydowska." Although it was under tight Nazi control, it managed to uplift the spirits of Polish Jews. 56 Jewish illegal periodicals – in Yiddish, Polish, and Hebrew – appeared from October 1940; their production was possible due to help from the Polish underground. The underground Jewish press began to report on the mass murder and resistance of Jews in March 1942. Based on Rosenthal's personal recollections, describes the Nazi action in the Warsaw ghetto in April 1942 intended to silence the underground press, when many Jewish printers were murdered. *DR*

0573 Rossino, Alexander B.: Nazi Anti-Jewish Policy during the Polish Campaign: The Case of the Einsatzgruppe von Woyrsch. *German Studies Review* 24, 1 (Feb 2001) 35-53.

Describes the formation, composition, and activities of the Einsatzgruppe zur besonderen Verwendung (Special Purpose Operational Group) headed by Udo von Woyrsch, in September 1939 in East Upper Silesia and Western Galicia. The group was deployed in the framework of "Operation Tannenberg," aimed at eliminating the Polish ruling and educated classes and the former Polish Silesian insurgents of 1919-21. In addition, the Einsatzgruppe of von Woyrsch, with the utmost brutality, killed hundreds of Jews in the area between Katowice and the San River. Examination of the orders received by von Woyrsch, as well as events of that time, shows that the Einsatzgruppe's activities were part of the Nazi policy of forced expulsion of Jews in 1938-39. The group terrorized the Jews in order to make them leave the areas of German interest. The terror was part of a campaign to ethnically reshape southern Poland for subsequent German colonization. *DR*

0574 Rumer-Zarayev, Mikhail: *Soshestvie v ad: Varshavskoye getto: Paralleli istorii* [Descent into Hell: The Warsaw Ghetto: Historical Parallels]. Hannover: Alexpress, 1999. 226 pp.

A history of the Warsaw ghetto and the uprising, interspersed with Rumer-Zarayev's recollections of his life in Moscow – during Stalin's regime and in the post-Soviet period. Describes Jewish life in Nazi-occupied Warsaw before the establishment of the ghetto, the ghetto, first attempts to form an underground organization, the deportations, and the uprising. Dwells on the personalities of leaders in the ghetto: from Czerniaków and Gancwajch to Yitzhak Zuckerman and Zivia Lubetkin. Compares the USSR, where memory of the Holocaust was suppressed, to Poland of the 1960s which preserved it. *DR*

0575 Sakowska, Ruta, ed.: *Archiwum Ringelbluma: Konspiracyjne Archiwum Getta Warszawy. Vol. 2: Dzieci – tajne nauczanie w getcie warszawskim* [Ringelblum Archives: Clandestine Archives of the Warsaw Ghetto. Vol. 2: Children – Clandestine Teaching in the Warsaw Ghetto]. Warszawa: Żydowski Instytut Historyczny, 2000. xvi, 413 pp. In Polish and Yiddish. Vol. 1 appeared in 1997.

A collection of documents from the Ringelblum Archives. The volume contains, among others, compositions (solicited by "Oneg Shabbat") written by children about the fate of their families during the war; testimonies of teachers about the children; documents illustrating activities of the clandestine schools, such as theatrical performances, concerts, or meetings celebrating a "Month of the Child" (Sept.-Oct. 1941); fragments of textbooks for teaching literature; and a text written by Nusen Koniński, "Oblicze dziecka żydowskiego" ("Visage of a Jewish Child") in November 1941 about the situation of children in the ghetto and their education (pp. 358-377). Contains some documents related to Janusz Korczak. Includes facsimiles of documents. *HV*

0576 Schwarberg, Guenther: *In the Ghetto of Warsaw: Heinrich Joest's Photographs*. Trans.: George Frederick Takis. Goettingen: Steidl, 2001. 137 pp. Originally published as "Das Getto" (Goettingen: Steidl, 1989).

This edition contains the photographs of the Warsaw ghetto made by Heinrich Joest – an officer in the German army in September 1941 –

together with descriptions of life in the ghetto. It does not contain the introductory essay which appears in the original German edition. *SSC*

0577 Stachura, Peter D.: *Poland in the Twentieth Century*. Basingstoke, Hants: Macmillan, 1999. xv, 180 pp.

A collection of eight essays, two of which deal with the Jews. Ch. 3 (pp. 41-57), "The Polish-Jewish Symbiosis in the Second Republic, 1918-39," argues that the extent of Polish antisemitism in this period is often exaggerated. While admitting its increase, especially from 1935, states that Jews enjoyed wide freedoms and that evidence of a high living standard among Jews belies allegations of discrimination. Notes that antisemitism had its counterpart in "polonophobia." Ch. 6 (pp. 97-111), "Poles and Jews in the Second World War," argues against allegations regarding the wartime conduct of Poles toward Jews – e.g. that Polish antisemitism paved the way for the Holocaust, that the Poles collaborated with the Nazis, and that they were apathetic to the plight of the Jews. States that the Holocaust was the result of German racist antisemitism, and that there were relatively few instances of Polish collaboration. Holds that antisemitism, as well as polonophobia, peaked after the war due to several factors; among them, attributes antisemitism to the enthusiasm with which Jews greeted the Soviet occupation in 1939-41, and to the postwar imposition of the communist system, in which a small Jewish elite was seen as wielding considerable power. *YC*

0578 Steinbacher, Sybille: Auschwitz: Die "Musterstadt" und das Vernichtungslager. *Dachauer Hefte* 17 (Nov 2001) 42-55.

Traces the history of Auschwitz – the town and the concentration camp – during the annexation of this part of Poland to Nazi Germany. The town was half Jewish, half Polish, and Germanization did not seem feasible; thus, at first, Jews were not deported but segregated and exploited. The situation changed when IG-Farben chose Auschwitz as the site for its Buna factory, and planned to turn it into a model German town. To this end, it had to be free of Jews; they were transferred to other towns in the region. Meanwhile the camp changed from a prison camp for political opponents and prisoners of war to an extermination camp for Jews. Describes the good life led by the SS-men and their families in the camp as well as by the new German settlers, mainly IG-Farben employees, in the town. Though many knew about the mass murder in the camp, most lived comfortably with their knowledge or suspicions: Germanization of the East required extermination of inferior races. *RW*

0579 Tomaszewski, Jerzy: Jews as the Citizens of the Polish Republic: Some Legal Aspects. *Justice* 28 (Sum 2001) 14-18.

From the beginning of its independence, Poland declared legal equality for all its citizens. However, having granted citizenship to every ethnic Pole, the Republic was reluctant to recognize the citizenship of many non-Poles, among them 600,000 Jews (in 1921). The majority of their cases were settled only by 1928. Another problem was lack of uniformity of Polish law; Russian or Austrian laws still in use had clauses discriminating against Jews, which were annulled only by 1931. The nationalists, especially the National Democrats, meanwhile tried to shift legislation in an anti-minorities direction. After Piłsudski's death in 1935, nationalist influence grew and laws were enacted which impaired the rights of Jews. *DR*

0580 Turski, Marian: The Warsaw Ghetto Uprising: Why So Late? *Justice* 28 (Sum 2001) 10-13.

Presents several widely known reasons why the uprising did not occur before April 1943, including lack of information on the mass killings and inability of the ghetto Jews to believe it, Nazi terror which paralyzed the will to act, fear that such an act would precipitate the annihilation of the Jews, and the lack of weapons and military training. Notes two other main issues which are generally not mentioned: the starvation conditions, due to which the majority could not even think of revolting because they had no energy; and the fact that as long as families were together, until the summer of 1942, the young people had to take care of their parents and grandparents; after the Great Deportation, they had nothing more to lose. *DR*

0581 Tych, Feliks; Hensel, Juergen, eds.: *Bund: 100 lat historii, 1897-1997* [Bund: 100 Years of History, 1897-1997]. Warszawa: Żydowski Instytut Historyczny, 2000. 352 pp.

A collection of papers delivered at a conference in Warsaw in November 1997. Partial contents: Blatman, Daniel: Narodowa ideologia Bundu w zderzeniu z antysemityzmem i Holocaustem (1933-1947) [The National Ideology of the Bund in the Face of Antisemitism and the Holocaust (1933-1947] (213-229); Sakowska, Ruta: Grupa Ringelbluma a Bund [Ringelblum's Group and the Bund] (271-282); Dunin-Wąsowicz, Krzysztof: Bund i polski ruch socjalistyczny w latach wojny i okupacji hitlerowskiej, 1939-1945 [The Bund and the Polish Socialist Movement during the War and Nazi Occupation, 1939-1945] (283-287). *HV*

0582 Ungar, William; Chanoff, David: *Destined to Live*. Lanham, MD: University Press of America, 2000. xix, 361 pp.

Memoirs of Wilo (Wolf) Ungar, a Jew born in 1913 in Krasne, near Tarnopol. In September 1939 he was drafted into the Polish Army; notes the anti-Jewish mood of Poles after the Polish defeat. He fled to Lvov, where the Nazi invasion caught up with him in July 1941. Describes Nazi measures against the Jews and life in the ghetto. Ungar's wife and newly-born child perished during the Nazi "action" in June 1942. He lived on "Aryan papers" in the vicinity of Lvov and later fled to Warsaw. From there he moved to Nowy Dwór, northwest of Warsaw, and before March 1943 was interned in the ghetto there. He fled and returned to Lvov. Having been exposed as a Jew, he narrowly escaped being sent to the Janowska camp; he was hidden by a Ukrainian woman until the liberation. He tried to enlist in the Polish army in Lvov in 1944, but was rejected as a Jew. After the war he immigrated to the USA. *DR*

0583 Weinstock, Nathan, ed.: *Chroniques du désastre: Témoignages sur la Shoah dans les ghettos polonais*. Genève: Metropolis, 1999. 468 pp.

A collection of testimonies translated from Yiddish and presented by Nathan Weinstock. Most of the texts appeared previously in the "Revue d'Histoire de la Shoah – Le Monde Juif." Contents: Lichtensztajn, Yisroel: La grande déportation de Varsovie, juillet-septembre 1942 [Appeared in the "Revue d'Histoire de la Shoah" 162 (1998).] (31-55); Perle, Yehoshua: Khurbm [!] Varshe: L'anéantissement de la Varsovie juive [Appeared in the "Revue d'Histoire de la Shoah" 164 (1998), edited together with Micheline Weinstock.] (56-131); Frank, Shlomo: Journal du ghetto de Lodz (extraits) [Appeared in "Le Monde Juif" 154 (1995).] (133-193); Meed, Vladka: Des deux côtés du mur (extraits) [From her memoirs, published in Yiddish in 1948.] (194-303); Weinstock, Nathan: Sur Shmuel Zygielbojm: Témoignages, récits autobiographiques, correspondance [Appeared in "Le Monde Juif" 158 (1996) and 159 (1997).] (305-438). *HV*

0584 Wiehn, Erhard Roy, ed.: *Totengebet: 60 Jahre Beginn des Zweiten Weltkriegs und der Schoáh in Polen*. Konstanz: Hartung-Gorre, 1999. 84 pp.

A collection of excerpts from books published by Hartung-Gorre. Partial contents: Wiehn, Erhard Roy: Anfang der Schoáh in Polen [Published in Wiehn's "Kaddisch – Totengebet in Polen" (1984) and in his "Ghetto Warschau" (1993).] (13-20); Kalter, Joachim: Dann kam der 1. September 1939 [From his book "Eine juedische Odyssee" (1997).] (21-30); Stein, Jehuda L.: Beginn der Jahre der Verfolgung [From his book "Die Steins" (1999).] (31-34); Mlawski, Joseph: Ein Zufall hat das Leben gerettet [An interview with Mlawski, published in Wiehn's "Kaddisch – Totengebet in Polen" (1984) and in his "Ghetto Warschau" (1993).] (35-68); Goldkorn, Josef: Der Tag des Synagogenbrandes [In Będzin. From his book "Im Kampf ums Ueberleben" (1996).] (69-73); Friedmann, Adam: Der Bart des orthodoxen Juden [Published in Wiehn's "Kaddisch – Totengebet in Polen" (1984) and in his "Ghetto Warschau" (1993).] (74-77). *ID*

0585 *Więź* 42, 7 [489] (July 1999).

This issue is entitled "Polacy i Żydzi w upiornej dekadzie" ("Poles and Jews in the Ghastly Decade"). Pp. 4-22 contain discussions on Jan Gross' book "Upiorna dekada" ("The Ghastly Decade"), by Włodzimierz Borodziej, Helena Datner, Andrzej Friszke, Dariusz Stola, Jacek Borkowicz, Agnieszka Magdziak-Miszewska, and Wojciech Wieczorek. The following are articles by students who were asked to comment on Gross's book: Słotwiński, Łukasz: Książka, którą czyta się bez przyjemności [A Book Which Is Read without Pleasure] (37-39); Sarnecka-Drużycka, Joanna: Prawdziwi Żydzi, upiorni Polacy [True Jews, Ghastly Poles] (39-41); Lau, Dominik: Pokonać stereotyp [To Overcome a Stereotype] (41-44); Buras, Piotr: W poszukiwaniu języka pamięci [Looking for Expression of Memory] (44-49); Miszewska, Anna: Jesteśmy ignorantami [We Are Ignorant] (50-52). *HV*

0586 Wilhelmus, Wolfgang, ed.: *Flucht oder Tod: Erinnerungen und Briefe pommerscher Juden ueber die Zeit vor und nach 1945*. Rostock: Ingo Koch Verlag, 2001. 263 pp.

A collection of 38 brief memoirs of Holocaust survivors who were originally from Pomerania; most escaped before 1939, others survived in ghettos and concentration camps. Pp. 243-253 contain letters written by Jews interned in the ghettos of Szczecin, Bełżyce, and Trawniki. *ID*

Romania

0587 Bauer, Markus: Paradies, Fegefeuer, Hoelle: Exil und Holocaust in Rumaenien. *Exilforschung* 19 (2001) 145-167.

Romania attracted relatively few racial or political refugees from Nazi Germany; names some of the more prominent ones. Although antisemitism was pervasive in Romania, and although the Iron Guard was strong, until 1939 the refugees enjoyed intellectual freedom; they could obtain anti-Nazi publications from abroad and also publish their own. Mentions, also, the ships carrying refugees from Central Europe to Romania. During the Soviet occupation of parts of Romania, many Jewish leaders were deported. With the start of the anti-Soviet war in alliance with the Germans, there were pogroms in Iaşi and Bucharest. From the Nazi-occupied provinces, the Jews were deported to Transnistria, where over 200,000 perished. Jews in central Romania were drafted for forced labor. The Jewish community tried to aid the deportees and to pressure the regime. Surveys some of the extensive German-language literature written by survivors of the camps in Transnistria. *RW*

0588 Gut, Ion Teodor: La situation politique des Juifs de Bihor pendant le Diktat de Vienne. *Transylvanian Review* 9, 3 (Fall 2000) 61-66.

Describes the situation of the Jews during World War II in the Bihor region (Romania) which was occupied by Hungary. Surveys consecutive anti-Jewish measures of the Horthy regime from 1938 to 1944. In the most important town in the region, Oradea, a ghetto was established in May 1944 where all the Jews of the Bihor region were concentrated. During July 1944 the Jews were deported to Auschwitz. *HV*

0589 Iancu, Carol, ed.: *La Shoah en Roumanie: Les Juifs sous le régime d'Antonescu (1940-1944): Documents diplomatiques français inédits.* Montpellier: Université Paul Valéry-Montpellier III, 1998. 189 pp.

Ch. 1 (pp. 11-47) examines the fate of Romanian Jewry under the Antonescu regime (1940-44). States that incriminating evidence about Romanian persecution of Jews remained classified until recently due to the negationist campaign orchestrated by the right-wing parties after 1989. Studies the basis for populist nationalism appearing in the local press and notes that despite Ceaucescu's criticism of antisemitism, the reputations of those responsible for it emerged unscathed. Presents two explanations for this: a counter-reaction to the commemoration ceremonies and documentation attesting to Romanian collaboration with the Nazis. The rest of the book comprises 100 documents found in the archives of the French Foreign Ministry which cast new light on antisemitic legislation in Romania and other persecution of the Jews. *TS*

0590 Iancu, Carol, ed.: *Shoah în România: Evreii în timpul regimului Antonescu (1940-1944): Documente diplomatice franceze inedite*. Trans.: Cosmina Ghebaur. Iaşi: Polirom, 2001. 204 pp. Originally published as "La Shoah en Roumanie: Les Juifs sous le régime d'Antonescu (1940-1944): Documents diplomatiques français inédits" (Montpellier: Université Paul Valéry-Montpellier III, 1998).

0591 Popovici, Traian: *Spovedania = Testimony*. Trans.: Viviane Prager. [Bucureşti]: Fundaţia Dr. W. Filderman, [2000?]. 98 pp. In Romanian and English.

Memoirs of a Romanian official, who in 1941-42 served as the mayor of Czernowitz. Describes the ghettoization and deportation of the city's Jews to Transnistria in fall 1941. Considers Antonescu responsible for Romania's anti-Jewish policies, but states that the governor of Bukovina, Calotescu, after taking up his position in September 1941, was even harsher, making persecution of the Jews a matter of top priority. Popovici states that he himself opposed the ghettoization and later the deportation of the Jews from Czernowitz. He requested that Antonescu exempt "useful" Jews from deportation; the intervention succeeded, and ca. 20,000 Jews were exempted. *DR*

0592 Voinea, Andrei; Kovács, János: *Clepsidra cu bolovani: Amintiri despre munca obligatorie a evreilor 1940-1944* [Hour-Glass with Boulders: Memoirs of Jewish Forced Labor in 1940-1944]. Bucureşti: Hasefer, 1999. 131 pp.

Voinea (b. 1922), from the village of Beliu in Transylvania, describes his experiences in a series of labor camps in this province between 1940-44. He relates that he began to write the book together with his comrade János Kovács, who died before it was finished. *RW*

0593 Voinea, Andrei; Kovács, János: *Sanduhr aus Steinen: Juedische Zwangsarbeiter in Rumaenien 1940-1944*. Trans.: Mirjam Bercovici-Korber. Ed.: Erhard Roy Wiehn. Konstanz: Hartung-Gorre, 2000. 140 pp. Originally published as "Clepsidra cu bolovani: Amintiri despre munca obligatorie a evreilor 1940-1944" (Bucureşti: Hasefer, 1999).

The editor's introduction, "Zur Schoáh in Rumaenien" (pp. 15-20), chronicles Romanian persecution of Jews since 1940, including the Transnistrian labor and extermination camps. The essay by William Totok, "Der revisionistische Diskurs" (pp. 91-134), discusses the rehabilitation of Ion Antonescu in post-1990 Romania, and surveys Romanian literature denying mass killings of Jews and depicting Antonescu as their savior. *RW*

Slovakia

0594 Breuer, Alexander: *Vojak č. 151* [The 151st Soldier]. Trans.: Dominik Bartosiewicz. Bratislava: Slovenské Národné Múzeum – Múzeum Židovské Kultúry, 2001. 83 pp. Translated from the English manuscript. Originally published as "He-hayal ha-151" (Kibbutz Dalia: Ma`arekhet, 1999).

0595

מנהיגות בעת מצוקה: קבוצת העבודה בסלובקיה 1942-1944. קיבוץ דליה: מערכת, 2001. 166 עמ'.

[*Leadership in a Time of Distress: The Working Group in Slovakia 1942-1944*. Kibbutz Dalia: Ma`arekhet, 2001. 166 pp.]

A collection of essays on the rescue activities of the underground Jewish organization Pracovná Skupina (Working Group) in Slovakia during the Holocaust. *RE*

0596 Nižňanský, Eduard: Akcja "Nisko": Obóz dla Żydów w Vyhniach (w Słowacji) [Operation "Nisko" and the Camp for Jews in Vyhne, Slovakia]. *Kwartalnik Historii Żydów* 200 (2001) 566-574.

In the "Nisko" action, transports were directed to Nisko from Vienna, Ostrawa, and Katowice at the end of 1939. Discusses one of the transports from Ostrawa (317 men) which did not reach Nisko; the Jews were first interned in Sosnowiec, and then 275 of the men were transferred to Vyhne in Slovakia in February 1940. These Jews were mainly Polish citizens who lived in Czechoslovakia before World War II. Describes the conditions of life in the camp, which were not very harsh; e.g. the Jews could work outside the camp, they could leave the camp for family visits, a school and kindergarten were organized there, etc. In 1942, Vyhne became a labor camp. The prisoners remained there and were not deported to the death camps. This camp existed up to August 1944, when the Slovak uprising broke out. *HV*

0597 Rajský, Milan: *Nebol som hrdina* [I Was No Hero]. Bratislava: Kontakt Plus, 1999. 152 pp.

A partially-fictionalized memoir by a Slovak Jew who was a teenager when the war began. Discusses the anti-Jewish policy of the Slovakia government and the antisemitic attitudes of the populace. Describes his failed attempt to flee to Hungary at the beginning of 1942, followed by imprisonment in the Nováky camp. He was freed several months later, after his family received a presidential exemption due to the need for agricultural workers. After his parents' deportation in August 1942, he took refuge with a non-Jewish neighbor, posing as his son. He worked as a minor economic official in several places, being forced to flee again and hide in March 1945 to avoid enlistment in the Slovak army. *JT*

0598 Slneková, Veronika: Dislokácia židovských obyvatelov v rámci mesta, v rámci slovenska a arizácia židovského domového majetku v Trnave v rokoch 1939-1945 [Dislocation of the Jewish Inhabitants within the City and within Slovakia, and the Aryanization of Jewish Real Estate in Trnava in 1939-1945]. *Acta Judaica Slovaca* 7 (2001) 99-116.

A study of the residential movements of Jews in Trnava during the war period, before the deportations. Describes the progressive limitation of rights of the Jewish homeowners in Trnava and the moving of Jewish inhabitants from the city center to suburbs by the local authorities. There were 2,467 Jews living in the city in December 1938, owning 265 houses and inhabiting more than 800 flats, according to a register completed by the local authorities on 20 March 1941. From March to November 1941, "temporary supervisors" were assigned to 154 houses. Moreover, on 1 November 1941, all the real estate owned by Jews, with few exceptions, was confiscated by the Slovak state. On 16 May 1941, the Jews had been banned from renting flats in certain streets in the city center and, as a result, 1,309 Jews were forced to move to outlying neighborhoods. In the fall of 1941, Trnava became part of a plan to dislocate Jews from Bratislava, but it was cancelled in September 1942. *JT*

Spain

0599 Boecker, Manfred: *Antisemitismus ohne Juden: Die Zweite Republik, die antirepublikanische Rechte und die Juden: Spanien 1931 bis 1936.* Frankfurt a.M.: Peter Lang, 2000. 392 pp. Based on the author's diss. – Universitaet Muenster, 1998.

After tracing the tradition of anti-Judaism in Spain, describes antisemitism in some of the anti-liberal, anti-republican factions and their press organs in the Second Republic. This antisemitism was directed at the "Jewish-Masonic-Bolshevist world conspiracy," which was blamed for the fall of the monarchy and identified with the Republic and its reformist and anticlerical policies. The conspiracy allegedly targeted Spain, seen as the champion of Christendom against the satanic powers; it threatened Spanish national identity, conceived as "cosubstantial" with Catholic orthodoxy. The preservation of this identity justified both the expulsion of 1492 and opposition to Jewish immigration in the 1930s. "The Protocols of the Elders of Zion" were widely disseminated; other influences came from French Catholic antisemitism rather than from Nazi Germany, whose racism was contrary to Catholic doctrine. During the revolution and under the Franco regime until 1945, antisemitism and conspiracy theory became part of official doctrine. *RW*

0600 Rodríguez Jiménez, José Luis: "Los Protocolos de los Sabios de Sión" en España. *Raíces* 38 (Spr 1999) 27-40.

Antisemitic propaganda became prevalent in Spain only after 1931, with the fall of the monarchy and the establishment of the Second Republic. Catholic integrationists and the extreme right appealed to the "theory of the international conspiracy of Masonry and Judaism" to explain the crisis in Spanish society and to set in motion active opposition to the Republic. The divulgation of "The Protocols of the Elders of Zion" played a key role in the campaign. The first edition appeared in 1927, followed by several other editions during the 1930s; the most widespread was that of 1932, published in the fascist weekly "Libertad." The editions were usually accompanied by prologues which connected the "Protocols" to "Spanish reality," explaining that the institutional crisis, the devaluation of Spanish currency, and the spread of liberalism were all the result of a Jewish plot to conquer the world, beginning with the Russian Revolution, continued in France, and now introduced into Spain, using Masonry as its political instrument. *MZ*

0601 Rother, Bernd: *Spanien und der Holocaust*. Tuebingen: Max Niemeyer, 2001. vi, 359 pp.

Although one cannot say that the Spanish government under Franco supported the murderous intentions of Nazi Germany, it also cannot be said that Spain granted asylum to the persecuted Jews. Antisemitic attitudes, rooted in traditional Catholicism, were prevalent in Spain. So long as Jews with Spanish citizenship who lived under Nazi occupation were only persecuted, but were not in danger of losing their lives, the Spanish government did not intervene. Only when Germany ordered their repatriation in 1943 and threatened that otherwise they would deport them, was Spain forced to act. Then, Jews were allowed to enter Spain only in transit to another country; they were helped by Jewish aid organizations. Draconic rules for inspection of citizenship were applied in order to assuage anti-Jewish elements in the Spanish government. As it was difficult to find countries willing to accept Jews, many remained without protection, especially in France. Altogether, ca. 20,000-35,000 Jews crossed the Spanish border legally or illegally. In France, Spain recognized 500 as Spanish citizens; 2,000 remained without refuge. In Greece, out of 700 Spanish Jews, 367 were repatriated. Altogether, Spain's rescue actions were limited, contrary to its claims after 1945. *ID*

Sweden

0602 Einhorn, Lena: *Handelsresande i liv: Om vilja och vankelmod i krigets skugga* [Commercial Traveler Dealing with Lives: On Will and Fickleness in the Shadow of War]. Stockholm: Prisma, 1999. 521, [16] pp.

Relates the activities of Gilel (Hillel) Storch, a wealthy businessman from Latvia. When the Nazis occupied Latvia, Storch was in Sweden for business purposes, and remained there. He managed to get his wife and daughter out of Latvia. Gradually, Storch became more and more active in rescue activities for Jews from all over Europe. He had contacts with officials and representatives of the Foreign Office, as well as with Raoul Wallenberg and Folke Bernadotte. Eventually, he even met Himmler. Gives detailed information about all the persons with whom Storch had contact, including important Swedish officials such as Gösta Engzell, head of the Legal Division of the Foreign Office. Includes numerous letters and documents. *RSL*

0603 Jarlert, Anders: Die Anwendung der Nuernberger Gesetze in der Schwedischen Kirche 1935-1945. *Kirchliche Zeitgeschichte* 14, 1 (2001) 159-174.

Based on a lecture held at the Unversity of Heidelberg, July 2000. States that according to a European convention of 1902, marriages of foreigners in the signatory states were to be regulated by the laws of the state of which the partners were citizens. During the Nazi period, Sweden enforced this convention scrupulously: the Foreign Office issued a directive that no "non-Aryan" was to marry an "Aryan" if one or both were refugees from Germany. Since in Sweden marriage licenses were the province of the Church, it was the local ministers who had to carry out this regulation. Some objected and sought subterfuges in order to conduct such marriages despite the prohibition, but most complied. *RW*

Switzerland

0604 Bloch, Rolf: Kann man aus der Geschichte lernen? *Internationale Kirchliche Zeitschrift* 91, 3 (July-Sept 2001) 209-221.

After discussing the question theoretically, takes as an example the Swiss refugee policy during the Holocaust. The Bergier Commission, established by the Swiss government in 1996, concluded that although the closing of the borders was not in violation of either Swiss or international law, the authorities, who were well aware that the Jewish refugees were in danger of losing their lives, should have used judgment and given them asylum. The main reason they did not was latent antisemitism, a fear of "Verjudung," which determined policy since 1919. Some Swiss who remember this time deny that they knew of any antisemitism; but this is contradicted by official documents as well as Jews' personal recollections. Concludes that it is indeed possible to learn from history: to suppress antisemitism and xenophobia, individually and collectively. The international community has applied this lesson by expanding international law on human rights and setting up courts to enforce it. *RW*

0605 Fischer, Elmar: Abwehr im Innern: Zur schweizerischen Fluechtlingspolitik im Zweiten Weltkrieg. *Zeitschrift fuer Geschichtswissenschaft* 48, 3 (2000) 214-238.

Swiss immigration policy toward refugees from Nazi persecution was driven by fear of flooding by foreigners, particularly Jews. Until 1938, refugees were allowed to enter the country for three months. To ensure that they would then move on, they were not permitted to work, to lease apartments or acquire real estate; neither could they obtain residence rights by the transfer of businesses from Germany. From September 1938 the border was closed to refugees who had a J stamped in their passports. Refugees found in the country illegally were interned under harsh conditions. The Swiss tried to find other countries willing to take them and also provided vocational training. After the occupation of France, the authorities, aware of the fate awaiting the refugees, did not have the heart to turn them back at the border. But when the stream of refugees increased after June 1942, the police chief, Heinrich Rothmund, ordered the border closed hermetically. Only public and parliamentary protest made him modify this order for special cases. Refugees who slipped through continued to be interned in camps. Immediately after the war they were again pressured to leave. *RW*

0606 Teitelbaum, Raul; Sanbar, Moshe: *Holocaust Gold: From the Victims to Switzerland: Trails of the Nazi Plunder*. Trans.: Amy Teibel. Tel Aviv: Moreshet, 2001. 261 pp. Originally published as "Zehav ha-Shoah: Netivot ha-bizah ha-natsit min ha-korbanot li-Schweiz" (Tel Aviv: Moreshet, 2000).

0607 Wagner, Meir: *The Righteous of Switzerland: Heroes of the Holocaust*. Eds.: Andreas C. Fischer, Graham Buik. Hoboken, NJ: Ktav, 2001. xli, 269 pp. An English version of his book "Die Gerechten der Schweiz: Eine Dokumentation der Menschlichkeit" (New York: Eli Meir, 1999).

The English edition includes statements and speeches by various persons (pp. xvii-xxxviii), the stories of two more Swiss citizens nominated by Yad Vashem for recognition as Righteous of the Nations, and other lesser-known accounts of Swiss citizens who helped save Jews during the war. Some of them are published here for the first time. *SSC*

Ukraine

0608 Boll, Bernd: "Aktionen nach Kriegsbrauch": Wehrmacht und I. SS-Infanteriebrigade 1941. *Zeitschrift fuer Geschichtswissenschaft* 48, 9 (2000) 775-788.

Describes the directives and the chains of command under which the SS in the Southern Army Territory (Ukraine), under SS- and Police Chief Friedrich Jeckeln, placed its 1st Infantry Brigade at the disposal of the army, and chronicles the brigade's activities. It was understood that the brigade was not to serve at the front but to secure the rear by "cleansing" it of allegedly dangerous elements. In the summer and fall of 1941 the brigade reported to the army command on the massacre of a total of over 12,000 civilians, mostly Jews. *RW*

0609 Finkelshtein, Yurii Ye.: *Simon Petlyura*. Rostov-na-Donu: Feniks, 2000. 349-510 pp. Bound with A.I. Kozlov, "General Denikin."

Pp. 349-510 contain a biography of Petlyura (1879-1926), the Supreme Ataman of the Army of the People's Republic of Ukraine (UNR) in 1918-20. Deals mainly with the question of his antisemitism and his responsibility for the pogroms perpetrated by his army in 1919. The other armies that fought in the Civil War in Ukraine were also guilty of pogroms, but Petlyura's army, with more than 1,000 pogroms, surpassed them all. Deplores attempts to whitewash Petlyura in recent historiography. Points out that Petlyura was not antisemitic in his youth, but the atrocities of his army against the Jews and the necessity to maintain good terms with his "boys" made him internalize antisemitism. Never in 1919-20 did Petlyura punish pogromists; his Order #131, condemning pogroms, was intended for Western consumption, not for his men. Describes the pogrom in Proskurov (now Khmelnytskyi) in February 1919. Contrasts Major Klodnytskyi, who fearlessly prevented a pogrom in Khmelnik, with Petlyura. Concludes that the defeats suffered by the armies that fought the Reds in 1918-20 led them to seek compensation in pogroms. The pogroms signified the moral disintegration of the armies and were an additional factor in their final defeat. *DR*

0610 Heer, Hannes: Einuebung in den Holocaust: Lemberg Juni/Juli 1941. *Zeitschrift fuer Geschichtswissenschaft* 49, 5 (2001) 409-427.

Attempts to trace the events in Lvov (Lemberg) in the first week of the German occupation. The NKVD, before the Soviet retreat, had murdered some 3,500 political prisoners in the city's three prisons. The Nazis used this atrocity, allegedly perpetrated by Jewish Bolshevists, on the one hand for propaganda purposes and on the other to set off a pogrom against the Jews, who were driven into the prisons supposedly to lay out the corpses, but actually to be exposed to the attacks of soldiers and civilians. 4,000 Jews were killed and many more injured. The pogrom was apparently led by Ukrainian units attached to the German army and under German command, but probably German units also took part; tries to identify these. German soldiers stationed in Lemberg or passing through were led through the prisons to see the atrocities, in order to increase their motivation for the war against the Jewish Bolshevists. Four days later, Einsatzgruppen began with more systematic massacres of Jews. *RW*

0611 Hesse, Klaus: NKWD-Massaker, Wehrmachtsverbrechen oder Pogrommorde? Noch einmal: Die Fotos der "Tarnopol-Stellwand" aus der

"Wehrmachtsausstellung." *Geschichte in Wissenschaft und Unterricht* 51, 12 (Dec 2000) 712-726.

The Wehrmacht exhibition of the Hamburg Institut fuer Sozialforschung was severely criticized for showing photos of victims of massacres as though these were the work of the Wehrmacht, when, according to the critics, they had in fact been perpetrated by the NKVD just before the Soviet withdrawal. One of the sites of disputed photos is Tarnopol. Analyzes these photos, and related ones not exhibited, and points to internal evidence that most of them, taken a few days after the arrival of the Wehrmacht in Tarnopol, show victims not long dead, among them soldiers of the Red Army; they must have died in the pogrom carried out by Ukrainians against Jews and communists in the first days of the German occupation. The many Wehrmacht soldiers seen looking on do not prove that any of them participated in the pogroms, but do make it seem likely. *RW*

0612 Hoffman, Henryk: *Z Drohobycza do Ziemi Obiecanej* [From Drohobycz to the Promised Land]. Lublin: Norbertinum, 1999. 135 pp.

Memoirs about the author and his family in interwar and wartime Drohobycz. Pp. 65-86 deal with the Holocaust period and life in the Drohobycz ghetto. Hoffman's father, as a physician, worked also outside of the ghetto. After the liquidation of the ghetto in 1943 the family lived in a labor camp. During winter 1944 they left the camp and survived in the forests until the end of the war. After the war they settled in Poland. Mentions, also, postwar antisemitism in Poland. After the antisemitic campaign of 1968, Hoffman immigrated to Israel with his wife and son. *HV*

0613 Lerner, Sima: *Skitaniya po dorogam zhizni: Vospominaniya* [Wanderings along the Roads of Life: Memoirs]. Tel-Aviv: Krugozor, 2000. 128 pp.

Most of the book was published previously as "Moi rok: Vospominaniya" (Tel Aviv: Krugozor, 1998). Some chapters on Lerner's postwar life are added here. Pp. 100-106 deal with Lerner's postwar encounter with antisemitism in Kharkov, Ukraine, where her daughter lived in the early 1980s. *DR*

0614 Lichtenstein, Heiner: 33.000 Morde an zwei Tagen: Zum 60. Jahrestag des Massakers von Babij Jar. *Tribuene* 159 (2001) 128-132.

Chronicles briefly the massacre of the Jews of Kiev at Babii Yar on 29-30 September 1941 by Sonderkommando 4a under Paul Blobel, quoting testimony by two participants in the massacre. Discusses, also, the subsequent murder on the same site of many more Jews as well as victims of other nationalities; and the exhumation and burning of the ca. 100,000 corpses in August-September 1943, again under the command of Blobel. Notes that the Soviet authorities prevented any mention of Jews as the main victims; a monument to them was erected only in 1991. *RW*

0615 Mallmann, Klaus-Michael: Der qualitative Sprung im Vernichtungsprozess: Das Massaker von Kamenez-Podolsk Ende August 1941. *Jahrbuch fuer Antisemitismusforschung* 10 (2001) 239-264.

In Kamenets-Podolskii (Western Ukraine), 23,600 Jews, most of them deported by Hungary, others residents of the town, were massacred. Argues that, contrary to statements of other historians assigning responsibility to the RSHA through its Einsatzgruppen, the massacre was carried out on the orders of the Police and SS Chief of the southern section of the rear, Friedrich Jeckeln, in full agreement with the Wehrmacht commander, Karl von Roques. Von Roques, like all the commanders in the area, shared the antisemitism of the national-conservative German elite. The army was worried by the problem of feeding the Hungarian deportees; Jews were expendable, so practical and ideological factors came together. The massacre was carried out by Jeckeln's staff, which was already experienced in the killing of Jews, and by Police Batallion 320, for which this was the first time. The police commander announced that no-one would be forced to participate, but only one man refused. This batallion went on to carry out massacres elsewhere. *RW*

0616 Nidergofer, Boris: *Doroga smerti* [The Road of Death]. Tel-Aviv: "Krugozor," 2001. 160 pp.

Memoirs of a Jew who was born in 1934 in Sadgora, northern Bukovina. Under Romanian rule in 1941, he, his mother, and his grandfather were deported to Transnistria. After a long trek, during which Nidergofer's grandfather died, he and his mother were sent to the Zatishye camp, near Zhmerinka. Describes his tactics of survival in the camp. Relates negatively to the Jewish Council, which was appointed by the Romanians. Nidergofer and his mother were liberated by the Soviets in spring 1944; in 1979 they immigrated to Israel. *DR*

0617 Rosen, Abraham; Sarig, H.; Bernstein, Y., eds.: *Kaminits-Podolsk & Its Environs: A Memorial Book of the Jewish Communities in the Cities of Kaminits-Podolsk, Balin, Dunivits, Zamekhov, Zhvanets, Minkovitz, Smotrich, Frampol, Kupin and Kitaygorod Annihilated by the Nazis in 1941*. Trans.: Bonnie Schooler Sohn. Bergenfield, NJ: Avotaynu Foundation, 1999. xiii, 234 pp. Originally published as "Kaminits-Podolsk u-sevivatah" (Tel-Aviv: Organization of Jews from Kaminits-Podolskii and Surroundings in Israel, 1965).

Relates the history of the Jews in southern Podolia (southwest Ukraine) in the early 20th century and in the Holocaust. Contains brief historical essays on towns in the area, as well as personal narratives and eyewitness accounts. Describes the pogroms of 1905 and of the Civil War period (1918-20). Jewish self-defense proved to be effective in Balyn in 1905, in Orynin in 1919, and in Zhvanets in 1918-19; it was ineffective in Kupyn in 1918. Dwells on Petlyura's treatment of the Jews in Kytai-Horod and other places. Pp. 65-69, by Moshe Deutsch, describe the ghetto of Kamenets-Podolskii, where ca. 20,000 local, Bessarabian, and Hungarian Jews were killed in August 1941; with this action, the ghetto was liquidated. *DR*

0618 Vanukevich, Anatolii: *Kholokost na Poltavshchine (1941-1943 gg.)* [The Holocaust in the Poltava Region (1941-1943)]. Poltava: [The Author], 1999. 19 pp.

Briefly describes the Nazi genocide of Jews in Poltava and in the towns of the Poltava region: Piryatin, Kremenchug, Lubny, Zolotonosha, Khorol, Lokhvitsa, and some others. Estimates that 22,340 Jews were killed by the Nazis. Presents a list of Nazi criminals – perpetrators of the genocide in the region. Pp. 17-19 contain a note by Lev Vaingort describing the monument to the victims erected in Poltava in 1970. *DR*

0619 Volovik, Larisa, ed.: *Materialy naukovoyi konferentsiyi, Kharkiv 16 hrudnya 1998* [Materials of the Scholars' Conference, Kharkov 16 December 1998]. Kharkiv: Kharkivskyi Prosvitnytskyi Tsentr "Holokost," 1999. 52 pp.

Partial contents: Leonidov, Leonid: Tri perioda unichtozheniya yevreyev v Ukraine [Three Periods of the Destruction of Jews in Ukraine] (4-8); Volovik, Larisa: Spaseniye kak odna iz sostavlyayushchikh Kholokosta [Rescue as One of the Components of the Holocaust] (11-20); Kovba, Zhanna: "Bili plyamy" istoriyi Holokostu halitskykh yevreyiv ["Blank Spots" in the History of the Holocaust of Galician Jews] (24-33); Epshtein, Arkadii I.: Antisemitizm i ksenofobiya na razvalinakh Soyuza [Antisemitism and Xenophobia on the Ruins of the [Soviet] Union] (37-41); Feller, Marten D.: Deyaki sotsialno-psykholohichni peredumovy dukhovnoho padinnya [Some Social and Psychological Preconditions for Inner Degradation. On the methods by which the Nazis instilled antisemitism into the popular mind.] (41-51). *DR*

0620 Zabarko, Boris, ed.: *Zhivymi ostalis tolko my: Svidetelstva i dokumenty* [Only We Survived: Eyewitness Accounts and Documents]. Kiev: Instytut Yudayiky, 1999. 575 pp.

A collection of brief memoirs by Holocaust survivors from Ukraine. Some of the accounts appeared previously in the Ukrainian press or elsewhere. Mentions events in Bar, Berdichev, Bershad, Chernovtsy, Chudnov, Kamenets-Podolskii, Kiev, Korets, Mogilyov-Podolskii, Odessa, the Pechory labor camp, Shargorod, Tulchin, Uman, and other places in Ukraine, as well as in Pinsk, Belarus. Two of the eyewitnesses are survivors of Auschwitz. Many of the survivors emphasize that they were rescued by Ukrainians, sometimes – by Germans. *DR*

USSR

(see also Belarus; Ukraine)

0621 Altman, Ilya A.; Gerber, Alla Ye.; Poltorak, David I.: *Istoriya Kholokosta na territorii SSSR: Uchebnoye posobiye dlya srednei obshche-obrazovatelnoi shkoly* [A History of the Holocaust in the Territory of the USSR: A Textbook for the General Secondary School]. Moskva: Fond "Kholokost," 2001. 79 pp.

Describes Nazi anti-Jewish policies in 1933-39 and the subsequent destruction of European Jewry. Discusses the Holocaust in the USSR (in its 1941 borders), dwelling on the Nazi occupation regime, anti-Jewish propaganda, ghettoization, mass shootings of Jews by the SS, Jewish participation in the resistance, and reactions of the non-Jewish population and of the Soviet government to the Holocaust. *DR*

0622 Arad, Yitzhak: Plunder of Jewish Property in the Nazi-Occupied Areas of the Soviet Union. *Yad Vashem Studies* 29 (2001) 109-148. Appeared in Hebrew in "Yad Vashem" 29 (2001).

The Nazi plunder of Jewish property in the USSR was different from that in other occupied countries. There was no private property in the USSR, so only homes, furniture, clothes, valuables, artworks, etc. could be confiscated; most of the Jews were killed on the spot and their property remained at the disposal of the local authorities. Describes the organization of the plunder: orders and directives of the German authorities which regulated the expropriation of the Jews and the transfer of the spoil. Besides the "official plunder," unofficial pillage also took place: marauding German soldiers, indigenous police, and the general populace stole much of the Jewish property. Among those who took part in the official plunder and benefited from it were German authorities, Einsatzgruppen and other SS formations, German and indigenous police, as well as various institutions and individuals in Germany. *DR*

0623

ארד, יצחק: שוד רכוש היהודים בשטחי ברית-המועצות הכבושים בידי גרמניה הנאצית. יד ושם 29 (2001) 85-116.

[Arad, Yitzhak: The Plunder of Jewish Property in the Areas of the Soviet Union Occupied by Nazi Germany. *Yad Vashem* 29 (2001) 85-116.] Appeared in English in "Yad Vashem Studies" 29 (2001).

0624 Glushkin, Oleg, ed.: *Krovotochashchaya pamyat Kholokosta* [The Bleeding Memory of the Holocaust]. Kaliningrad: Regionalnyi Obshchestvennyi Fond "Shofar," 2001. 145 pp.

Pp. 18-86, "Utselevshiye v adu" ("Those Who Survived the Hell"), contain recollections of Holocaust survivors now residing in Kaliningrad. Deals with events in Glubokoye, Minsk, Ilyintsy, Narodichi, etc. Pp. 89-120, "Kholokost: Kyonigsbergskiye khroniki" ("Holocaust: Chronicles of Koenigsberg"), deal with Nazi policy and the Holocaust in Koenigsberg and Tilsit. Pp. 123-143, "Palmnikenskaya tragediya" ("The Tragedy of Palmnicken"), contain a short essay on the Nazi mass murder of former prisoners of the Stutthof concentration camp on 31 January 1945 in Palmnicken (now Yantarnyi) and excerpts from various eyewitness accounts of this event. *DR*

0625 Golbraikh, Yefim: *Eto nasha sudba* [This Is Our Fate]. Tel Aviv: Krugozor, 1999. 396 pp. Part two of a three-volume work; the first volume was published in 1995.

A collection of 35 personal accounts by Jewish survivors of World War II born in the USSR, Poland, or Romania, most of them former servicemen in the Soviet army or in partisan groups, most of them now residing in Israel. Some of the narratives deal with the Holocaust: in Korets (Western Ukraine), Dąbrowa Górnicza (Poland), Beshenkovichi (Belarus), Transnistria, and other places. Four memoirs relate the fate of Jewish servicemen in German captivity. Many of the witnesses experienced antisemitism in the Soviet Army and in partisan units, as well as in the postwar USSR. *DR*

0626 Golbraikh, Yefim: *Posledniye uchastniki* [The Last Participants]. Rishon le-Tsion: "BK-2000," 2001. 256 pp. Part three of a three-volume work; the first volume was published in 1995; the second in 1999.

A collection of 39 personal accounts by former World War II veterans, mainly Jewish, now residing in Israel. Four of them experienced the Nazi occupation in Slonim (Belarus), Melnytsya and Yarun (Ukraine), and Transnistria; one is a Georgian Jew who survived German captivity. *DR*

0627 Pohl, Dieter; Angrick, Andrej: *Einsatzgruppen C and D in the Invasion of the Soviet Union*. London: Holocaust Educational Trust, 2000. 34 pp. (Holocaust Educational Trust Research Papers, 1, 4).

Contents: Bloxham, Donald: The Einsatzgruppen: An Overview (1-6); Pohl, Dieter: Einsatzgruppe C [Appeared in German in "Die Einsatzgruppen in der besetzten Sowjetunion 1941/42" (1997).] (7-15); Angrick, Andrej: Einsatzgruppe D [Ibid.] (16-26). Pp. 27-32 contain the notes to the above articles. *DR*

0628

רפל, יואל, עורך: שלושה מיליון: השואה בשטחי ברית-המועצות. משואה 29 (2001).

[Rappel, Yoel, ed.: Three Million: The Holocaust in the USSR. *Massuah* 29 (2001).]

תוכן חלקי: ארד, יצחק: שואת יהודי ברית-המועצות (44-9); רדליך, שמעון: הועד היהודי האנטי-פשיסטי והיהדות הסובייטית: פעילות קהילתית – מדינית – תרבותית במלחמה ולאחריה (66-45); שנייר, אהרון: גילויי אנטישמיות בברית-המועצות בתקופת השואה (107-83); דובסאן, וואדים: הגטאות בשטחים הכבושים בפדרציה של רוסיה (1941-42): חיי היום-יום (124-109); פרנקל, אלכסנדר: לזכר השואה: המצב בפדרציה הרוסית (132-125); רומנובסקי, דניאל: השואה בעיני האזרח הסובייטי: סקר המבוסס על ראיונות עם יהודים ולא-יהודים בצפון-מזרח בילורוסיה וצפון-מערב רוסיה (158-133); רולניקיטה, מאשה: עדיין אין הכרעה בענין כתב-היד (171-159); לויטס, פליקס: השואה באוקראינה: על-פי חומר ארכיוני מאוקראינה (189-173); ספקטור, שמואל: הפלוגה הפרטיזנית היהודית במחוז ויניצה שבאוקראינה (198-191); סמילוביצקי, ליאוניד: הנסיונות להינצל בשטחי בילורוסיה (220-199); זייץ, דן: גטו מינסק ותולדותיו לאור התיעוד החדש (249-221); לוין, דב: 60 שנה ל"ניצוץ" (265-251); רזניק, ניסן: ווילנה הגרמנית נאצית: עדות (271-267); בייסקי, משה: מסקנות הוועדה הציבורית לנושא יער בולגריה (286-273).

Partial contents: [Arad, Yitzhak: The Holocaust of Soviet Jewry [Appeared in "Yad Vashem" 21 (1991).] (9-44); Redlich, Shimon: The Jewish Anti-Fascist Committee and Soviet Jewry: Communal – Political – Cultural Activity during and after the War (45-66); Shneyer, Aron: Antisemitic Manifestations in the USSR in the Holocaust Period [Also on anti-Semitism in the Russian Army.] (83-107); Dubson, Vadim: The Ghettos in the Occupied Territories of the Russian Federation (1941-42): Daily Life [Appeared in Russian in "Vestnik Yevreiskovo Universiteta" 21 (2000).] (109-124); Frenkel, Aleksandr: In Memory of the Holocaust: The Situation in the Russian Federation [On memorialization of the Holocaust in the 1990s.] (125-132); Romanovsky, Daniel: The Holocaust in the Eyes of "Homo Sovieticus": A Survey Based on Northeastern Belorussia and Northwestern Russia [Appeared in English in "Holocaust and Genocide Studies" 13 (1999).] (133-158); Rolnikaite, Mariya (Mashe): No Decision as Yet Concerning the Manuscript [On difficulties faced by Rolnikaite in the publication of her memoirs in the 1960s.] (159-171); Levitas, Feliks: The Holocaust in Ukraine: Based on Archival Material from Ukraine (173-189); Spector, Shmuel: The Jewish Partisan Group in the Region of Vinnytsya in Ukraine (191-198); Smilovitskii, Leonid L.: Attempts to Survive in the Territories of Belarus [Also on the rescue of Jews by local Christians and even by German soldiers.] (199-220); Zhits, Dan: The Minsk Ghetto and Its History in Light of New Documentation [Also on the Judenrat and the resistance organization.] (221-249); Levin, Dov: 60 Years of "Nitzotz" [On the illegal wartime periodical of Irgun Brit Zion in the Kovno ghetto.] (251-265); Reznik, Nisan: German Nazi Vilna: Testimony (267-271); Bejski, Moshe: Conclusions of the Public Committee on the Topic of the Bulgaria Forest [On objections to the planting of a forest in Israel in memory of King Boris III, since he cooperated with Nazis to a large extent.] (273-286).] *RE*

0629 Sigal, Yakov: Ya smerti skazal: "Podozhdi!.." [I Said to Death: "Wait!.."]. Detroit: The Author, 2001. 221 pp.

Memoirs of a Jew born in 1921 in Uman, Ukraine. In 1941 he was drafted into the Soviet Army, and in October was captured by the Germans. Sigal managed to flee from the train transporting the POWs to a camp in Germany and found himself in Western Belarus, near Smorgon. Passing as a Ukrainian, he survived in various villages in the area. In the fall of 1942 he joined the partisan unit "Dima," one of whose commanders was David Keimakh, a Jew. Even in this unit, Sigal witnessed antisemitism directed at a Jewish soldier (Sigal himself continued to pass as a non-Jew). *DR*

0630 Trostanetskii, Yevsei; Shpitsburg, Moshe, eds.: *Voina sovsem ne feierverk* [War Is Certainly Not Fireworks]. Tel Aviv: Izrailskii Soyuz Voinov i Partizan – Invalidov Voiny s Natsistami, 1999. 232 pp.

A collection of brief accounts by Jews – former Red Army servicemen who fought in World War II and Holocaust survivors. The memoirs of the Holocaust survivors include a description of events in Grodno, Odessa, Vilna, Bobruisk, Mena in Ukraine and some other places, as well as stories of Jews who survived German captivity as POWs. *DR*

Yugoslavia

0631 Friedman, Tuviah: *Die Vernichtung der Juden in Jugoslawien durch die Nazis in den Jahren 1940-45: Dokumentensammlung*. Haifa: Institute of Documentation in Israel for Investigation of Nazi War Crimes, 2000. Various pagings.

Contains a biography of Harald Turner, an SS officer serving as military governor of Serbia, who gave the orders for the extermination of the Jews of Serbia; Turner's correspondence referring to this subject; documents on the Holocaust collected by the Council of Jewish Communities in Serbia; documents of the German Foreign Office on cooperation with foreign countries on the deportation of their Jews; testimony given at Yugoslav war crimes trials; Croatian anti-Jewish legislation of 1941; documents dealing with Kurt Gerstein; and miscellaneous other documents on the Holocaust. *RW*

0632 Pečarić, Josip: *Serbian Myth about Jasenovac*. Trans.: Ivana Pečarić. Zagreb: Stih, 2001. 494 pp. Pt. I was originally published as "Srpski mit o Jasenovcu" (Zagreb: Dom i svijet, 1998).

Pt. I (pp. 9-233) is a response to Milan Bulajić's "'Jasenovacki mit' Franje Tuđmana" (1994). Pt. II (pp. 237-478) is a reply to Bulajić's response to part I. Contests the widely accepted estimate of the number of victims at Jasenovac – ca. 600-700,000. Affirms that Jasenovac was a labor camp, and that the bulk of its victims were Serbian Chetnik prisoners and postwar Croatian prisoners held by Tito. Asserts that the main sites of the perpetration of the genocide of Yugoslavia's Jews were Sajmište (near Belgrade) and other Serbian camps. States that Bulajić wrote his book in order to slander the Croats and brand them as a genocidal nation, while it is the Serbs themselves who were always antisemitic and genocidal. *DR*

0633 Pečarić, Josip: *Srpski mit o Jasenovcu: Skrivanje istine o beogradskim konc-logorima*. Zagreb: Dom i svijet, 1998. 216 pp. Unseen. A two-volume edition appeared in Zagreb: Element, 2000 (200; 200 pp.). See the abstract for the English translation: "Serbian Myth about Jasenovac" (Zagreb: Stih, 2001).

0634 Ristović, Milan: *U potrazi za utočištem: Jugoslovenski jevreji u bekstvu od holokausta 1941-1945* [In Search of Refuge: Yugoslavian Jews Fleeing the Holocaust 1941-1945]. Beograd: Službeni list SRJ, 1998. 382 pp. With an English summary (pp. 345-350).

Before the occupation of Yugoslavia in April 1941, many Jews found refuge in the country. Thousands of Jews entered Yugoslavia on tourist visas, challenging the authorities' attempts to restrict this influx and to expel those who had entered. After April 1941, thousands of Jews were forced to flee the country, or at least the area occupied by Germany. Most of those who managed to survive did so in the area occupied by Italy. Dwells on hardships endured by refugees under Italian rule (part of whom were interned in camps in Italy), as well as by those who tried to find refuge in neutral or Allied countries or who looked to the Vatican for protection. Discusses the mission of the Yugoslavian Zionist leader

Martin Weltmann who went from Palestine to Istanbul in an attempt to rescue Jews from Yugoslavia. Relates, also, to the role of the Yugoslavian government-in-exile in rescuing Jews. *DR*

1919-1945: Far East

India

0635 Bhatti, Anil; Voigt, Johannes H., eds.: *Jewish Exile in India, 1933-1945.* New Delhi: Manohar, in association with Max Mueller Bhavan, 1999. 195 pp.

Papers presented at a symposium held at the Centre of German Studies, Jawaharlal Nehru University, March 1995. Partial contents: Voigt, Johannes H.: Persecution of the Jewish People: Prelude to the Holocaust [On antisemitism in Nazi Germany.] (13-21); Oesterheld, Joachim: British Policy towards German-Speaking Emigrants in India, 1939-1945 [Focuses on the British policy of immigration restrictions and internment of enemy aliens vis-à-vis German and Austrian Jews seeking asylum in India.] (25-44); Siddiqi, Majid Hayat: Jews and Central European Nationals in Exile in Colonial India between the Two World Wars (45-54); Sareen, Tilak Raj: Indian Responses to the Holocaust [Including that of Nehru.] (55-63); Dengle, Rajendra: "Ranangan" or Response in Marathi Literature to the Theme of Jewish Emigration [On the novella "Ranangan" (1939), by Vishram Bedekar, which relates the encounter of the Maharashtrian protagonist with Jews fleeing from Nazi Germany on board a ship carrying them to Bombay.] (172-185). *DR*

0636 Kumaraswamy, P.R.: India and the Holocaust: Perceptions of the Indian National Congress. *The Journal of Indo-Judaic Studies* 3 (Apr 2000) 117-125.

India, which never knew persecution of Jews, was also indifferent to the "Jewish problem" in Europe. In 1933-45, and after World War II, the Indian National Congress maintained silence on the suffering of the European Jews; it adopted no formal resolution on the Holocaust. Nehru did not object to the admission of Jewish refugees to India, mainly due to pragmatic considerations. There were several reasons for the silence maintained by the INC: from the 1920s it opposed Zionism as an ally of British imperialism and supported Arab nationalists in Palestine; the INC was interested in Hindu-Muslim unity, which made it downplay the issue of the persecution of Jews in Europe; and the influence of anti-Jewish Congress leaders such as Bose. With all that, the crucial fact is that India was neither an accomplice nor a beneficiary of the Holocaust. *DR*

1919-1945: Islamic World

Algeria

0637 Msellati, Henri: *Les Juifs d'Algérie sous le régime de Vichy, 10 juillet 1940 – 3 novembre 1943*. Paris: L'Harmattan, 1999. 302 pp.

A history of the Jews of Algeria under the Vichy government which was in power there from July 1940 until November 1943. Algerian Jews, French subjects since 1870 and numbering about 110,000 at the beginning of the war, were spared the fate of European Jewry but were subjected to the Vichy anti-Jewish legislation. They were stripped of their French nationality, deprived of their rights, and robbed of their property. The anti-Jewish laws remained in effect a year after the Allies landed in Algeria in November 1942, proving false the claims by France that it had adopted the racist policy under German pressure. Discusses the effects of the humiliations suffered by the Jews during the war on their political stance in the Algerian war of independence in 1960, when most of them sided against extremist defenders of French Algeria. *EO*

1919-1945: Latin America

General

0638

בן-דרור, גרסיאלה: דמיון ושוני ביחסן של מקסיקו וברזיל להגירה יהודית בצל השואה. ילקוט מורשת 67 (אפר' 1999) 107-127.

[Ben-Dror, Graciela: Similarities and Differences in the Attitudes of Mexico and Brazil to Jewish Immigration in the Shadow of the Holocaust. *Yalkut Moreshet* 67 (Apr 1999) 107-127.]

Describes the political factors which influenced the attitudes of Mexico and Brazil toward Jewish immigration, especially during the 1930s when Jews could still leave Europe. Focuses on the government of Lazaro Cárdenas in Mexico and that of Getúlio Vargas in Brazil. Points out that the relatively progressive Mexico absorbed only a few hundred Jews despite Cardenas' tendency to let Jews enter his country, probably because of the influence of xenophobia and antisemitism amongst the masses on the government. Brazil absorbed a much larger number of Jews, despite restrictive immigration laws in that country and an antisemitic bureaucracy. *LFo*

Argentina

0639

בן-דרור, גרסיאלה: הכנסייה הקתולית והיהודים: ארגנטינה 1933-1945. ירושלים: מרכז זלמן שזר לתולדות ישראל; המרכז הבינלאומי לחקר האנטישמיות ע"ש וידאל ששון, האוניברסיטה העברית, 2000. 320 עמ'.

[Ben-Dror, Graciela: *The Catholic Church and the Jews: Argentina 1933-1945*. Jerusalem: Zalman Shazar Center for Jewish History; Vidal Sassoon International Center for the Study of Antisemitism, Hebrew University, 2000. 320 pp.] Based on the author's diss. – Hebrew University of Jerusalem, 1993.

Uruguay

0640 Raicher, Rosa Perla: Los judíos refugiados del nazismo, 1933-1941. *Judaica Latinoamericana* 4 (2001) 63-77.

Discusses the attitude of the Uruguayan authorities toward Jewish refugees between 1933-41. In response to the wave of refugees, the rightist government of Gabriel Terra (1933-38), in an atmosphere of growing antisemitism, enacted restrictive immigration laws which resulted in the rejection of hundreds of Jewish immigrants. Later, the more liberal government of Alfredo Baldomir (1938-43) publicly declared its willingness to help refugees, but in fact he continued with the policy of his predecessor, by means of administrative and bureaucratic obstacles. Despite the restrictions, almost 10,000 Jewish refugees entered the country between 1933-41. *MZ*

1919-1945: North America

Canada

0641 Stingel, Janine Lianne: *Social Discredit: Anti-Semitism, Social Credit, and the Jewish Response*. Montreal: McGill-Queen's University Press, 2000. xv, 280 pp. Based on the author's diss. – McGill University, Montreal, 1997.

Discusses the Social Credit Party which ruled Alberta from 1935-71, highlighting its propaganda of antisemitism based on a conspiracy theory that blamed Jews for Canada's, especially Alberta's, economic and political problems. Social Credit antisemitism was rhetorical and ideological rather than aggressive or discriminatory. Reaction to Nazi policies affected Canadian public opinion and undermined antisemitism in North America. For this reason, Alberta Prime Minister and Social Credit Party head Ernest Manning repudiated antisemitism in 1949 and purged persistent antisemites. The quiet diplomatic approach of the Canadian Jewish Congress was unsuccessful in opposing Social Credit. Eventually the Congress became more assertive but it could not claim credit for ending Social Credit's antisemitism. *YC*

USA

0642 *American Jewish History* 89, 1 (Mar 2001).

This issue is entitled "America and the Holocaust: New Perspectives – II." Contents: Wyman, David S.: The Bergson Group, America, and the Holocaust: A Previously Unpublished Interview with Hillel Kook/Peter Bergson [Pp. 8-32 present excerpts of interviews with Kook conducted in 1973 and 1978.] (3-34); Zucker, Bat-Ami: Frances Perkins and the German-Jewish Refugees, 1933-1940 [Perkins was the Secretary of Labor in this period; her office was in charge of immigration and naturalization issues until 1940; a humanitarian, she did her utmost to oppose the nativist restriction and antisemitism of other departments in the administration.] (35-59); Herman, Felicia: Hollywood, Nazism, and the Jews, 1933-41 [Discusses the question why Hollywood did not produce anti-Nazi films in the 1930s; contends that Jewish organizations in the U.S. did not want overt condemnation of Nazism or overt defense of Jewry to be depicted on screen. Nevertheless, three films produced in the 1930s can be considered anti-Nazi.] (61-89); Medoff, Rafael: New Evidence concerning the Allies and Auschwitz [Documents show that there is no question as to whether the Allies could have bombed Auschwitz; they clearly could have but did not wish to.] (91-104); Sheramy, Rona: "There Are Times When Silence Is a Sin": The Women's Division of the American Jewish Congress and the Anti-Nazi Boycott Movement [From 1934 to 1941 the Women's Division assumed a pivotal role in the day-to-day operation and local supervision of boycott work.] (105-121); Ansell, Joseph P.: Arthur Szyk's Depiction of the "New Jew": Art as a Weapon in the Campaign for an American Response to the Holocaust [On his drawings, posters, and caricatures depicting Jews as both victims and fighters, and his close collaboration with the Bergson group.] (123-134). *SSC*

0643 Bendersky, Joseph W.: The Absent Presence: Enduring Images of Jews in United States Military History. *American Jewish History* 89, 4 (Dec 2001) 411-436.

Examines antisemitism in the U.S. military from the end of World War I through World War II and beyond. Focusing on the Military Intelligence (MI), shows that antisemitic attitudes and conspiracy theories about Jews (e.g. as Bolsheviks) prevailed among top MI officers. The heads and teachers of the War College believed contemporary racial theories that prejudiced them against Jews; their attitudes affected their juniors, who later held key posts on the general staff and as military attachés in Europe. Thus, more than one generation of antisemites in the military were in a position to and did influence policy regarding Jewish immigration (i.e. the basic denial of shelter to refugees from Hitler's Germany), rescue of Jews during the Holocaust, the admission of Jewish DPs after the war, and policy toward the establishment of Israel. *YC*

0644 Brackman, Harold David: Just Another Klansman? *Midstream* 47, 1 (Jan 2001) 19-24.

With the rise of Hitler, African American publications equated Nazi antisemitism with American racism, while White newspapers rejected this link. At the same time, some Black opinion makers resented the attention paid to Jewish victims, and sympathy with Hitler was expressed by antisemitic rabble-rousers such as Sufi Abdul Hamid. After Pearl Harbor, Blacks ceased to sympathize with Hitler due to Nazi anti-Black racism and Black-Jewish cooperation in the "Popular Front" against fascism. In the U.S., the destruction of European Jewry was not seen as Hitler's main crime in either the Black or the White press. The reactions in the Black press to Nazi genocide related mainly to the Black experience in the U.S., e.g. comparing Nazi racism to that of the Ku Klux Klan. The "nonreporting" of the treatment of Black POWs and civilians by the Nazis facilitated a postwar trend to view Germany, in contrast to America, as a paradise for non-Whites. *YC*

0645 Herschenhous, Nicole: The Response of the Providence Jewish Community to Events in Prewar Germany, 1935 to 1939. *Rhode Island Jewish Historical Notes* 13, 1 (Nov 1999) 39-54.

Explores the response of the American Jewish community to the Nazi persecution of Jews by studying the reactions of Providence Jews to prewar events reported by the local newspaper, the "Jewish Herald." From 1935 to 1939, the focus of the community's actions shifted from attempts to affect the Nazi regime (e.g. with a boycott) to concern for the refugees from Germany. Providence Jews failed to conduct a concerted action: the participation in the boycott was only partial, and the Jews of Providence did not try to influence the U.S. government to change its immigration policy. Among the reasons for their inaction were difficult economic conditions in the post-Depression years and belief that the conditions in

Germany would improve, but the chief reason was the belief that the U.S. government would take some action on behalf of the refugees. *DR*

0646 Klein, Gary: When the News Doesn't Fit: The "New York Times" and Hitler's First Two Months in Office, February/March 1933. *Journalism & Mass Communication Quarterly* 78, 1 (Spr 2001) 127-149.

Despite its intensive coverage of events in Germany in February-March 1933, the "New York Times" tended to misinform and confuse its readers by discounting, ignoring, and even withholding disturbing facts that most other major publications included in their analyses. The paper deviated from its mainstream counterparts in both the degree and duration of its rosy view of Hitler's Germany. In particular, it tended to cover-up and downplay the Nazi persecution of Jews. Behind this deliberate policy were the assimilationist views of the paper's Jewish publishers, Ochs and Sulzberger, who did not want to draw attention to Jewish issues in the "New York Times" and implemented measures to "de-Jewify" the paper. Towards the end of World War II and immediately after it, the paper grossly underplayed its reporting on the Holocaust and did not acknowledge that most of the victims were Jews. *DR*

0647 Luconi, Stefano: The Response of Italian Americans to Fascist Antisemitism. *Patterns of Prejudice* 35, 3 (July 2001) 3-23.

Disputes the view that Mussolini's antisemitism was rejected by Italian Americans. Attributes Italian American antisemitism, which was manifest until Pearl Harbor, to general antisemitism in the U.S. and to friction between Italians and Jews over jobs, political patronage, housing, and relief benefits, especially during the Depression. Furthermore, identification with Italian fascism helped Italian Americans to solidify their identity. For these reasons, ethnic Italians in the U.S. did not sympathize with Jewish victims of antisemitism in Italy. *YC*

1945-1990: General

Responses to the Holocaust

0648 Baer, Ulrich Christian, ed.: *"Niemand zeugt fuer den Zeugen": Erinnerungskultur und historische Verantwortung nach der Shoah*. Frankfurt a.M.: Suhrkamp, 2000. 277 pp.

A collection of essays, all but one published previously. *RW*

0649

באואר, יהודה: היכן היה האלוהים? היהדות החרדית ויחסה לשואה. יהדות חופשית 24-23 (סתיו 2001) 38-29.

[Bauer, Yehuda: Where Was God? Haredi Jewry and Its Attitude to the Holocaust. *Yahadut Hofshit* 23-24 (Fall 2001) 29-38.] Translation of an excerpt from his book "Rethinking the Holocaust" (2001).

Discusses reactions of Orthodox Jewry to the Holocaust, including the view that God "hid His face" or reduced His presence in order to allow people to choose between good and evil, Irving Greenberg's view that God is just but not omnipotent, and the Haredi (ultra-Orthodox) view that the Jews were punished for their sins. Focuses on the reaction of Rabbi Menahem Mendel Schneersohn, expressed in some of his writings in the 1970s and in a letter to Chaika Grossman in August 1980. He contended that, since God is just, the Holocaust must have been good for the Jews spiritually, although it was bad for them physically. He compared the Holocaust to an operation in which one part of the body is amputated in order to save a person. However, when asked why so many Orthodox Jews were killed, especially in Poland, he stated that humans cannot understand the ways of God. Bauer notes that the death of one million children also cannot be explained, and that Walter Zvi Bacharach advocated continued belief in God but also anger with Him. *RE*

0650 Bensoussan, Georges: *Auschwitz en héritage? D'un bon usage de la mémoire: Suivi de Brève histoire de la destruction des Juifs d'Europe*. Paris: Mille et une Nuits, 1998. 207 pp.

Clarifies the need of postwar generations to remember the Holocaust in order to prevent its repetition. Underscores the role of the bureaucratic political machine in his analysis of the banality of evil. Contends that the emphasis on the need for commemoration should be complemented by a historical study of the events that culminated in the genocide of millions. Pp. 55-77, "Enseigner quoi?," examines the role that an organized curriculum can play in viewing Auschwitz as a symbol of systematic destruction. Pp. 155-181 contain a brief history of the Holocaust. *TS*

0651 Deoriti, Alessandra; Paolucci, Silvio; Ropa, Rossella, eds.: *Le storie estreme e la storia: I racconti della Shoah*. Chiaravalle: Orecchio di Van Gogh, 1999. 312 pp.

A collection of papers stemming from a course conducted in 1997-98 at the Istituto per la Storia della Resistenza e della Società Contemporanea nella Provincia di Bologna. *MZ*

0652 Morgan, Michael Lewis, ed.: *A Holocaust Reader: Responses to the Nazi Extermination*. New York: Oxford University Press, 2001. vi, 378 pp.

An anthology of articles and excerpts from books by various writers, Jewish and non-Jewish, religious and secular, reflecting a wide spectrum of intellectual responses to the Holocaust. The book is intended for use in teaching about the Holocaust. *DR*

0653 Stone, Dan: Ontology or Bureaucracy? Hannah Arendt's Early Interpretations of the Holocaust. *European Judaism* 32, 2 (Fall 1999) 11-25.

Contrasts Arendt's early postwar writings with "Eichmann in Jerusalem." In the former she stressed Nazi ideology more than Nazism's modern characteristics (use of technology, functional rationality, and bureaucracy) which she later focused on as the basis for the Holocaust. In both periods she stressed the lack of connection between Nazism and Western, especially German, philosophical and political traditions. Her letters to Karl Jaspers ca. 1946 offered a different, ontological perspective concerning what it meant to be human and the Nazi attempt "to eradicate the concept of the human being." This idea, which is present in "The Origins of Totalitarianism," is dropped in her book on Eichmann. There, the "banality of evil" implies that evil cannot be radical; questions about the nature of humanity are avoided. Concludes that the latter reflects Arendt's attempt to protect her philosophical and emotional stability. Avoidance of the question whether Nazism was anti-humanist or the culmination of humanism not only allows Arendt to salvage a belief in human goodness, it also denies any link between Nazism and the Western philosophical tradition that had become Arendt's "home" in exile. *YC*

0654 Stone, Dan, ed.: *Theoretical Interpretations of the Holocaust*. Amsterdam: Rodopi, 2001. viii, 239 pp.

A collection of essays which examine, theoretically, various representations of the Holocaust in Western culture. *DR*

0655 Trautwein, Dieter: *Oskar Schindler, ...immer neue Geschichten: Begegnungen mit dem Retter von mehr als 1200 Juden*. Frankfurt a.M.: Societaets-Verlag, 2000. 333 pp.

A personal account of the author's meetings with Schindler and his examination of Schindler's life and personality, with a view to understanding his motives for rescuing more than 1,200 Jews during the Holocaust. In the framework of a panel discussion, organized by Trautwein in 1967 in Hannover, German adolescents learned about Schindler's rescue activities, which were then barely known. Schindler was honored in Israel in 1962 at a reception organized by the Jews whom he rescued, and, after the celebration of his 60th birthday in Israel, he gradually began to gain recognition in Germany. Quotes extensively from those meetings and others, including interviews with Schindler on television and programs about him. *ID*

0656 *The Yale Journal of Criticism* 14, 1 (Spr 2001).

This special issue of the journal, entitled "Interpretation and the Holocaust," is edited by Laura Frost et al., and focuses on representations of the Holocaust in literature and films. *SSC*

Nazi War Criminals and War Crimes Trials

0657

בשביל הזיכרון 41 (אפר'-מאי 2001).

[*Bishvil ha-Zikaron* 41 (Apr-May 2001).]

This issue is entitled "40 Years since the Eichmann trial." *SSC*

0658 Bloxham, Donald: *Genocide on Trial: War Crimes Trials and the Formation of Holocaust History and Memory*. Oxford: Oxford University Press, 2001. xix, 273 pp. Based on the author's diss. – University of Southampton, 1998.

Discusses the American and British roles in the Nuremberg war crimes trials. Focuses not on the punishment aspect but rather on the aims of deterrence and education. Stresses the role of the trial medium and of the Allies' values in creating views about the essence of Nazism that affected the subsequent historiography of the Holocaust. Notes the trials' focus on concentration camps and the Wehrmacht as exemplifying the Holocaust and the involvement in it of the German people. Contends that some Holocaust historians have been "overinformed" by Nuremberg historiography. Concludes that for decades the murder of Jews hardly impinged on postwar consciousness. This is partially explained by the way the World War II story was used by the Allies to further their goal of reeducating Germans, as well as punishing war criminals. *YC*

0659 Douglas, Lawrence: *The Memory of Judgment: Making Law and History in the Trials of the Holocaust*. New Haven: Yale University Press, 2001. xiii, 318 pp.

Examines how the law has struggled to find an appropriate idiom to represent and judge the traumatic history of the Holocaust. Deals mainly with the Nuremberg Trials and with the trials of Eichmann, Zundel, and Demjanjuk. Focuses not on whether the trials did justice to the defendants but on whether the trials did justice to the unprecedented crimes of the Holocaust. In contrast to Hannah Arendt, argues that a didactic goal is an important aspect of such trials. One way that the law has dealt with the radicalness of Nazi atrocities has been to establish the concepts of genocide and crime against humanity. Notes that the Nuremberg and Eichmann trials were powerful, socially necessary responses to extreme crimes, while the Demjanjuk and Zundel trials were disasters of didactic legality, where history was obfuscated. Yet even in the latter cases, the law emerged victorious. Together, all these trials comprise a range of efforts, at different points in time, to protect collective memory. Discusses, also, the documentary film "Nazi Concentration Camps." *YC*

0660 Gerlach, Christian: The Eichmann Interrogations in Holocaust Historiography. *Holocaust and Genocide Studies* 15, 3 (Win 2001) 428-452.

Argues that Eichmann's interrogations, as well as those of other major Nazi perpetrators, cannot serve as the exclusive or main basis of any historiographical arguments. His statements made at the trial were self-exculpatory and a part of his defense strategy. Eichmann's main goal was to present himself as a cog in the wheel rather than an initiator of mass murder. Examines some of Eichmann's statements; e.g. shows that the purpose of his travels to Minsk in 1941-42 was not "inspection" but preparation for forthcoming deportations of Jews from Central Europe to this ghetto. Dismisses Eichmann's self-portrait as a cog; although he was not the architect of the genocide, as the Israeli prosecution tried to present him, he was also not an automaton fulfilling orders from above. Criticizes scholars who took Eichmann's self-exonerating statements at face value, especially Hannah Arendt, who depicted Eichmann as a "desk killer" and thereby influenced many other historians. *DR*

0661 Lundgren, Svante: Den banala debatten: Hannah Arendt i Jerusalem [The Banal Debate: Hannah Arendt in Jerusalem]. *Nordisk Judaistik* 22, 2 (2001) 131-156.

Discusses the anger and debate aroused by Arendt's book "Eichmann in Jerusalem." She was viewed as a self-hating Jew because she accused the Jewish leaders during the Holocaust of complying with Nazi orders, thereby facilitating the mass murder. Her views on Eichmann's personality were considered to be wrong and her way of writing was considered inappropriately lacking humility in dealing with the Holocaust. States that although much of the criticism was unfair, Arendt's undiplomatic style of writing contributed to the negative reactions which ensued. *SSC*

0662 Tobias, Jim G.; Zinke, Peter: *Nakam – juedische Rache an NS-Taetern*. Hamburg: Konkret Literatur Verlag, 2000. 173 pp. Another edition appeared in Berlin: Aufbau Taschenbuch Verlag, 2003 (179 pp.).

Immediately after World War II, groups consisting of former partisans, survivors of concentration camps, and soldiers of the "Jewish Brigade" planned revenge against Nazi perpetrators. Focuses on the group called Nakam (Revenge) which was affiliated with the "Jewish Brigade" from Palestine. Based on interviews with former members of the group now living in Israel, recounts some of their revenge actions in Germany and Austria until they received the order to immigrate to Palestine in summer 1946. The interviewees were: Dov Shenkal (Bernhard Schenkel), Chaim Miller, Ollie Giveon, Leipke Distel, Olek Hirsch, Joseph Harmatz. *ID*

0663 Wojak, Irmtrud, ed.: *"Gerichtstag halten ueber uns selbst...": Geschichte und Wirkung des ersten Frankfurter Auschwitz-Prozesses*. Frankfurt a.M.: Campus, 2001. 356 pp. (Fritz Bauer Institut: Jahrbuch 2001 zur Geschichte und Wirkung des Holocaust).

Partly based on lectures delivered at a colloquium of the Fritz Bauer Institut, 1999. Partial contents: Wojak, Irmtrud: "Die Mauer des Schweigens durchbrochen": Der erste Frankfurter Auschwitz-Prozess, 1963-1965 [Analyzes the legal problems involved, using the example of one of the accused, Josef Klehr.] (21-42); Miquel, Marc von: "Wir muessen mit den Moerdern zusammenleben!" NS-Prozesse und politische Oeffentlichkeit in den sechziger Jahren (97-116); Dirks, Christian: Selekteure als Lebensretter: Die Verteidigungsstrategie des Rechtsanwalts Dr. Hans Laternser (163-192); Plato, Alice von: Vom Zeugen zum Zeitzeugen: Die Zeugen der Anklage im ersten Frankfurter Auschwitz-Prozess (1963-1965) (193-215); Braese, Stephan: "In einer deutschen Angelegenheit": Der Frankfurter Auschwitz-Prozess in der westdeutschen Nachkriegsliteratur (217-243); Loewy, Ronny: "Zeugin aus der Hoelle" [On a film directed by Zica Mitrovic, a joint West German/Yugoslav production, based on the Auschwitz trial.] (265-275); Yablonka, Hanna: Nazi-Prozesse und Holocaust-Ueberlebende, Israel 1950-1967 (277-292); Laster, Kathy: Eine Buehne zur Selbstdarstellung der Kultur: Kriegsverbrecherprozesse in Australien [On the trial in 1990-94 of

Timofeyevich Polyukovich, a Ukrainian accused of assisting in the massacre of the Jews of Sarnicki. He was acquitted.] (293-316). *RW*

0664

יבלונקה, חנה: מדינת ישראל נגד אדולף אייכמן. תל-אביב: משכל, 2001. 357 עמ'.

[Yablonka, Hanna: *The State of Israel vs. Adolf Eichmann*. Tel-Aviv: Miskal, 2001. 357 pp.]

Pt. 1 (pp. 43-172) discusses various aspects of the Eichmann trial in 1961, including Israeli reactions to Eichmann's capture, especially those of David Ben-Gurion; technical and investigative preparations for the trial; the witnesses, most of whom were Holocaust survivors; the appointment of the defense attorney and the judges; and the sentencing. Pt. 2 (pp. 175-240) deals with effects of the trial on various groups in Israel: Holocaust survivors, youth, the Sephardi community, and the educational system. Pt. 3 (pp. 243-287) discusses various issues which were raised and debated as a result of the trial, as well as whether it was actually a show trial. Contends that it was a trial about the past, for the future. *RE*

1945-1990: Europe

Austria

0665 Embacher, Helga: Vom Hilfsfonds zur Versoehnungskonferenz: Oesterreichs Umgang mit Entschaedigungszahlungen und Rueckstellung. *Kirche und Israel* 15, 2 (2000) 144-153.

Traces the history of Austria's reluctant and scanty payments to Holocaust victims (called "aid funds" or "reconciliation funds," never "reparations," since Austria did not hold itself responsible). These were attributed by the press and public opinion to extortion by Jewish emigrants who had supposedly spent the war in luxury abroad while Austrians suffered. Similar antisemitic complaints arose during the WJC's campaign against Waldheim (called a Jewish conspiracy) and the recent demand for compensation to forced laborers (despite the fact that most of the recipients will be non-Jews). *RW*

0666 Wassermann, Heinz Peter: Antisemitismus in Oesterreich nach 1945: Eine empirische Bestandsaufnahme. *Jahrbuch fuer Antisemitismusforschung* 10 (2001) 156-183.

Analyzes polls measuring antisemitic attitudes from 1945 to the present, finding that there are great differences in this respect among the

Austrians, and that these are not dependent on age. Concludes that social antisemitism has stayed disturbingly constant, that denial of Austria's responsibility in the Holocaust has decreased, and that racist antisemitism is negligible. *RW*

Belarus

0667 Basin, Iakov: Smert korolya Lira [The Death of King Lear]. *Mishpokha* 5 (1999) 90-97.

Based on an interview with the former actress of the Belorussian State Jewish Theater Yudif Aronchik, conducted in 1990 and published in "Rodnik," as well as the recollections of a photographer who took the last photographs of Mikhoels, reconstructs the atmosphere in Minsk before and immediately after the murder of Mikhoels in January 1948. Attempts to reconstruct the circumstances of the murder; surmises that Mikhoels was killed at the former farmstead Loshitsa near Minsk, and that Mikhoels put up resistance to his murderers. Despite the fact that the authorities annihilated anyone who might have known that Mikhoels was killed on Stalin's order, everybody in Mikhoels' circle felt that it was not an accident but a political murder. *DR*

0668 Nordshtein, Mikhail: Podnyatsya i vystoyat [Getting Up and Holding Out]. *Mishpokha* 6 (1999) 114-119.

Lev Ovsishcher, Yefim Davidovich, and Naum Alshanskii were officers in the Soviet Army, veterans of World War II. During and after the war they were confronted with antisemitism. In the 1970s, all three requested official permission to immigrate to Israel but were refused. They protested publicly against the official antisemitism and, from 1975 on, chaired the rallies in Minsk held on the site of the mass murder of the ghetto Jews during the Nazi occupation. The authorities tried to thwart the rallies and to demolish the monument commemorating the Jewish victims. A criminal case was opened against Davidovich. In the late 1980s, Ovsishcher and Alshanskii immigrated to Israel; Davidovich died. An appeal to the president of Belarus in 1995 asking for posthumous rehabilitation of Davidovich was refused. *DR*

0669 Puzhalovskii, Aleksandr: "Pyataya grafa lichnovo listka uchyota kadrov stala na puti moyom" ["The Fifth Item of the Workers Registration Sheet Stood in My Way"]. *Mishpokha* 9 (2001) 70-73.

Presents a sample of letters sent in the late 1940s-early 1950s to the authorities (local and national) by citizens of Belarus. Four of the letters describe antisemitic incidents and cases of anti-Jewish discrimination regarding employment; they also complain about the reluctance of the police and the local authorities to act against this antisemitism. Four other letters complain about the "dominance" of Jews, and their "nationalism" and "subversive activities." Comments that almost all of the letters (both antisemitic and against antisemitism) viewed by the author were not responded to by the authorities. *DR*

0670 Smilovitskii, Leonid L.: The Non-Jewish Reaction to the "Doctors' Plot" in Belorussia in Light of New Documents (January-March 1953). *Shvut* 9 [25] (2000) 67-92.

In January-March 1953, at the time of the "Doctors' Plot" affair, Soviet authorities attempted to launch a mass antisemitic campaign. The press, and party propagandists who were sent to workplaces and to educational institutions in both urban and rural areas, were instrumental in instigating an anti-Jewish mood in the population. Analyzes reports on mass meetings and rallies that took place at factories, collective farms, medical and educational institutions, etc. in Belorussia in this period. The accusations made by participants in these meetings ranged from comments on Jewish unreliability and parasitism to accusations of Jewish collaboration with the American and British imperialists to unleash a Third World War. Many speakers demanded brutal punishments for the "guilty" doctors and a mass deportation of the Jews. Belorussian authorities attempted to replace some Jewish medical administrators and physicians with non-Jews. The campaign ended in April 1953 as abruptly as it had begun. *DR*

0671 Smilovitskii, Leonid L.: The Struggle of Belorussian Jews for the Restitution of Possessions and Housing in the First Postwar Decade. *East European Jewish Affairs* 30, 2 (2000) 53-70.

After World War II the authorities of the Belorussian SSR were reluctant to return prewar Jewish property, especially former Jewish dwellings, to their legal owners. There was a measure of antisemitism in the hostility of the Belorussian property owners toward the Jewish claimants, especially since some of the former had acquired the property as a reward for their services to the Germans or through plunder. *DR*

0672 Vyaliki, Anatol: Razhrom habreiskai namenklyatury: "Habreiskaye pytanne" w Belarusi w 40-50-ya hady [The Destruction of the Jewish Nomenclatura: "The Jewish Question" in Belarus in the 1940s-50s]. *Arche* 3 (2000) 206-211.

In the 1930s, Jews constituted a large part of the Belorussian ruling strata (in Soviet political slang, "nomenclatura"). A relatively large part of this Jewish faction survived World War II; in 1945, the proportion of Jews in the political and economic leadership of the republic remained principally the same. The situation changed drastically in 1946-47, with the Stalinist campaign against "rootless cosmopolitans"; numerous Jews were purged from their positions. The "Jewish nomenclatura" suffered a major purge in 1951-52 at the initiative of the Belorussian party leadership. In the wake of the "Doctors' Plot" affair in Moscow, popular antisemitism in Belorussia increased greatly; the local authorities encouraged this trend. Among the "people's demands" was one to remove all the Jews from administrative positions. *DR*

France

0673 Fresco, Nadine: *Fabrication d'un antisémite*. Paris: Seuil, 1999. 795 pp.

A biography of the founder of Holocaust "negationism," Paul Rassinier (1906-1967). In different periods of his life he was a pacifist, communist, socialist, and anarchist. During the Second World War, Rassinier was deported to concentration camps in Germany as a member of the resistance. After the war he returned to his hometown Belfort (Alsace) in a bid to become mayor of the town, but a Jewish communist won the election. Rassinier left Belfort in 1947, settled near Paris, and then began his negationist activities. Concludes that the antisemitic atmosphere and the pressure of the times in which he lived were not enough to make Rassinier an antisemite. His lack of success in the political sphere engendered in him strong resentments which had an impact on his negationist ideology. Analyzes his postwar writings (including description of his experiences in the concentration camp Dora) and states that Rassinier did not describe realities but created them. His negationist ideas claimed that reality was fiction (e.g. that the Holocaust did not take place) and fiction was reality (e.g. the "Jewish conspiracy" needed the existence of a Holocaust for its own goals, the creation of the State of Israel). *HV*

0674 Guland, Olivier: *Le Pen, Mégret et les Juifs: L'obsession du "complot mondialiste."* Paris: La Découverte, 2000. 223 pp.

Examines the place of antisemitism in the history of the Front National, from its creation in 1972 until its scission in 1999. Includes interviews with Le Pen, Bruno Mégret, and other leaders of the movement. Shows that an obsession with "the Jewish conspiracy" lies at the heart of the FN's ideology, characterized by a strong anti-democratic impetus and racism aimed mainly at Arabs from the Maghreb. Sheds new light on the "detail" (an expression used by Le Pen regarding the Holocaust) and Carpentras affairs. Describes French-Jewish reactions to the rise and increasing power of the FN. Concludes that the extreme right in France has marked the mental climate of the country and warns against attributing the negative effects of globalization to hidden powers that run the world, thus feeding into old antisemitic fantasies. *EO*

0675

אופן, ישראל ז'ק: מקור הכחשת השואה בצרפת והתפשטותה. ילקוט מורשת 71 (אפר' 2001) 83-107.

[Offen, Yisrael Jacques: The Origins of Holocaust Denial in France and Its Dissemination. *Yalkut Moreshet* 71 (Apr 2001) 83-107.]

The first publications denying the Holocaust appeared shortly after the war in France. The works of Maurice Bardèche and Paul Rassinier set the groundwork for Holocaust denial, but they were largely ignored until the Six-Day War (1967). Thereafter, the Jews were perceived as being strong, and the world was free of guilt feelings towards them. During the 1970s the works of Bardèche and Rassinier became more popular and Holocaust denial was accepted throughout the world as a legitimate academic pursuit, taken up by scholars such as Robert Faurisson, who claims that the Jews are attempting to control France. States that both the Right and Left in France use Holocaust denial for political purposes. Both use it in the struggle against Zionism and the State of Israel, but the Right also uses it to support its belief in a Jewish world conspiracy and to condemn communism. *RE*

Germany

0676 Knoch, Habbo: *Die Tat als Bild: Fotografien des Holocaust in der deutschen Erinnerungskultur*. Hamburg: Hamburger Edition, 2001. 1120 pp.
Examines the postwar reactions of the German public to photographs relating to the Holocaust which appeared in the press. Photographs of war crimes, cadavers in liberated concentration camps, and survivors on the brink of death were provided by the Allies in 1945 and placed the Germans close to the crimes both geographically and mentally (i.e. in their consciousness) at a time when the crimes were being vehemently denied. In a second stage, during the founding years of the BRD, Germans considered themselves victims of Nazism, as well as war victims, and the Nazi past was silenced. Between 1955-65 photographs of Nazi crimes reappeared in daily newspapers, journals, textbooks, films, and exhibitions. At this stage, Auschwitz designated the ultimate narrative of war crimes, the interpretation of which codified the culture of remembrance through the 1990s. Includes some photographs. *ID*

0677 Leo, Annette: Der Friedhof, die Schule und die Erinnerung. *Jahrbuch fuer Antisemitismusforschung* 10 (2001) 67-85.

Over the course of several weeks in 1988, five East Berlin teenagers, from broken homes and with a record of minor delinquency, entered the Jewish cemetery opposite the school attended by three of them in the Schoenhauser Allee, toppling headstones and shouting neo-Nazi slogans. They did this night after night, after first getting thoroughly drunk, in a sort of contest of strength and daring. Analyzes the (in)effectiveness of communist indoctrination in the school and describes the brainwashing to which the pupils were subjected after the apprehension of their delinquent classmates. The East German government, which was just then seeking a rapprochement with "world Jewry," was extremely embarrassed by the incident and turned it into a show trial, meting out heavy prison sentences to the accused. *RW*

0678 Neiss, Marion: Friedhofsschaendungen in Deutschland: Ein Ueberblick. *Jahrbuch fuer Antisemitismusforschung* 10 (2001) 55-66.

The desecration of Jewish cemeteries has a continuous history beginning in the Middle Ages. The Nazis desecrated ca. 90% of the Jewish cemeteries in Germany; many were levelled completely and converted into parks and building sites. Since 1945, and especially since 1990, cases of cemetery desecration have increased in number and intensity. Not only are gravestones overturned, they are dynamited; antisemitic and neo-Nazi slogans are spray-painted on gravestones and walls. Although the police take these acts very seriously, finding the perpetrators is difficult. Argues that cemetery desecration is not gratuitous vandalism; it is aimed at obliterating the memory of the Jewish presence in Germany and of the Holocaust. *RW*

0679 Schuppener, Georg, ed.: *Juedische Intellektuelle in der DDR: Politische Strukturen und Biographien.* Leipzig: Arbeitskreis Hochschulpolitische Oeffentlichkeit beim StuRa der Universitaet Leipzig, 1999. 382 pp. (Hochschule Ost, Jg. 8,1-2).

The following articles deal with antisemitism: Pasternack, Peer; Schuppener, Georg: Juedisch und intellektuell: Zu den verdoppelten Schwierigkeiten eines DDR-geschichtlichen Themas (8-20); Jonas-Maertin, Esther; Mertens, Lothar: Intellektuelle in den juedischen Gemeinden in der Fruehphase der DDR (47-61); Herzog, Andreas: War die DDR antisemitisch? Kritische Anmerkungen zu den Studien einiger Historiker (62-74); Hollitzer, Siegfried: Vom Evangelisch-Lutherischen Ausschuss fuer Mission unter Israel zur Juedisch-christlichen Arbeitsgemeinschaft Leipzig: Streiflichter [On Protestant action against antisemitism in East Germany.] (75-89); Brentjes, Sonja; Kuczynski, Thomas: "Es gibt ebensowenig eine juedische Oekonomie wie eine deutsche Physik oder eine marxistische Biologie": Interview mit Thomas

Kuczynski am 1.10.1998 [Kuczynski denies having had any experience of antisemitism in the GDR.] (90-105). *RW*

0680 Wiedemann, Wilfried: Memorial Sites in Lower Saxony: Reminders of German Crimes during the Nazi Era. *Midstream* 47, 4 (2001) 5-7.

Memorials in the German federal states have developed independently and in different ways. Discusses two examples in Lower Saxony – Bergen-Belsen and the former Israelite School of Horticulture in Ahlem. In Bergen-Belsen the British liberators commissioned a German architect to design the memorial; the result was the covering up or removal of all of the camp buildings. The site of the mass graves was landscaped like a park. No information was available as to what had happened at the site. Only in the 1980s was pressure increased to improve the situation, and in 1990 the new building for the documentation center was completed. In Ahlem, the memorial site was opened in 1987, dealing with the history of the site as a place of Jewish culture and as a place of the persecution of many people. Notes the importance of citizens' action groups for the culture of remembrance in Germany. *SSC*

Great Britain

0681 Caven, Hannah: Horror in Our Time: Images of the Concentration Camps in the British Media, 1945. *Historical Journal of Film, Radio and Television* 21, 3 (Aug 2001) 205-253.

Examines the harrowing images from Bergen-Belsen shown in newspapers and newsreels in Britain in the summer of 1945. Discusses who photographed these images and for what purpose. The bulk of these images were produced by the men of the British Army Film and Propaganda Unit (AFPU), who were among the first troops to enter the camp and remained there for several weeks. Describes the shock they experienced in the first days in the camp, as well as their guiding principles in choosing the images to be shot: they emphasized, in particular, the humanity of their subjects. The AFPU's photographs and film footage shocked the people who viewed them. Besides arousing resentment against the Germans, they raised the question of what the British government had known and why it did nothing. The AFPU's camp images convinced skeptics that reports on Nazi camps were neither atrocity propaganda nor "Jewish stories." *DR*

0682 Levin, Itamar: *His Majesty's Enemies: Great Britain's War against Holocaust Victims and Survivors*. Trans.: Natasha Dornberg, Judith Yalon-Fortus. Westport, CT: Praeger, 2001. xviii, 238 pp. Originally published as "Oyevei hod malkhuto: Britania, ha-Shoah u-rekhush ha-korbanot" (Tel-Aviv: Ministry of Defence, 1999).

0683 Miller, Rory: *Divided against Zion: Anti-Zionist Opposition in Britain to a Jewish State in Palestine, 1945-1948*. London: Frank Cass, 2000. xii, 275 pp.

Examines the ideological tenets and activities of three British-based anti-Zionist bodies active in 1944-48: the Jewish Fellowship, the Arab Office, and the Committee for Arab Affairs. Contends that many of the era's leading non-Jewish anti-Zionists, while claiming not to be antisemitic, displayed an attitude which was derogatory to both Judaism and Jewry. Some of them also compared Zionism to Nazism. *DR*

0684 Stetz, Margaret D.: Rebecca West and the Nuremberg Trials. *Peace Review* 13, 2 (June 2001) 229-235.

Rebecca West (born Cicely Fairfield, 1892-1983), a British writer and journalist, was dispatched to the Nuremberg Trials as a correspondent. The trials impressed her; she returned to this topic several times and summed up her impressions of the trials and of postwar Germany in the 1955 volume "A Train of Powder." Her narrative became part of the master narrative about Nuremberg which prevailed in Britain for decades. West's essays focused on Germany and the Germans. Paradoxically, although she noted popular German antisemitism which survived the defeat of 1945, she never related to the sufferings of Jews as the targets of organized genocide. Thus, she shares responsibility for the fate of those refugees who were turned away by the British government and for the high level of ignorance about the Holocaust in Britain of the 1950s. *DR*

Italy

0685 *La memoria della legislazione e della persecuzione antiebraica nella storia dell'Italia repubblicana*. Milano: FrancoAngeli, 1999. 119 pp.

A collection of papers presented at a meeting of IRSIFAR (Istituto Romano per la Storia d'Italia dal Fascismo alla Resistenza), Rome, October 1998. Partial contents: Miccoli, Giovanni: Tra rimozione e memoria: Antisemitismo e Shoah nel mondo cattolico (9-26); Schwarz, Guri: Identità ebraica e identità italiana nel ricordo dell'antisemitismo fascista (27-43); Levi, Fabio: La memoria della persecuzione a Torino (45-54); Koch, Francesca; Lunadei, Simona: Il 16 ottobre nella memoria cittadina (55-70); Garribba, Pupa: "Occorre vedere e non guardare in aria" (71-76); Farinelli, Fiorella: La difficile staffetta della memoria (77-82); Gioia, Annabella: Nuovi sguardi sull'antisemitismo italiano (83-94). *MZ*

0686 Roth, Philip; Levi, Primo: Primo Levi et Philip Roth parlent boutique. *La Nouvelle Revue Française* 559 (Oct 2001) 1-19.

Presents a conversation held by Roth with Primo Levi in September 1986 in Levi's home in Torino. They discussed Levi's life experiences, his profession as a chemist and director of a factory, his career as a writer, the impact of Auschwitz on his writings, his condition of being a survivor, and his feelings of being "other" in Italy. *HV*

Poland

0687 Bauman, Janina: *Nigdzie na ziemi* [Nowhere on Earth]. Warszawa: Żydowski Instytut Historyczny, 2000. 258 pp.

Memoirs dealing with the postwar period. Describes, among other topics, the engagement of herself and her husband (the sociologist Zygmunt Bauman) in the communist cause in Poland, and their disappointment with this ideology. Pp. 216-258 present a personal account of the antisemitic campaign in Poland in 1968, which led to the family's emigration. *HV*

0688 Blatman, Daniel: Polish Jewry, the Six-Day War, and the Crisis of 1968. *The Six-Day War and World Jewry,* ed. Eli Lederhendler. Bethesda, MD: University Press of Maryland, 2000. Pp. 291-310.

Poland's Jewish community dwindled throughout the period of 1946-66; the anti-Jewish climate in Poland after 1956 caused a wave of Jewish emigration. The 1960s were marked by a crisis in Gomułka's leadership, which gave rise to a nationalist opposition, the Moczar group, in the Polish Communist Party. Mieczysław Moczar attacked Gomułka with anti-Jewish and anti-Zionist arguments, accusing him of failure to deal with "Zionists" and accusing the Jews of imposing a Stalinist system on Poland. Gomułka opposed Moczar's nationalism, but he also enlisted antisemitism and anti-Zionism to his cause. Beginning in June 1967, he accused the Jews of being an anti-Polish and "reactionary" force. Slogans equating Zionism with Nazism were also introduced into the propaganda. Thus, the Six-Day War served as a catalyst for the virtual expulsion of the Jews from Poland in 1968-69. The propaganda campaign, which began with guidelines from Moscow in June 1967 and continued with a futile attempt to cope with Poland's internal problems, metamorphosed into an antisemitic campaign. Jewish life in Poland was suppressed and most of the Jews left the country. *DR*

0689 Grabski, August: Sytuacja Żydów w Polsce w latach 1950-1957 [The Situation of the Jews in Poland in the Years 1950-1957]. *Biuletyn Żydowskiego Instytutu Historycznego* 196 (Dec 2000) 504-519.

Describes the political context of the founding of the Towarzystwo Społeczno-Kulturalne Żydów (Socio-Cultural Association of the Jews) in Poland in 1950. Presents reasons why many Jews were not pleased with the Association, whose leaders expounded anti-Zionist propaganda in favor of internationalism. After the death of Stalin, when the general political atmosphere improved, many Jews decided to emigrate due to disappointment with life under a communist regime and because of antisemitism in the years 1956-57. The Communist Party leaders condemned antisemitism, but this was seen as a move which served the internal struggles inside the party. In 1957, Hersz Smolar, the chairman of TSKŻ, wrote a memorandum to the Central Committee of the Communist Party entitled "Kwestia żydowska w chwili obecnej" ["The Jewish Question Now"] (appended on pp. 515-519). Suggests that this document led to the sharp condemnation of antisemitism by Premier Józef Cyrankiewicz and to the issuing of a circular by the Central Committee in April 1957 against antisemitism. *HV*

0690 Stola, Dariusz: "The Anti-Zionist Campaign Was to Persecute Someone for His Political Opinion." *Justice* 29 (Fall 2001) 22-25.

In 1967, in the wake of the Six-Day War in the Middle East, the First Secretary of the Polish Communist Party, Władysław Gomułka, launched an anti-Zionist campaign which went beyond the campaign common to all communist countries except Romania. He branded Zionists and, in fact, all Polish Jews as a fifth column. In 1968, after the student rebellion in Warsaw, Gomułka once more mentioned the "Zionist threat" in order to revile the students. The main objectives of Gomułka's last propaganda campaign were to purge the party's top echelons of undesirables and to reconsolidate the party leadership in his own hands. The purge quickly spread from the party to other spheres. In 1968-70 Polish Jews were "allowed" to emigrate. Nothing of what Gomułka did in this period was linked to any real transgressions of Zionism, but the anti-Zionist slogan helped him to attain his political goals. *DR*

0691 Szlajfer, Henryk: Konfrontation der Stereotypen: Zum Verhaeltnis zwischen Polen und Juden. *Das Juedische Echo* 50 (Oct 2001) 223-236.

Discusses the persistence of antisemitism in postwar Poland, and its nature – religious, racist, or cultural, asserting that all these form a whole. The Poles' perception that the Holocaust was perpetrated by the Germans, while they themselves were victims, and their loss of sovereignty to the communists after the war, "saved" them from having to confront their past. Thus antisemitism could survive unchecked, transmitted through the family, the school, and the Church, encouraged by the regime, and reinforced by the identification of Jews with communists. It was an antisemitism not based on familiarity with actual Jews; most of the Holocaust survivors left the country. Jews, in turn, viewed Poles stereotypically as antisemites and active partners of the Nazis. Since its democratization, Poland and the Polish Church are starting to catch up with the rest of the world in combating antisemitism. Suggests that Jews would do well to support them by giving up their own negative stereotypes. *RW*

0692 Wolak, Arthur John: *The Polish Government's Anti-Semitic Campaign, 1967-68.* M.A. thesis – California State University, Dominguez Hills, 2000. 108 pp. Unseen.

Examines the causes of the 1967-68 anti-Zionist campaign in Poland, which led to the effective eviction of the majority of Poland's Jewish population. While the 1967 Arab-Israeli war served as the pretext for the persecution of Poles of Jewish ancestry, argues that the overt persecution of Jews, which accelerated after the student protests of March 1968, primarily served domestic political aims. The anti-Zionist campaign was intended to appeal to broad antisemitic sentiments in Polish society, while helping the leader of Poland's Communist Party, Gomułka, to appease an influential rival faction and still retain Moscow's favor. Besides examining domestic economic and political problems that contributed to the Polish regime's use of antisemitism as a political weapon, attention is also given to Soviet anti-Jewish tactics and policies for their suspected influence on Poland's antisemitic campaign.

Ukraine

0693 Briman, Shimon: Zimnii psikhoz: Kharkov i "delo vrachei" 1953 g. ["A Winter Psychosis": Kharkov and the "Doctors' Plot Case" of 1953]. *Istoki* 8 (2001) 25-42.

Traces effects of the "Doctors' Plot" affair in Kharkov, a city where Jews were over-represented in medicine since the 19th century. The anti-Jewish campaign in Kharkov began in January 1953 with restrictions against physicians of Jewish origin. The party leadership, and not the NKVD, initiated these measures. After that, measures were taken against intellectuals and administrative workers of Jewish origin. The antisemitic campaign was supported by the party press. Stalin's death marked the end of the "Doctors' Plot" affair. Some of the initiators of the anti-Jewish campaign in Kharkov did a turnabout and condemned the "provocation."

Some of them were later conspicuous in carrying out anti-Jewish policies in the 1950s-60s. *DR*

0694 Goldshtein, Mikhail I.: The "Doctors' Plot" in the Poltava Region (Excerpts from the Poltava Press and Archives). *Shvut* 9 [25] (2000) 93-101.

The beginning of the "Doctors' Plot" affair in January 1953 was accompanied by an antisemitic press and radio campaign in the Poltava region of Ukraine. Gives examples of anti-Jewish publications in the local newspaper "Zorya Poltavshchyny." Presents four top secret reports to the Poltava regional authorities and to authorities in Kiev describing the popular reaction to the beginning of the "Doctors' Plot" affair, its abrupt end, as well as the dismissal of the Minister of Interior Beriya in July 1953. The documents testify to a great measure of fear among the Poltava Jews and to strong anti-Jewish sentiments among non-Jews. After Stalin's death many people believed that Jews murdered Stalin; they alleged that Beriya was Jewish and that the release of the doctors "by Beriya" was the result of a Jewish plot. *DR*

USSR

(see also Belarus; Ukraine)

0695 Altshuler, Mordechai: The Synagogue in the Soviet Union on Passover 1953. *Jews in Eastern Europe* 46 (2001) 58-76.

Discusses the uniqueness of the Soviet antisemitic campaign that began on 13 January 1953. It affected every Soviet Jew, with its threats (including the threat of deportation) and with individual experiences of suspicion and hate. The Stalinist campaign began with the announcement of the "Doctors' Plot" that instilled fear of Jewish medical personnel throughout the USSR; it aroused resentment against Jews and a desire to punish them. The synagogues, the only Jewish institutions allowed to exist, albeit in reduced numbers, were threatened with closure and, in the reigning atmosphere, Jews feared to attend. However, official statistics reveal that synagogue attendance increased dramatically during Passover 1953, after Stalin's death, when the official press announced the dropping of the case against the Jewish physicians. Many non-religious Jews, including some who held important jobs in the Soviet economy, attended synagogue to express relief at the deliverance of their people. *YC*

0696 Averbukh, Semyon L.: *Nasytilis my prezreniyem...* [We Are Gorged with Contempt...]. Kiev: Glav. Spetsializirovannaya Red. Lit-ry na Yazykakh Natsionalnykh Menshinstv Ukrainy, 2000. 444 pp.

A collection of essays, most of them written in the early 1990s, dealing mainly with the "anti-Zionist" campaign unleashed in the USSR in 1967-90 and its aftermath in Russia and Ukraine in the 1990s. The goal of this campaign was to incite mass hatred toward Israel and the Jews. Gives statistics on "anti-Zionist" publications in the USSR in the 1960s-80s. Discusses specific authors, such as Yurii Ivanov, Vladimir Begun, Yevseyev, Korneyev, Romanenko, and others. Antisemitism disguised as anti-Zionism continued to be part of the ideology of nationalist and communist groups in Russia and Ukraine, of which Pamyat was the most conspicuous. Focuses, also, on the group of nationalist and antisemitic intellectuals which arose in 1985, known as "Zayedinshchiki," especially on its most well-known representative, Shafarevich. *DR*

0697 Kostyrchenko, Gennadii V.: *Tainaya politika Stalina: Vlast i antisemitizm* [Stalin's Secret Policy: State Power and Antisemitism]. Moskva: Mezhdunarodnye Otnosheniya, 2001. 779 pp.

Stalin, outwardly a communist internationalist, secretly pursued an anti-Jewish policy from the late 1930s to 1953. He tried to suppress grassroots pogromist antisemitism, but he implemented state antisemitism. The elimination of Jews from the political elite and the suppression of Jewish culture in the 1930s was not motivated by antisemitism, but was part of a broader suppression campaign. Official antisemitism ripened in the late 1930s, when state ideology turned from internationalist to Russian nationalist. In the 1940s-50s, Stalin drifted into rabid antisemitism. However, as a pragmatic politician, unable to declare his anti-Jewish stance, he was inconsistent in his ethnic politics. Traces the purge of Jews from the political, cultural, industrial, etc. elites in the late 1930s-early 1950s. Highlights the roots of Soviet official anti-Zionism. Describes postwar anti-Jewish affairs, such as that of the Jewish Anti-Fascist Committee and the "Doctors' Plot". Denies that Stalin planned a mass resettlement of Jews: such a measure demanded radical and dangerous changes in the official ideology. Stalin's antisemitism stemmed not from racism but from his "ethnic mentality" and his fear of a "Zionist influence" on him and his politics. *DR*

0698 Madievskii, Samson A.: 1953: La déportation des Juifs soviétiques était-elle programmée? *Cahiers du Monde Russe* 41, 4 [Oct-Dec 2000] (2001) 561-568. Appeared in English in "Midstream" 46, 3 (Apr 2000).

0699 Madievskii, Samson A.: Czy Żydom radzieckim zagrażała w 1953 roku deportacja? [Were Soviet Jews Threatened by Deportation in 1953?]. *Biuletyn Żydowskiego Instytutu Historycznego* 194 (June 2000) 215-221. Appeared in English in "Midstream" 46, 3 (Apr 2000).

0700 Madievskii, Samson A.: Standen die sowjetischen Juden im Jahre 1953 vor einer Deportation? *Freiburger Rundbrief* 8, 1 (2001) 13-20. Appeared in English in "Midstream" 46, 3 (Apr 2000).

0701 Madievskii, Samson A.: Were Soviet Jews to Be Deported? *Midstream* 46, 3 (Apr 2000) 18-21.

Discusses the pros and cons of the contention that Stalin had a plan to deport Soviet Jews that was thwarted by his death. Focuses on the opposing views of Vladimir Naumov and Gennadii Kostyrchenko. The former argues that such a deportation was indeed being planned; the latter maintains that there is no clear proof of it. Naumov's arguments include the antisemitism of Stalin himself and antisemitic attitudes among the Soviet people. Fears of antisemitism among Soviet Jews gave credence to the possibility of a plan for deportation, especially in the aftershock of the Holocaust that encompassed part of Soviet Jewry, but cannot confirm its actuality. This article is based on a discussion which took place at a symposium at the Universitaet Eichstaett in 1998, in which Kostyrchencko, Naumov, and Madievskii took part. *YC*

0702 Rubenstein, Joshua; Naumov, Vladimir P., eds.: *Stalin's Secret Pogrom: The Postwar Inquisition of the Jewish Anti-Fascist Committee*. Trans.: Laura Esther Wolfson. New Haven: Yale University Press, in association with the United States Holocaust Memorial Museum, 2001. xxvii, 527 pp. Originally published as "Nepravednyi sud: Poslednii stalinskii rasstrel: Stenogramma sudebnovo protsessa nad chlenami Yevreiskovo antifashistskovo komiteta," ed. V.P. Naumov (Moskva: Nauka, 1994).

Pp. 1-64 contain an introduction written for the English-language edition, "Night of the Murdered Poets," by Joshua Rubenstein, drawing on newly-released documents in Moscow archives and on interviews with relatives of the defendants. It sets the trial in the historical and political context of Stalin's antisemitic campaign. *SSC*

1945-1990: Islamic World

General

0703

בודנסקי, יוסף: איסלאמיות אנטישמית כמכשיר פוליטי. תרגום: ברוך קורות. תל-אביב: תמוז, 2000. 208 עמ'.

[Bodansky, Yossef: *Islamic Anti-Semitism as a Political Instrument*. Trans.: Baruch Korot. Tel-Aviv: Tammuz, 2000. 208 pp.] Originally published as "Islamic Anti-Semitism as a Political Instrument" (Shaarei Tikva, Israel: Ariel Center for Policy Research; Houston, TX: Freeman Center for Strategic Studies, 1999).

0704

לוין, איתמר: שקיעה במזרח: חיסול הקהילות היהודיות במדינות ערב ושוד רכושן. תל-אביב: משרד הביטחון - ההוצאה לאור, 2001. 366 עמ'.

[Levin, Itamar: *Decline in the East: The Destruction of the Jewish Communities in the Arab Countries and the Theft of Their Property*. Tel-Aviv: Ministry of Defence, 2001. 366 pp.]

Discusses persecution of the Jews and expropriation of Jewish property in Iraq (pp. 17-138), Egypt (pp. 139-225), and Syria (pp. 226-290). In Iraq, in the 1940s, many wealthy Jews were persecuted by the government for economic gain. In 1950 Jews were allowed to leave the country, with almost no possessions, if they gave up their citizenship. Although their assets were frozen from 1951, ca. 120,000 Jews immigrated to Israel at the time. In Egypt, many Jews were arrested with the establishment of the State of Israel in May 1948. In 1949-50, ca. 20,000 Jews left the country legally or illegally; most of their possessions were confiscated. In Syria, persecution began in 1947 and worsened in the 1950s, when Jewish property was confiscated and most of the Jews left the country without possessions. Also mentions Jews in Lebanon, who were not persecuted, but left the country in the 1950s-60s. Discusses efforts made by the Israeli government from the 1950s to claim reparations for the Jews of Iraq, Egypt, and Syria. *RE*

0705 Nordbruch, Goetz, ed.: *Die Araber und die Shoa: Ueber die Schwierigkeit dieser Konjunktion*. Trier: Kulturverein AphorismA, 2000. 88 pp.

Pp. 9-33 present an essay by Azmi Bishara, "Die Araber und die Shoa" [based on a lecture held at the University of Innsbruck in the winter semester of 1992/93; first published in "Der Umgang mit dem Holocaust" (1994); appeared in Hebrew in "Zmanim" 53 (1995)], in which he surveys Arab historiography of the Holocaust in connection with the Palestine-Israel conflict. States that the Arab public tends to deny or minimize the extent of the Holocaust because it cannot believe that Jews, if they knew such suffering, would inflict it on others – the Palestinians. Arab historians, even when describing Nazi racist crimes, keep silent on the Holocaust. Like revisionist Israeli historians, they see Zionism as concerned only for itself, not for Diaspora Jews, to the point where it exploited antisemitism and the Holocaust and even collaborated with the Nazis. In the eyes of many historians, Zionism is itself racism, and support for the State of Israel is the Western world's attempt to rid itself of its moral debt to the Jews at the expense of the Arabs. Pp. 34-39 contain five brief items by other persons (including Edward Said and Roger Garaudy) on this topic. Pp. 40-73 contain reactions to Bishara's views by historians and scholars. *RW*

Egypt

0706

לוין, איתמר: רכושם השדוד של יהודי מצרים. נתיב 14, 1 (ינ' 2001) 42-48.

[Levin, Itamar: The Stolen Jewish Property in Egypt. *Nativ* 14, 1 (Jan 2001) 42-48.]

Reports on the persecution of Jews in Egypt in 1948, when Jewish property was confiscated and many Jews were arrested, and in 1956, when many Jews were forced to leave the country virtually penniless. Ca. 50,000 Jews lived in Egypt before the Sinai Campaign (1956), and their property was estimated to be worth ca. $100 million at the time. In 1956, Jewish businesses were confiscated and salaried Jews lost their jobs. The Secret Police insisted that the Jews leave; those who hesitated were beaten. After 1957, Jews who were British or French citizens received partial compensation for their losses, but others received no compensation at all. After the Six-Day War of 1967, almost all the remaining Jews who left the country were forced to sign documents giving up their possessions. *RE*

Iraq

0707

לוין, איתמר: רכושם השדוד של יהודי עירק. נתיב 13, 6 (נוב' 2000) 46-52.

[Levin, Itamar: The Stolen Property of Iraqi Jews. *Nativ* 13, 6 (Nov 2000) 46-52.]

Based on material held in various archives in Israel, discusses the plunder of Jewish property in Iraq. There were ca. 130,000 Jews living in Iraq in 1947, most of them involved in business and services. Estimates that the value of their property was ca. $560 million at the time (now ca. $4 billion). After May 1948, the Jews suffered from discrimination: they were not allowed to leave the country, their homes were looted by police, many were arrested and some were executed, all those who had government jobs were dismissed, and they were not allowed to deal in imports. Thousands of Jews escaped from the country illegally in 1948; in 1949 they were permitted to leave, but lost their citizenship. In 1951 the Iraqi government froze Jewish assets, but did not confiscate them, in order to avoid compensation. The government took possession of the Jewish properties gradually, ostensibly in order to pay for their care. By the end of 1951, ca. 120,000 Jews had left the country, most of them with no possessions. The assets of the remaining Jews were frozen in 1967; they were forced to live in poverty until they also fled. *RE*

1945-1990: Israel

0708 Ben-Amos, Avner; Bet-El, Ilana: Holocaust Day and Memorial Day in Israeli Schools: Ceremonies, Education and History. *Israel Studies* 4, 1 (Spr 1999) 258-284.

Traces the development of commemorative ceremonies in Israel for two thematically similar days, both occurring in the spring, one week apart: the Holocaust Remembrance Day (Yom Hashoah) and the Memorial Day for Fallen Israel Defense Forces Soldiers. Yom Hashoah was introduced later than Memorial Day – only in 1951 – and underwent a long evolution, in the course of which it gained greater importance and its ritual came close to that of Memorial Day. Examines the ceremonies as part of a series of holidays and remembrance days which start with Passover and end with Jerusalem Day six weeks later. Focuses on similarities and differences in the two ceremonies' micro-narratives. *DR*

0709

בן-עמוס, אבנר; בית-אל, אילנה: טקסים, חינוך והיסטוריה: יום השואה ויום הזכרון בבתי ספר בישראל. חינוך והיסטוריה: הקשרים תרבותיים ופוליטיים, עורכים רבקה פלדחי, עמנואל אטקס. ירושלים: מרכז זלמן שזר לתולדות ישראל, 1999. עמ' 457-479.

[Ben-Amos, Avner; Bet-El, Ilana: Ceremonies, Education and History: Holocaust Day and Memorial Day in Israeli Schools. *Hinukh ve-historia: Heksherim tarbutiim u-politiim* [Education and History: Cultural and Political Contexts], eds. Rivka Feldhay, Immanuel Etkes. Jerusalem: Zalman Shazar Center for Jewish History, 1999. Pp. 457-479.] Appeared in English in "Israel Studies" 4 (1999). Appeared in Hebrew also in "Defusim shel hantsahah: Asufat ma'amarim" (Tel-Aviv: Ministry of Defence, 2000).

0710 Dror, Yuval: Holocaust Curricula in Israeli Secondary Schools, 1960s-1990s: Historical Evaluation from the Moral Education Perspective. *The Journal of Holocaust Education* 10, 2 (Fall 2001) 29-39.

Surveys and criticizes the way the Holocaust has been taught in Israel, as evidenced in curricula and materials located in the pedagogical center of Yad Vashem. Asks whether, in recent decades, the Holocaust might have received too much attention, particularly in terms of Israel's civil religion and politicization of the Holocaust. Advocates balancing the emotional with the cognitive aspects of the Holocaust, avoiding focusing only on the Holocaust in Jewish history, integrating personal Holocaust testimonies, eschewing demagogic use of the Holocaust, and raising consciousness about other genocides and about the moral responsibility of the individual to oppose discrimination and oppression. *YC*

0711 Enemark, Birgitte: Holocaust Remembrance and Education in the State of Israel 1948-2000. *Nordisk Judaistik* 22, 2 (2001) 107-130.

Traces the development of Holocaust consciousness in Israel and its impact on Israeli education. In Israel's first decade Holocaust education focused on heroism. With the Eichmann trial, the voices of the victims began to be heard. Identification with Holocaust victims developed along with Israeli feelings of vulnerability after the 1973 Yom Kippur War. Thereafter two new high school programs were developed. The Holocaust became a compulsory subject in schools in 1982. The Likud's rise to power led to Holocaust politicization and fragmentation, with rightists using Holocaust rhetoric for nationalistic purposes and leftists for universalistic ones. Textbooks included new themes, e.g. the spiritual struggle of religious Jews to preserve their values in the face of Nazism, and the comparison of the Holocaust to other genocides. Trips to death camps in Poland are controversial, since they often lead to nationalistic conclusions. Although identification with the victims is high, some warn of the dangers of Israel's obsession with the Holocaust. *YC*

0712 Sheffi, Na`ama: *The Ring of Myths: The Israelis, Wagner and the Nazis*. Trans.: Martha Grenzeback. Brighton: Sussex Academic Press, 2001. x, 182 pp. Originally published as "Taba`at ha-mitosim: Ha-Yisraelim, Wagner ve-ha-Nazism" (Haifa: University of Haifa Press; Tel-Aviv: Zmora-Bitan, 1999).

1945-1990: North America

USA

0713 Feldman, Egal: *Catholics and Jews in Twentieth-Century America*. Urbana, IL: University of Illinois Press, 2001. xiii, 323 pp.

Discusses, briefly, Catholic-Jewish relations in the U.S. from 1890 to the end of the 20th century, against the background of the theology of contempt that permeated Christian attitudes toward Jews and Judaism from early Christianity. Highlights Catholic insensitivity toward the suffering Christianity has inflicted on Jews. In terms of the Holy Land, Catholic anti-Zionism reflected theological resentment of Jews. Focuses on the post-1945 period and issues related to Catholic repentance for its failings regarding the Jews, particularly during the Holocaust. Such a change, involving revision of traditional Christian teaching, was initiated with Vatican II, which revolutionized Catholic attitudes (including in regard to conversion) and opened possibilities for Catholic-Jewish dialogue. The latter was complex, with continuing Catholic insensitivities, e.g. regarding the significance of Israel to the Jews. American Catholics also played a role in introducing revisions in Catholic education and liturgy, and fighting anti-Jewish prejudice. *YC*

0714 Kun, Josh: The Yiddish Are Coming: Mickey Katz, Antic-Semitism, and the Sound of Jewish Difference. *American Jewish History* 87, 4 (Dec 1999) 343-374.

Argues that the music and words of band leader, clarinetist, and Yiddish-English parodist Mickey Katz represented unassimilated notes of Jewishness or "Yidditude" that in 1950s America, at least publicly, often received a cold reception from Jews who wished to assimilate. The latter rejected Katz's "aurality of Jewish difference" and his parodies that mocked American values and heroes (like the cowboy). The entertainment industry followed suit in marginalizing this figure, whose musical Yinglish revealed and reveled in a double Jewish marginality (vis-à-vis the majority culture and other minorities). Antisemites resented both Jews wearing white masks and the unassimilable Katz, who made fun of their White American values. *YC*

0715 Staub, Michael E.: "Negroes Are Not Jews": Race, Holocaust Consciousness, and the Rise of Jewish Neoconservatism. *Radical History Review* 75 (Fall 1999) 3-27.

From the mid-1950s to the early 1970s, leaders and intellectuals of the American Jewish community managed to reinterpret the memory of the Holocaust. While, at the early stage, Holocaust images evoked analogy with anti-Black politics in the USA and encouraged Jewish radicals to support the African-American struggle for social justice, in the late 1960s-70s those images helped the conservatives to discourage such support. In the 1960s a neo-conservative Holocaust consciousness emerged, whose leitmotif was a focus on Jewish problems and survival in the post-Auschwitz world, isolating them from the problems of others. Many conservatives also expressed concern over growing Black antisemitism. Despite the conservatives' contempt for "Jewish masochism," which expressed itself in involvement with the Black liberationist struggle and a call for a universalist interpretation of the "lessons of the Holocaust," Jewish left-wing radicalism did not disappear in the 1970s. *DR*

1991-2001: General

0716 *Antisemitizmus na konci 20. storočia: Zborník referatov z konferencie (Nitra, 15.-17. máj 2000)* [Antisemitism at the End of the 20th Century: A Collection of Papers from a Conference (Nitra, 15-17 May 2000)]. Bratislava: Judaica Slovaca, 2001. 259 pp.

Partial contents: Vago, Raphael: Antisemitizmus a politika v postkomunistickej strednej a východnej Európe [Antisemitism and Politics in Post-Communist Central and Eastern Europe] (15-26); Gbúrová, Marcela: Antisemitizmus v kontexte teoretickej reflexie národného a občianskeho princípu [Antisemitism in the Context of Theoretical Reflection on National and Civic Principles] (27-32); Stauber, Roni: Od revizionizmu k popieraniu holokaustu: David Irving ako prípadová štúdia [From Revisionism to Holocaust Denial: David Irving as a Case Study] (53-71); Benz, Wolfgang: Židia v Nemecku po roku 1945: Medzi antisemitizmom a filosemitizmom [Jews in Germany after 1945: Between Antisemitism and Philosemitism. Deals, mainly, with the 1990s.] (73-80); Halpert, Marta S.: Činnosť Ligy proti hanobeniu (ADL) vo svetle politickej situácie v Rakúsku [The Anti-Defamation League (ADL) in Light of the Political Situation in Austria] (81-85); Shafir, Michael: Človek, ktorého s radosťou nenávidia: Slimačí domček medzi holokaustom a gulagom [The Man They Love to Hate: The Snail's House between Holocaust and Gulag. On Norman Manea. Appeared in English in "East European Jewish Affairs" 30 (2000).] (87-117); Mindlin, Vladimir: Prehlad antisemitských publikácií a prejavov na Ukrajine [A Survey of Antisemitic Publications and Periodicals in Ukraine] (119-138); Cała, Alina: Polsko-postoje k Židom v svetle sociologických výskumov [Polish Attitudes toward Jews in Light of Sociological Research] (139-144); Jelinek, Yeshayahu Andrej: Antisemitizmus v postkomunistickej východnej Európe a štát Izrael [Antisemitism in Post-Communist Eastern Europe and the State of Israel. Relates to Israeli views of the phenomenon.] (145-159); Fatran, Gila: Životnosť antisemitských javov [The Viability of Antisemitic Manifestations. In Slovakia. Appeared in English in "Jews and Antisemitism in Public Discourse of the Post-Communist European Countries" (2001).] (161-176); Mešt'an, Pavol: Niektoré špecifické podmienky reprodukcie negatívneho stereotypu Žida na Slovensku po roku 1989 [Some Specific Conditions for the Reproduction of the Negative Stereotype of the Jew in Slovakia since 1989] (177-189); Spitzer, Martin: Antisemitizmus a kresťanstvo na Slovensku koncom 20. storočia [Antisemitism and Christianity in Slovakia at the End of the 20th Century] (191-195); Baranova, Daniela: Postoje evanjelickej cirkvi k antisemitizmu na Slovensku v 20. storočí [Attitudes of the Evangelical Church toward Antisemitism in Slovakia in the 20th Century. Both during and after the Holocaust.] (197-209);

Poláková, Eva: Možnosti Múzea židovskej kultúry pri výchove mládeže k tolerancii [Possibilities of the Museum of Jewish Culture in Educating the Young to Tolerance] (211-223); Slnekova, Veronika: Antisemitizmus – holokaust: Výskum a školská prax (správa o výskume a výučbe holokaustu na katedre histórie FF UKF v Nitre) [Antisemitism – Holocaust: Research and Teaching Practice (Report on Holocaust Research and Teaching in the Department of History of Constantine the Philosopher University in Nitra)] (225-236); Čuchranová, Darina: Informácia versus intolerancia [Information versus Intolerance. On Christian-Jewish relations in Slovakia.] (237-243). *SSC*

0717 Dahmer, Helmut: Antisemitism and Xenophobia: How to Solve the Riddle of the Sphinx. *Australian Journal of Jewish Studies* 15 (2001) 72-87.

A paper delivered at the conference "Psychoanalysis among the Disciplines," University of Michigan, 1994. Reflects on the nature of antisemitism and ways to overcome it in contemporary Western society. In the context of psychoanalysis, interprets antisemitism as part of a person's inherited cultural tradition. The social sciences and psychoanalysis, operating together, need to expose the genesis of this antisemitic tradition – the way in which historical experiences of feudal and capitalist societies with Jewish minorities came to be transformed into social mythology. After the Holocaust, antisemitism became a form of xenophobia. Calls on psychoanalysts to engage in critical examination and relativization of this difficult cultural inheritance, and to develop a transnational ("cosmopolitan," "internationalist") identity. Antisemitism and xenophobia should be held in contempt and outlawed. *DR*

0718 Eden, Shevach: The Work and Recommendations of the Polish-Israeli Textbooks Committee. *Polin* 14 (2001) 306-314.

In 1991 a cultural agreement was signed between Poland and Israel, one of whose resolutions was to establish national committees in both countries to examine history and geography textbooks and their treatment of the two nations, with the aim of rectifying mistakes that could lead to prejudices and distortions of the truth. Discusses the functioning of the committees between 1992-95. One prime area of concern dealt with by the scholars who examined the books was the indifference of the majority of Poles to the fate of the Jews in World War II. Topics suggested for inclusion in Polish history textbooks were the treatment of surviving Jews (e.g. the Kielce pogrom), the role of the Polish Catholic Church in spreading antisemitism, and continuing antisemitic stereotypes in a Poland largely without Jews. Members of the Polish committee raised questions about anti-Polish stereotypes in Israel (e.g. identifying Poles with Nazis) and faults in the program of Israeli student trips to Poland.

The committees drafted recommendations for changes in the textbooks to further mutual understanding. *YC*

0719 *Gesher: Journal of Jewish Affairs* [Special edition] (Oct 2001).

This issue, entitled "The Jewish People at the Threshold of the New Millennium," presents brief papers comprising a written symposium, published on the eve of the 11th Plenary Assembly of the World Jewish Congress. The following papers deal with antisemitism and/or the Holocaust: Gelber, Yoav: Holocaust Denial in Israel [On Israeli revisionist history which relativizes the Holocaust, obscures its uniqueness, and links it to the Jewish-Palestinian conflict.] (37-40); Gottschalk, Alfred Yechiel: Don't Blur the Uniqueness of the Holocaust (44-46); Porat, Dina: The Church and the Jews: Stereotypes Derive from Education and from Memories of the Past [On points of friction in current relations between the Jewish people and the Catholic Church.] (64-68); Wistrich, Robert Solomon: Indeed, a War between Two Civilizations [In the context of the Islamic assault against the West, discusses Islamic Judeo-phobia as the most virulent, active, and totalitarian form of antisemitism that exists today.] (100-104); Katz, Steven Theodore: A Century of the Unexpected, the Unthinkable, and the Impossible – 1900 to 2000 [Pp. 117-120 deal with the Holocaust.] (112-122). *SSC*

Responses to the Holocaust

0720

באואר, יהודה: השואה: המיוחד והאוניברסלי. חינוך לתרבות בחברה רב-תרבותית, עורכת מרים ברלב. ירושלים: האוניברסיטה העברית, בית-הספר לחינוך, 2000. עמ' 159-168.

[Bauer, Yehuda: The Holocaust: The Unique and the Universal. *Hinukh le-tarbut be-hevrah rav-tarbutit* [Cultural Education in a Multi-Cultural Society], ed. Miriam Barlev. Jerusalem: Hebrew University, School of Education, 2000. Pp. 159-168.]

Based on a lecture delivered at the Bundestag in Bonn, 27 January 1998. Enumerates several reasons for the uniqueness of the Holocaust: the Jews were murdered simply for having been born Jewish; the intention was to murder all Jews throughout the world, while other genocides were restricted to specific areas; the ideology behind the genocide was based on an irrational racial antisemitism, including belief in a Jewish world conspiracy, while other genocides had more realistic motivations; and the Nazis were also determined to humiliate the Jews before they murdered them. Although the Holocaust was unique, the threat of its recurrence is universal, because extremism can lead to genocide. Germany bears responsibility for the culture which allowed for the rise of Nazism – since the German intelligentsia supported racism and antisemitism, it was easy to convince the general public to do so. The Jews became a target for the German public's general discontent and a symbol of the scorned Western culture. However, some Germans rejected Nazism and some even risked their lives to save Jews. Both Germans and Jews must learn from the Holocaust and try to prevent its recurrence. *RE*

0721 Brown, Erica: "Turning All the Past to Pain": Current Trends in the Memorialization of the Holocaust. *The Torah u-Madda Journal* 9 (2000) 108-132.

Current commemoration of the Holocaust is aimed, inter alia, at instilling Jewish identity in secularized Jews. There are several types of Holocaust memorialization: Holocaust museums and memorials, survivor accounts, historical and fictional accounts – all of these are problematic in some respects. Moreover, the attempts by theologians and historians to make the Holocaust a defining event of Jewish history and to separate it in kind and magnitude from other tragedies has resulted in a number of thorny issues for observant Jews. Contemporary writers and rabbis (e.g. Peter Novick, J.B. Soloveitchik) criticize making a new master Jewish story from the Holocaust or making a "Torah" out of its history. The traditional Jewish memorialization of the Temples' destruction is based on creating empty spaces as a memorial to loss. A similar form of memorialization can apply to the Holocaust. The empty spaces in this case are the prewar Jewish communities extinguished by the Holocaust. *DR*

0722 *Cahier International sur le Témoignage Audiovisuel* 6 (Mar 2001).

Partial contents: Pedon, Eric; Walter, Jacques: De la découverte de photographies à l'esthétisation du témoignage: Chronique couleur du ghetto de Lodz (7-17); Beyrak, Nathan: Oral Documentation of the Holocaust of Lithuanian Jewry: A Project of the U.S. Holocaust Memorial Museum / The Jeff and Toby Herr Collection [Includes excerpts from testimonies.] (19-31); Burtin, Jérôme: Shoah, comédie et représentation(s) [Discusses comedy in Holocaust films made between 1940-98.] (33-53); Gotthardt Strage, Alberta: The Use of Audio-Visual Testimony at the Permanent Holocaust Exhibition of the Imperial War Museum in London (55-59); Zarka, Josette: Six ans de malheur: De l'enfance à l'adolescence sous les persécutions nazies [On the difficulties of child survivors in recounting their experiences.] (61-71); Giacomozzi, Carla; Paleari, Giuseppe: Augenzeugenberichte aus den Lagern: RAI Educational erwirbt 50 Videoaufzeichnungen [On video films made for the Italian television station in 2000 about Italians deported in World War II; some of the interviewees were Jews.] (73-83); Stroscio, Renzo: Témoignages des survivants de l'Holocauste: L'expérience de la libération (91-94). *TS*

0723 Cotler, Irwin: The Holocaust, 'Thefticide' and Restitution: A Legal Perspective. *Cardozo Law Review* 20, 2 (Dec 1998) 601-623.

Contends that the plunder of Jewish property during the Nazi period, disregarding who or which state committed it or benefited from it, is a criminal act falling under international law, rather than a civil violation, because it was part and parcel of the genocide of the Jewish people, one of its stages which can be characterized as "thefticide." Thus, restitution is a moral and juridical imperative resulting from international responsibility for crimes defined at the Nuremberg Trials. Lists "national myths," on the basis of which some states have tried to evade responsibility for the plunder (e.g. they were victims themselves, it was committed by the occupiers). Presents nine basic principles of state responsibility for these crimes which call for restitution or reparations. *DR*

0724 Davies, Ian, ed.: *Teaching the Holocaust: Educational Dimensions, Principles and Practice*. London: Continuum, 2000. xiii, 178 pp.

A collection of essays dealing with teaching the Holocaust in Europe and the U.S. *DR*

0725 Eley, Geoff, ed.: *The "Goldhagen Effect": History, Memory, Nazism – Facing the German Past*. Ann Arbor: University of Michigan Press, 2000. 172 pp.

A collection based on papers delivered at a symposium held at the University of Michigan in November 1997. They discuss aspects of the public reception of Goldhagen's book, concentrating on reactions in the USA, Israel, Germany, Austria, and elsewhere in Europe. *DR*

0726 Finkielkraut, Alain: *Une voix vient de l'autre rive*. [Paris]: Gallimard, 2000. 146 pp.

A collection of essays on the imperative to remember the Shoah. Echoing Vladimir Jankélévitch's anti-Hegelian and unhistorical position, insists that after Auschwitz history has lost its sense. Shares Jankélévitch's, Primo Levi's and Emmanuel Levinas' horror of being alive, which he sees as the opposite of the Hegelian belief in the future. Emphasizes the moral obligation to remember, to remain attentive to voices from the past. Warns against using the Holocaust in the service of some ulterior goal. Also cautions against casual or disturbing "friends of the Holocaust," who either freeze the memory of the Shoah or manipulate it. Reflects on different ways of listening to the past in essays dealing with Claude Lanzmann, Ernest Renan, René Char and Albert Camus, among others. Includes an essay about the function of art in Theresienstadt. *EO*

0727 Half, Lauren M.; Singer, Jefferson A.; MacKinnon, John R.: Jewish Identity, Hostility toward Germany, and Knowledge about the Holocaust in the United States and Germany. *Journal of Psychology and Judaism* 23, 3 (Fall 1999) 107-124.

Relates to the continuing impact of the Holocaust on Jewish identity in the U.S. and Germany. Presents results of a survey conducted in the late 1990s among 109 Jews living in the USA and 31 Jews living in Germany; the age range was 13 to 85. The results show that hostility toward Germany and knowledge of the Holocaust were related to Jewish identity among American Jews, but were not related to Jewish identity among German Jews. However, the Jews in Germany knew more about the Holocaust than did their American counterparts. *DR*

0728 Halivni, David Weiss: Prayer in the Shoah. *Judaism* 50, 3 (Sum 2001) 268-291.

An essay written to accompany the publication by Yad Vashem of a New Year prayerbook transcribed in 1944 by Naphtali Stern in the Wolfsberg labor camp. Halivni, then 16 years old, was present when the Nazis allowed this one-time public prayer in the camp. The essay includes theological observations as well as personal testimony. Based on biblical verses, stresses that, despite the claims of some prominent ultra-Orthodox rabbis, the Holocaust was not punishment for sins committed by the Jews. Views God's inaction when His people were being annihilated as a form of kabbalistic "tsimtsum," whereby God contracted himself, as it were, to allow room for human freedom, which was then used by the Nazis for ultimate evil. Focuses on one of the prayers, which was uttered by camp inmates with particular feeling, since it "asks God to eliminate the free will of the perpetrators and to take the reins of government back into His hands." The essay is preceded by Peter Ochs' "Introduction to David Weiss Halivni's 'Prayer in the Shoah'" (pp. 259-267), which provides biographical information about Halivni, including his Holocaust experiences and his combination of talmudic scholarship with modern religious commitment. See the response to Halivni by Berel Lang [in "Judaism" 52, 1-2 (2003) 51-59], in which he rejects Halivni's attribution of the Holocaust to God's "tsimtsum" since that too would be an intentional or, at least, a concurred-in action. Notes the difference between this view and the biblical one of a God of history. Halivni appears to accept the ominous implication that the Holocaust is somehow justified. Lang concludes that an adequate account of the Holocaust must either move outside the domain of religious discourse or revise the terms of that discourse. Suggests that God Himself is constrained or limited in history. *YC*

0729 Lang, Berel: *The Future of the Holocaust: Between History and Memory*. Ithaca, NY: Cornell University Press, 1999. xiv, 198 pp.

A collection of essays, most of them published previously. *DR*

0730 Lang, Berel: *Holocaust Representation: Art within the Limits of History and Ethics*. Baltimore: Johns Hopkins University Press, 2000. xiv, 175 pp.

A collection of essays (most of them published previously) and excerpts from books. *YC*

0731 Lerner, Michael, ed.: *Best Contemporary Jewish Writing*. San Francisco: Jossey-Bass, 2001. xxvii, 436 pp.

An anthology of essays, fiction, and poetry devoted to Jewish identity, spiritual renewal in Judaism, contemporary reactions to the Bible, the Holocaust, Israel, and Jewish culture. Partial contents: Rosen, Jonathan: The Trivialization of Tragedy [First appeared in "Dumbing Down: Essays on the Strip-Mining of American Culture" (New York: W.W. Norton, 1996). Discusses his opposition to Holocaust museums, specifically the U.S. Holocaust Memorial Museum.] (230-240); Bauman, Zygmunt: Hereditary Victimhood: The Holocaust's Life as a Ghost [Appeared in "Tikkun" 13 (1998).] (241-251); Gabel Peter: The Meaning of the Holocaust: Social Alienation and the Infliction of Human Suffering [Appeared in "Tikkun" 13 (1998).] (252-262); Slater, Jerome: Occupation and Antisemitism [Appeared in "Tikkun" 16 (2001). Disputes the view of Ruth Wisse and others, that attributes Palestinian opposition to Israel to antisemitism.] (284-296). *YC*

0732 Levy, Daniel; Sznaider, Natan: *Erinnerung im globalen Zeitalter: Der Holocaust*. Frankfurt a.M.: Suhrkamp, 2001. 254 pp.

Contends that commemoration of the Holocaust has become a model for critical historical retrospective reflection in the era of globalization. Memorials today commemorate the victims rather than perpetrators or heroes. Discusses museums and memorial sites, and discussions and debates, as examples of Holocaust commemoration in different countries, especially in Germany, the USA, and Israel, as well as the recent proliferation of reparations claims. All of these serve as an important institutional guidepost for cosmopolitical orientation in dealing with crimes against humanity worldwide and their consequences. *ID*

0733 Locke, Hubert G.: *Learning from History: A Black Christian's Perspective on the Holocaust*. Westport, CT: Greenwood Press, 2000. xv, 128 pp.

Compares historical experiences of Jews and of Blacks, including the Holocaust and slavery, and notes similarities and differences between these experiences. States that slavery cannot be equated with the Holocaust; although it was brutal, it was not genocidal. Opposes trivialization of the Holocaust. Maintains that the Black and Jewish communities must be more receptive to the traumatic experiences of the other. The Holocaust has significance for Christians, taking into account the Church's relative passivity toward the Nazi anti-Jewish policies; it also has significance for Blacks, who should realize that, had the Nazis won the war, they would have been killed as well. On the other hand, Jews should keep in mind that German racism began with anti-Black racism, and there is a continuity between the thesis on Black inferiority developed in the late 18th century and the Holocaust. Calls on Jews and Blacks to put an end to mutual recriminations – of antisemitism or racism – and unite against bigotry. *DR*

0734 Morris, Marla Beth: *Curriculum and the Holocaust: Competing Sites of Memory and Representation*. Mahwah, NJ: Lawrence Erlbaum, 2001. xii, 270 pp. Based on the author's diss. – Louisiana State University, 1999.

Working at an intersection of curriculum theory, post-Freudian psychoanalysis, and object relations theory, examines ways in which the Holocaust is represented in texts written by historians and by novelists. Argues that memory is the larger category under which history is subsumed, and that works by historians and novelists share many common features. Although the historian and the novelist are constrained by their disciplines, both operate out of their own memories, and for both, psychological transference, repression, denial, projection, and reversal shape personal memories and may therefore determine the ways in which they construct the past. The way the Holocaust is represented in curricula is the way it is remembered. Argues against redemptive representation of the Holocaust and for a dystopic curriculum in education as well as representation, which does not contain "rhetoric of hope" and is a "testament to despair and truthfulness." *DR*

0735 Presner, Todd Samuel: Subjunctive History? The Use of Counterfactuals in the Writing of the Disaster. *Storiografia* 4 (2000) 23-38.

Presents a philosophical argument on the representability of the Holocaust. After dealing with the concept of "modernity and the disaster," discusses the relation of "counterfactuals" (i.e. possibilities) to Holocaust studies. Stresses that realism is inadequate for representing events of modern disaster, which totally confound the experiences and expectations of death and time. Notes how Holocaust denial has been allowed to define the criterion of "truth" vs. "falsehood" as the basis for evaluating accounts of the Holocaust. Instead, recommends the value of "subjunctive" or possible history, i.e. what might have but did not happen, as conveyed in fiction. This leaves room for ethics, which "realistic" history avoids. Does not advocate the total abandonment of "realistic" history, but aims at depriving it of its apparent monopoly on "truth." *YC*

0736 Rosenberg, Alan; Watson, James R.; Linke, Detlef, eds.: *Contemporary Portrayals of Auschwitz: Philosophical Challenges*. Amherst, NY: Humanity Books, 2000. 355 pp.

A collection of papers from the first international conference of the Society for the Philosophic Study of Genocide and the Holocaust (Bonn, 1996). *YC*

0737 *Teaching History* 104 (Sept 2001).

This issue is devoted to the topic "Teaching the Holocaust"; the articles relate mainly to teaching the subject in Great Britain to Year 9 students (13-14 year-olds). *SSC*

0738 Tóth, Imre: *Zsidónak lenni Auschwitz után* [Being Jewish after Auschwitz]. Budapest: Pont Kiadó, 2001. 76 pp.

Tóth, a Hungarian Jew, was born in 1921 in Szatmárnémeti, Romania. In 1944 he was imprisoned as a communist, which saved him from deportation. In the 1950s he was a lecturer at Bucharest University. In 1958 he was expelled from the party as a "Zionist agent." In 1969 he left for Germany where he became a professor at Regensburg University. In this essay he tries to understand the problem of antisemitism and calls it the heritage of the European intellectual tradition. Explains the professional success of Jews and their prominence in the arts and sciences by the obstacles they had to overcome due to antisemitism. Asserts that the Holocaust resulted from emancipation, the attempt of Jews to integrate into society. The Holocaust also meant the suicide of German intellectual life, its descent into mediocrity. Believes that Auschwitz means the end of antisemitism – the horrors of the Shoah will eventually purge the souls of antisemitism. *AV*

0739 Weiss, Gershon: *The Holocaust Haggadah: Foundations of Our Faith*. Southfield, MI: Targum Press; Nanuet, NY: Feldheim, 2000. 306 pp. On title page also: In collaboration with Yehuda Sorscher.

Presents an ultra-Orthodox commentary on the traditional Passover Haggadah. On the basis of frequent comparisons between the suffering of the Children of Israel in Egypt under Pharoah and the persecution of the Jews under Hitler, deals with questions of religious faith. Finds models of heroism among the Jews in Egypt, who in their courage and spiritual autonomy showed all future generations, including those during the Holocaust, how to resist physical and spiritual oppression. Includes inspiring accounts of Jewish resistance, such as the Warsaw ghetto uprising that began on Passover. Referring to the Haggadah text, stresses an archetypal, recurring antisemitism "in every generation." Expresses the view that God will help the Jewish people survive and will ultimately redeem them. Includes the text of the Haggadah with an English translation. *YC*

Neo-Nazism and Holocaust Denial

0740 Chatwin, Margret: Der pure Hass: Rechtsextremismus im Internet. *Tribuene* 157 (2001) 122-133.

Surveys use of the web by neo-Nazis to spread their ideology in Germany, including antisemitism, Holocaust denial, and the call for a new Holocaust, and their attacks on Jewish web sites, flooding them with their own propaganda, jamming them with junk mail, and threatening opponents, sometimes in real life. Comments that neo-Nazi sites, which include games, music, and the lure of the exotic, appeal to the young; it is almost impossible to censor them, but essential to expose their lies. *RW*

0741 Cooper, Abraham; Brackman, Harold David: The Fight against Holocaust Denial. *Midstream* 47, 3 (Apr 2001) 2-4.

In March 2001, the Lebanese government banned an international gathering of Holocaust deniers scheduled to convene in Beirut, organized under the auspices of the Institute for Historical Review. Nevertheless, it must be borne in mind that Holocaust denial has become a widespread phenomenon in the Arab and Muslim world. In the USA, neo-Nazism and Holocaust denial flourish. Criticizes veteran war reporter John Sack who, in an article in "Esquire" (February 2001) covering an IHR convention, represents the deniers as harmless cranks and fighters for freedom of speech. *DR*

0742 Eaglestone, Robert: *Postmodernism and Holocaust Denial*. Cambridge: Icon Books, 2001. 75 pp.

Deborah Lipstadt, in "Denying the Holocaust" (1994), blames postmodernism for creating an intellectual climate favorable to Holocaust denial, and characterizes the latter phenomenon as "deconstructionist history at its worst." Eaglestone argues against this accusation. According to the postmodern view, there is no such thing as objective history. History is a compendium of texts on the past, each of them subjective and dependent on the worldview of its author. However, history as a genre of writing is subject to generic conventions. Holocaust denial is not a kind of history at all, because it does not obey the rules of the genre and has aims different from those of history. This is what the court had in mind in its judgment on the Irving vs. Lipstadt trial in 2000, when it characterized Irving as being not a historian but a racist and antisemite. Argues that the questions postmodernism asks of history and historians are in fact strong weapons in combating Holocaust denial. *DR*

0743 Evans, Richard John: *Lying about Hitler: History, Holocaust, and the David Irving Trial*. New York: Basic Books, 2001. xiv, 318 pp.

Presents an account of the libel trial brought by David Irving against Deborah Lipstadt and Penguin Books for allegedly sullying his reputation as a historian. Evans was one of the expert witnesses for the defense. Describes his scrupulous study of Irving's books and of vast material related to Irving's attitude toward Hitler, the Jews, the Holocaust, etc. The trial made clear that Irving had intentionally distorted documents, quotes, etc. to minimize the Holocaust and to deny Hitler's role in it; it also made clear Irving's antisemitism, both in his writings and his public activities. Irving lost his case, since the charges made in Lipstadt's book "Denying the Holocaust" were substantiated. Irving emerged as a confirmed Holocaust denier. Concludes that it took historians to unmask Irving's distortions; a jury trial would probably not have arrived at the same conclusion. Notes misunderstandings of the trial in the press. *YC*

0744 Goldberg, Jonathan: The Appeal Judgment in the Case of David Irving v. Professor Deborah Lipstadt and Penguin Books. *Justice* 29 (Fall 2001) 13-16.

In July 2001 the Court of Appeal refused Irving's request to appeal the judgment of the High Court of Justice in London of April 2000, in which the court justified statements about Irving made by Deborah Lipstadt. The Court of Appeal confirmed the conclusions of the trial judge that Irving's works are antisemitic and that he deliberately distorted his sources. The latter decision marks the end of this particular litigation. Meanwhile, Irving is continuing his struggle by means of the Internet. Having failed in the judicial field, he tries to gain publicity through the mass media. *DR*

0745 Gottfried, Ted: *Deniers of the Holocaust: Who They Are, What They Do, Why They Do It*. Brookfield, CT: Twenty-First Century Books, 2001. 112 pp.

A book for teenagers explaining the phenomenon of Holocaust denial, its various manifestations, and its dangers. Contrasts the lies of the deniers with facts that disprove claims such as that Hitler did not even know about the Holocaust, that there were no gas chambers at Auschwitz, that the Holocaust was no worse than slavery or the American internment of Japanese during World War II, etc. Stresses the dangers of allowing denial to rest unchallenged. Discusses the pros and cons of banning or filtering Internet hate sites. Notes connections between words and deeds, e.g. deniers and killers, and the link between antisemitism and other types of racism and persecution. *YC*

0746 Guttenplan, D.D.: *The Holocaust on Trial: History, Justice and the David Irving Libel Case*. London: Granta Books, 2001. xiv, [4], 328 pp. Appeared in the U.S. as "The Holocaust on Trial" (New York: W.W. Norton, 2001).

A journalist's account of the London trial of David Irving vs. Penguin Books and Deborah Lipstadt, held in January-April 2000, interspersed with reflections on Holocaust memory and denial. The case became a Holocaust trial, albeit without perpetrators and victims. It did not generate a new narrative of the Holocaust, a revisionist one, as Irving had hoped. Without endorsing Irving's views, defends his right to historical dissent and regrets that his reputation as an outstanding historian was undermined. Contends that the critics of Irving's views, including Lipstadt, misinterpreted his writing as an outgrowth of postmodern deconstruction; what Irving assaulted was not rationalism and history as a science, but the Jews. His main motive was antisemitism. Contends that the depoliticized myth of the "Nazi war against the Jews," characteristic of some historiographic works, made the deniers' task easier. Believes that the Jews are not immune to critical analysis of their history; however, Irving crossed the divide between legitimate criticism and prejudiced vilification of the Jews as a people. *DR*

0747 Longerich, Peter: Der Holocaust vor Gericht? Bericht ueber den Londoner Irving-Prozess. *"Gerichtstag halten ueber uns selbst...": Geschichte und Wirkung des ersten Frankfurter Auschwitz-Prozesses,* ed. Irmtrud Wojak. Frankfurt: Campus, 2001. Pp. 317-351.

Reports on David Irving's libel suit against Deborah Lipstadt, who accused him of misrepresentation of the facts in his denial of the Holocaust. Judge Gray's summation included detailed analysis of Irving's distortion of sources. The finding that Irving's allegation that Hitler did not give orders for the mass killing of Jews, and did not know of it, is incompatible with the evidence, and so is his denial that Jews were gassed in Auschwitz. In conclusion, the judge concurred with Lipstadt's characterization of Irving as a Holocaust denier and an antisemite, whose distortions are tendentious and intentional. *RW*

0748 Maegerle, Anton: Neonazis und Islamisten: Die unheilige Allianz zwischen Hakenkreuz und Halbmond. *Tribuene* 160 (2001) 218-232.

Traces connections between European, mainly German, neo-Nazi organizations and leaders who support the actions of Muslim terrorists, including their war on Israel and the attack on the World Trade Center in New York in September 2001. Neo-Nazi publications and Internet sites meld anti-capitalism, anti-Americanism, anti-Zionism, and antisemitism. Arab states, on their part, have given refuge to Nazis as well as neo-Nazis and deny the Holocaust, which they consider an invention used by the Zionists to justify their occupation of Palestine. *RW*

0749 Menasse, Eva: *Der Holocaust vor Gericht: Der Prozess um David Irving*. Berlin: Siedler, 2000. 191 pp.

After tracing Irving's career as a Holocaust denier and Deborah Lipstadt's as the Holocaust deniers' professional nemesis, chronicles Irving's libel suit against Lipstadt and her British publisher, Penguin, as reported by Menasse in the "Frankfurter Allgemeine Zeitung." Irving asserted that persecution of the Jews was initiated by Goebbels behind Hitler's back, and that typhus, not gas, caused mass death in the concentration camps. The defense exposed Irving's falsification of the sources, and accused him of association with neo-Nazis and of antisemitism. Also describes a tour of Auschwitz by a group of Viennese architecture students, led by Robert Jan van Pelt, the defense's expert witness on the structure and functioning of the gas chambers. Concludes that the trial raised public awareness of the Holocaust and disproved the allegations of Holocaust deniers once and for all, but did not put an end to them. *RW*

0750 *Summary of Judgment in the Case of David Irving vs. Penguin Books Ltd. and Deborah Lipstadt: Announced on April 11, 2000, Courtroom 36, Royal Courts of Justice, London, England.* [New York]: American Jewish Committee, [2000]. 66 pp.

Presents extracts of the full judgment in the libel case brought by David Irving against Deborah Lipstadt and Penguin Books. The summary first contains passages from Lipstadt's book "Denying the Holocaust," which Irving claimed were libelous. The judgment concludes that Irving falsified and misrepresented historical evidence, e.g. in regard to the "Kristallnacht" pogrom, in his portrayal of Hitler as a friend of the Jews until October 1943, and in regard to Hitler's alleged lack of knowledge of and/or responsibility for the Final Solution. Irving is declared to be a Holocaust denier and a racist who has been sympathetic to neo-Nazi and other antisemitic organizations. Irving's efforts to exculpate Hitler reflect deliberate distortion of evidence and perverse manipulation. Thus, the defendants were exonerated. *YC*

Anti-Zionism and Anti-Israel

0751 Gremliza, Hermann L., ed.: *Hat Israel noch eine Chance? Palaestina in der neuen Weltordnung*. Hamburg: Konkret, 2001. 239 pp.

A collection of essays on the Arab-Israeli conflict, analyzing its background and perspectives, including the role of antisemitism, the genesis of Arab anti-Judaism, and the radicalization of Islamic movements. The following relate to antisemitism: Améry, Jean: Der ehrbare Antisemitismus [An abridged version of an essay written thirty years ago.] (7-12); Gremliza, Hermann L.: Wir Palaestinenser (13-18); Schroeder, Ralf: Volkskrieg, Intifada, Jihad: Der nationale Kampf der Araber und Palaestinenser gegen Israel und die Juden (19-43); Wertmueller, Justus: Aufstand der Anstaendigen [On antisemitism in the Quran and ritual murder accusations.] (44-64); Tempel, Sylke: Beruf: Palaestinenser: Yassir Arafat und die PLO (65-76); Uwer, Thomas; Osten-Sacken, Thomas von der: Chronik der Al-Aksa-Intifada (77-101); Id.: Arabischer Antisemitismus (125-140); Kloke, Martin W.: Zwischen Scham und Wahn: Israel und die deutsche Linke 1945-2000 (207-236). *ID*

0752 *Justice* 29 (Fall 2001).

The section entitled "The Durban World Conference" relates to the conference against racism which was convened by the United Nations in Durban, South Africa in September 2001. It consisted of two events – the NGO (non-governmental) Forum and the Intergovernmental World Conference – which were supposed to pave the way for cooperation around the world in the fight against racism. Instead, it turned into an arena for promotion of hatred and violence, singling out Israel and the Jewish people as the enemies of mankind. The following papers relate to the events: Ben-Itto, Hadassa: "Failure to Provide Moral Leadership Will Result in a Grave Regression of International Human Standards": A Last Minute Appeal to the International Community by the International Association of Jewish Lawyers and Jurists on the Eve of the Durban World Conference against Racism [A statement made by Ben-Itto to the High Commissioner for Human Rights on behalf of the IAJLJ on 18 July 2001.] (4-6); Bayefsky, Anne: Anti-Semitism at the United Nations: The World Conference against Racism Becomes a World Conference for Racism [Highlights from her address at Durban as a representative of the IAJLJ.] (7-8); Lack, Daniel: Report on Durban World Conference: Few Satisfied and Many Dismayed [Lack was also a representative of the IAJLJ at the Durban Conference.] (9-12). *SSC*

0753 Melchior, Michael: Zionism, Racism, and the Distortions of the Durban Conference. *Midstream* 47, 6 (Sept-Oct 2001) 4-6.

A revised version of the text of a statement delivered on 3 September 2001 at the UN's World Conference against Racism in Durban by Ambassador Mordecai Yedid, head of the Israeli delegation. Condemns the antisemitism, "manipulated for political ends," that has emerged at the conference and taken the form of anti-Zionism. Notes the paradox that a conference against racism is characterized by a form of "racist atrocity," while another form, slavery, was one of the intended main targets of the conference. Stresses that condemnation of slavery is perhaps the main contribution of Judaism to civilization. Condemns antisemitism expressed in Palestinian textbooks and in the conference statement of Yasir Arafat, as well as Holocaust denial and anti-Zionism in Arab states, and antisemitic caricatures in the Arab press. Expresses regret that the racism directed against Jews prevented the conference from dealing with the desperate problems of millions of victims of racism and related oppression in the world. *YC*

0754 Puddington, Arch: The Wages of Durban. *Commentary* 112, 4 (Nov 2001) 29-34.

Discusses the UN-sponsored World Conference against Racism which opened in Durban on 31 August 2001, and the expressions of anti-Israel and antisemitic hatred which erupted there. Claims that this UN forum was perhaps unique in the virulence of its antisemitic language and its assumption that Israel deserved to be destroyed. A coalition of regimes, led by Arab countries, attempted to delegitimize Israel and demonize its people. Elements of Holocaust denial also surfaced in Durban. Racism and genocidal behavior perpetrated in other countries was ignored. Besides Israel, the main target was the U.S. A parallel conference for representatives of NGOs was also dominated by antisemitic attitudes, including calls for the reinstatement of the UN Zionism=racism resolution. Expresses disappointment that such groups as Amnesty International failed to repudiate the hate language directed against Israel and Jews. The NGO sessions were "an ongoing antisemitic carnival." Connects approval at Durban for acts of terrorism against Israel with support for the attack on the World Trade Center in New York on 11 September 2001. *YC*

0755 Puddington, Arch: Durban újtai [The Wages of Durban]. *Múlt és Jövő* 4 (2001) 3-9. Appeared in English in "Commentary" 112 (Nov 2001).

Christian-Jewish Relations

0756 Bader-Saye, Scott: Post-Holocaust Hermeneutics: Scripture, Sacrament, and the Jewish Body of Christ. *Cross Currents* 50, 4 (Win 2000-2001) 458-473.

Argues that if Christianity wishes to relinquish supersessionist and triumphalist readings of the Bible, new methodologies of interpretation are not sufficient, since they ignore the social context of reader and reading. In the contemporary West (at least America), where Christianity is not tied to the state, there is no longer a need for supersessionist attitudes toward Judaism to buttress Christian power. A radical return to Christianity's Jewish roots can alter Christian anti-Judaism. One way to do this is liturgical, to re-Judaize the sacrament of the Eucharist by recognizing the Jewishness of the body of Jesus and to reconnect the Last Supper with the Jewish Passover. Thus, Christianity could move from its ignoring or "forgetting" of Judaism to establishing a connection based on the ideas of liberation and the founding of a community tied to God, not the state. Furthermore, remembering that the one whom they worship was Jewish would make Christians less likely to mistreat his people. *YC*

0757 Madigan, Kevin: A Survey of Jewish Reaction to the Vatican Statement on the Holocaust. *Cross Currents* 50, 4 (Win 2000-2001) 488-505.

Contends that "We Remember" failed to achieve the goal proclaimed in its preface: "to heal the wounds of past misunderstanding and injustices." The statement, although assessed positively by some Jewish scholars and rabbis, aroused criticism amongst the majority of Jewish thinkers. The disappointment was especially strong given the fact that the French bishops' statement, "Declaration of Repentance," issued at Drancy in 1997, was perceived by both Christians and Jews as a genuine apology of the Church for its silence during the Holocaust. Discusses the main points in "We Remember" that evoked Jewish criticism. It was the attempt to justify Pius XII's silence which caused the most negative reaction. Criticizes the viewpoint of David Novak, who argues that the present-day Church cannot apologize for past behavior. *DR*

0758 Wistrich, Robert Solomon: The Vatican and the Shoah. *Modern Judaism* 21, 2 (May 2001) 83-107.

Since 1965, following the "Nostra Aetate" declaration, there has been a tendency on the part of the Vatican to redefine the Church's relations with the Jews. Pope John Paul II has been sincere and resolute in his wish to push forward Catholic-Jewish dialogue. His efforts found expression, in particular, in the document "We Remember: A Reflection on the Shoah" (1998). Although this is a significant document, it also has faults. It fails to accept responsibility for the Church's passivity during the Holocaust (unlike some local bishops' conferences); it removes any Christian responsibility for modern antisemitism, including that of the Nazis; it barely mentions individual clerics who tried to help Jews at great personal risk, and at the same time mentions Pope Pius XII in a positive light. Deplores the Pope's passivity during the Holocaust, as well as his political timidity and shortsightedness. Criticizes the Vatican's attempts to Christianize the Holocaust, expressed in clerical pronouncements following the beatification of Edith Stein in 1987. *DR*

1991-2001: Australia

0759 Blackford, Russell: Free Speech and Hate Speech. *Quadrant* 373 (Jan-Feb 2001) 10-17.

Discusses the conflict between the social values of free speech and speech used to express and/or incite hatred, focusing on Australian legislation in this area. Favors some limitation of absolute free speech, on the Canadian model. Argues that hate propaganda may justifiably be suppressed since racism can lead to bigotry, discrimination, and persecution. Suggests, however, that racial vilification statutes already extant in Australia should be interpreted narrowly, so that they do not interfere with the general principle of free speech. Stresses that Holocaust denial should not be legally protected, nor should hate messages of neo-Nazis and Nazi sympathizers. *YC*

0760 Gouttman, Rodney: Pernicious Vision: Challenge from within Australia's Extreme Right. *Australian Journal of Jewish Studies* 15 (2001) 46-63.

Describes the Citizens' Electoral Council (CEC), an extreme-right organization which was established in Queensland in the 1980s; in the 1990s it came under the influence of the ex-American Lyndon LaRouche. Unlike another organization of the LaRouche camp, the Australian League of Rights, the CEC is also a political party, participating in elections. Its program includes anti-globalization, racism, rejection of multiculturalism, "re-industrialization" of the Western economy, antisemitism disguised as anti-Zionism, and opposition to a worldwide financial "oligarchy." *DR*

0761 Jones, Jeremy: *Report on Antisemitism in Australia, October 1, 2000 – September 30, 2001*. Darlinghurst, NSW: [Executive Council of Australian Jewry], 2001. iv, 140 pp.

States that during the 12-month period covered here, an almost unprecedented level of antisemitic activity took place within Australia. These include acts of physical violence against Jews, vandalism of synagogues and community sites, and threats of violence communicated by mail, telephone, e-mail, or leaflets. A number of articles and letters in mainstream newspapers evinced antisemitic stereotyping. Discusses, also, actions taken against antisemitism. Believes that there is a significant under-current of racism in Australia, but that it is difficult to objectively assess the place of antisemitism in Australian racism. Pp. 66-140 contain facsimiles, mainly of newspaper items. *SSC*

1991-2001: Europe

General

0762 Germinario, Francesco: Critica del "mondialismo", stereotipi antisemiti e conspirazionismo storico nella cultura politica del radicalismo di destra. *La Rassegna Mensile di Israel* 65, 1 (Jan-Apr 1999) 115-140.

Opposition to globalization has been characteristic of right-wing radicalism (or the New Right) since the 1990s in Europe. Activists interpret globalization as an instrument created by Judaism to dominate the world. It is described by the New Right as a process of Judaization of all humankind, in which the identity of entire peoples (their country, tradition, and culture) would be destroyed in favor of the identity of the Jews – universalistic and cosmopolitan. These views recall the accusations of a "Jewish world conspiracy," at the core of which lies the alliance of international finance and Zionism. *MZ*

0763 Messmer, Matthias: *Antisemitismus in Russland, der Ukraine und Litauen – eine vergleichende Studie*. [Koeln]: Bundesinstitut fuer Ostwissenschaftliche und Internationale Studien, 1998. 40 pp.

States that after the collapse of the Soviet Union and the de-ideologization in all the former republics, antisemitism became increasingly instrumentalized by various groups, political parties, and even by the Church. The new post-totalitarian antisemitism is a conglomerate of different forms of hostility expressed in the past, including blood libels, conspiracy theories, and Jew-hatred propagated by the Nazis. It is used to divert public attention from political or economic malaise and to find a scapegoat for the country's ills. The "intellectual" antisemitism of the Gorbachev era merged in Russia and Ukraine with "popular" antisemitism, and formed "political" antisemitism. While antisemitism emerged in Russia in the course of the Russian search for identity, in Ukraine and in Lithuania the debate on antisemitism took place in the context of dealing with the events of World War II. In Ukraine the position of Jews is considerably better than in Russia. States that antisemitism may be used as an indicator of social and political conditions, or as a yardstick for how well East European countries are progressing towards democracy. *ID*

0764 Tuor-Kurth, Christina, ed.: *Neuer Antisemitismus – alte Vorurteile?* Stuttgart: W. Kohlhammer, 2001. 229 pp.

A collection based on a series of lectures organized by the Institut fuer Juedische Studien der Universitaet Basel, held in summer 2000. Contents: "Antisemitismus im Nachkriegseuropa": Bergmann, Werner: Aus der Geschichte gelernt? Konflikte ueber Antisemitismus in der Bundesrepublik Deutschland (1949-2000) [Appeared in "Kirche und Israel" 15 (2000).] (11-31); Hagemeister, Michael: Antisemitismus und Verschwoerungsdenken in Russland [In the 19th-20th centuries.] (33-52); Kreis, Georg: Antisemitismus in der Schweiz nach 1945 (53-63); Niederer, Ulrich: Organisierter Antisemitismus unter Schweizer SchriftstellerInnen? Ein Blick zurueck [On antisemitism in Swiss literature in the 1930s.] (65-77); Picard, Jacques: Antiuniversalismus, Ethnizismus, Geschichtspolitik (79-101). "Antisemitismus und Oeffentlichkeit": Wetzel, Juliane: Antisemitismus im Internet (105-116); Dichanz, Horst; Breidenbach, Barbara: Antisemitismus in den Medien: Beispiele und Analysen (117-136); Liatowitsch, Peter: Holocaust-Leugnung in der Schweiz und Art. 261bis StGB [Appeared in "Kirche und Israel" 15 (2000).] (137-153); Hettlage, Andrea; Hettlage, Robert: Antisemitismus – eine Vorurteilsspirale (155-175). "Antisemitismus und christliche Theologie": Ehrlich, Ernst Ludwig: Umgang mit den Juden im Religionsunterricht [Appeared in "Juedisches Leben und juedische Kultur in Deutschland" (2000).] (179-187); Raguse, Hartmut: Der strukturelle Antijudaismus in der Theologie Eugen Drewermanns (189-198); Stegemann, Ekkehard W.: Theologie zwischen Antisemitismuskritik und alten Vorurteilen [Appeared in "Kirche und Israel" 15 (2000).] (199-215). *ID*

Austria

0765 *Attitudes toward Jews and the Holocaust in Austria: A Public Opinion Survey: Conducted for the American Jewish Committee by Gallup Institute of Austria, March 1-April 17, 2001*. New York: American Jewish Committee, 2001. 32 pp.

2,000 Austrians, 14 years of age or older, were interviewed in the course of this survey. Results show that Austrians have general knowledge concerning the Holocaust but have difficulty in acknowledging Austria's role in events leading up to it. A majority of the respondents favor keeping the memory of the Holocaust alive and do not support Holocaust denial. Negative feelings toward Jews are expressed by a significant portion of Austrians. Most of them agree that Joerg Haider is a racist and antisemite; supporters of Haider's party are more likely than other Austrians to exhibit anti-Jewish sentiment. Pp. 11-32 present the questionnaire and tables with the results. Compares the results of this survey with those of a 1995 survey in Austria and with those obtained by the AJC in other countries. *DR*

0766 Peri, Anat: *Joerg Haider's Antisemitism*. Jerusalem: Hebrew University, Vidal Sassoon International Center for the Study of Antisemitism, 2001. 35 pp. (Analysis of Current Trends in Antisemitism, 18).

Haider's antisemitism is a typical example of postwar antisemitism in the German-Austrian cultural sphere. Haider, born after the war to parents who were both ardent Nazis, identifies deeply with his parents and their generation, and sees them as victims. Thus, relativization of the Holocaust lies at the core of his antisemitic views. He equates the Nazi mass murder of the Jews with Stalinist purges, expulsions of Germans from Eastern Europe after the war, etc. He often uses the strategy of comparing Jews to Nazis. In his public statements, Haider has attacked the "pretend Holocaust survivors" Ignaz Bubis and Robert Jungk as admirers of the Third Reich and collaborators with the Nazis. On the other hand, he praises "honest Jews" (e.g. Kreisky and Peter Sichrovsky) who make anti-Jewish pronouncements, and denounces "treacherous Austrians" (e.g. Anton Pelinka) who cooperate with Jews against Austria. In order to camouflage his antisemitism, Haider uses coded language, in which many terms (e.g. international, cosmopolitan) actually refer to Jews. *DR*

Belarus

0767

סמילוביצקי, ליאוניד: יהודי ביילרוס ב-1996-1998. יהודי ברית-המועצות במעבר 4 [19] (2000) 285-296.

[Smilovitskii, Leonid L.: The Jews of Belarus in 1996-1998. *Yehudei Brit ha-Mo'atzot be-Ma'avar* 4 [19] (2000) 285-296.]

Discusses Jewish life in Belarus in 1996-98; pp. 290-295 relate to antisemitism. Political instability and economic problems have caused an increase in antisemitic incidents, mainly vandalism in synagogues and Jewish cemeteries. Antisemitism is not officially sanctioned, but it is also not condemned by the government. Many antisemitic essays have appeared in popular journals and some of the editors, appointed by President Lukashenko, are openly antisemitic. The subject of the Holocaust is largely ignored in Belarus; it is barely mentioned in school textbooks. Members of the Jewish community have objected to the situation and attempted to rectify it. As a result of their activities, Lukashenko visited a Holocaust memorial in Minsk in May 1997, and declared that the Jews would not be ostracized in Belarus. *RE*

France

0768 Erhel, Catherine; Aucher, Mathieu; La Baume, Renaud de, eds.: *Le procès de Maurice Papon. Vol. 1-2.* Paris: Albin Michel, 1998. 953; 973 pp.

Presents excerpts of the official transcript of the Papon trial (October 1997-April 1998), the longest in the history of the French Assizes Court. Traces the major events in the trial, from the opening statements by the prosecution and the defense, testimony by historians and the plaintiffs, the closing statement by the accused, and the final verdict. Despite attempts by the defense to annul the trial on the basis of Papon's advanced age and ill health, Papon was pronounced guilty of crimes against humanity for his involvement in the deportation of over 1,600 Jews. *TS*

0769 Golsan, Richard Joseph, ed.: *The Papon Affair: Memory and Justice on Trial.* New York: Routledge, 2000. xiii, 279 pp. On title page also: Translations by Lucy B. Golsan and Richard J. Golsan.

A collection of essays, as well as articles, op-ed pieces, letters, and interviews from the French press, relating to Maurice Papon, the French official who was involved in deporting Jews under the Vichy regime and in the shooting of Algerian protesters in Paris in 1961. He was tried for crimes against humanity and sentenced to ten years in prison in 1998. *YC*

Germany

0770 Bergem, Wolfgang: So viel Vergangenheit war nie: Nationalsozialismus und Holocaust im Identitaetsdiskurs der Berliner Republik. *Deutschland Archiv* 34, 4 (2001) 650-658.

Discusses guilt for the Nazi past, and its cardinal crime, the Holocaust, as a formative element of the identity of unified Germany. Guenter Grass asserted that this guilt should deter Germans from unification; Martin Walser sparked a debate by arguing that the time had come to stop reproaching Germany for the Holocaust, to drop it from public discourse and ritual and leave it to the individual conscience. *RW*

0771 Bergmann, Peter E.: Daniel Goldhagen in Germany: An Exploration in German Historiography. *Historical Reflections/Réflexions Historiques* 26, 1 (Spr 2000) 141-159.

Discusses Goldhagen's "Hitler's Willing Executioners" (1996) and his highly successful German lecture tour after the book's translation into German. Within the context of the conflict between the "destructive" and "evasive" theories (which view the Holocaust as either the logical outcome of, or accidental to, German history), Goldhagen's work is located within the "destructive" camp. Although Goldhagen gives credit to contemporary Germany for having abandoned its Nazi past, in the continuing conflict between the two camps of German historiography he is clearly viewed as opposing those who try to help Germany evade responsibility for the Holocaust. Nevertheless, some aspects of Goldhagen's scholarly approach are criticized. *YC*

0772 Conard, Robert C.: Eye on Germany: The New Morality and the Politics of Memory. *Debatte* 7, 2 (Nov 1999) 175-184.

In his acceptance speech for the German publishers' Peace Prize in 1998, Martin Walser referred to the Holocaust as a "moral cudgel" used against the German people and spoke against the proposed Holocaust Memorial in Berlin. States that Walser was always a leftist intellectual and that his new discourse must not be misconstrued as his having joined the rightist camp. However, in his writings Walser has maintained that everyone has a right to his own memory, a view similar to that of Edgar Reitz. Walser expressed this stand in his last novel, "Ein springender Brunnen," which depicts the time of his youth without any mention of Auschwitz. The German nation has accepted both Walser's speech and his novel of 1998. Walser expressed the common state of mind – the postwar period is over, a new era has begun, and Germany's past must be treated like every other nation's past. *DR*

0773 *Furchtbare Antisemiten, ehrbare Antizionisten: Ueber Israel und die linksdeutsche Ideologie*. Freiburg i.Br.: ça ira, 2000. 155 pp. At head of title page: Initiative Sozialistisches Forum.

Reprints articles originally published by the ISF in connection with the occupation of the "alternative" Regionalradio Dreyeckland (Freiburg) by left-wing students objecting to its airing, during the Gulf War of 1991, of radical anti-Israel views clearly tinged with antisemitism. In the introduction (pp. 7-13), states that at present the Left no longer needs to disguise its antisemitism as anti-Zionism (which had been mostly a mark of solidarity with the communist East); expression of antisemitism has become acceptable over much of the political spectrum. *RW*

0774 Gregor, Neil: "Vergangenheitspolitik," CSU-Style: The Memory of Forced Labour in Nuremberg. *The Journal of Holocaust Education* 10, 3 (Win 2001) 83-104.

Discusses debates which took place in 1997-99 concerning the award of the status of honorary citizen to industrialist Karl Diehl in Nuremberg. The city council, led by the Christian Social Union, honored Diehl although it knew that he had exploited concentration camp inmates in his factories during World War II. Defending himself, Diehl alleged that he acted under compulsion and not out of conviction. The scandal highlighted the question of how honestly Nuremberg, and maybe Germany as a whole, has confronted its Nazi past. States that there is a need to recognize the crucial complicity of local economic elites in the crimes of the Third Reich. *DR*

0775 Hansen-Schaberg, Inge; Schmeichel-Falkenberg, Beate, eds.: *Frauen erinnern: Widerstand – Verfolgung – Exil, 1933-1945*. Berlin: Weidler Buchverlag, 2000. 248 pp.

A collection of papers, most of which were presented at a study day held in Berlin in 1999, focusing on ways to facilitate the anchoring of commemoration of the Holocaust in the collective memory of German society today. *ID*

0776 Heimrod, Ute; Schlusche, Guenter; Seferens, Horst, eds.: *Der Denkmalstreit – das Denkmal? Die Debatte um das "Denkmal fuer die ermordeten Juden Europas": Eine Dokumentation*. Berlin: Philo, 1999. 1298 pp.

Presents articles and speeches on German reactions to the Holocaust as reflected in the controversy over the central monument in Berlin; protocols of meetings of the citizens' initiative, the Berlin city council, and a series of symposia; conceptual discussions; and artists' proposals submitted to the panel of judges, documenting the history of the monument from its conception in 1988 up to, but not including, the Bundestag decision of 1999. *RW*

0777 Klausa, Ekkehard: Die deutsche Gedenkreligion des Holocaust. *Merkur* 53, 9-10 (Sept-Oct 1999) 911-921.

Notes that remembrance of the Holocaust has become, for Jews, a substitute religion and a source of identity. For Germans, the Holocaust is taboo in the sense of a sacred object that it is forbidden to touch. The very word "holocaust," meaning a burnt offering to God, is a ritualistic euphemism for mass slaughter. Such a religious approach is fitting in ceremonies of commemoration, but out of place in scientific as well as everyday discourse. Criticizes the legal prohibition of Holocaust denial, and the outcry against Martin Walser's protest against the demand that Germans continue to feel responsible for crimes committed by a past generation. Many Jews, also, have expressed discomfort with their role as the priests of the Holocaust religion, and wish for more normal relations with Germans. *RW*

0778 Luehe, Barbara von der: Zeitzeugen der "Shoah" im Offenen Kanal Berlin: Verfolgung und Exil in der Zeit des Nationalsozialismus als Gegenstand einer medienpraktischen Unterrichtseinheit: Ein Projekt an der Technischen Universitaet Berlin in Kooperation mit dem Offenen Kanal Berlin. *Exilforschung* 19 (2001) 254-274.

The course was organized for students preparing to work in the mass media or in oral history, to teach them how to conduct an interview with Holocaust survivors or refugees from Nazism without exploiting them or catering to easy emotionalism, as, in the organizers' view, commonly happens. The curriculum included background in German-Jewish history and on the Holocaust; the viewing of sample interviews from Lanzmann's "Shoah" and other Holocaust documentaries, and from the video archives of the Fritz Bauer Institut, the Spielberg Foundation, and the Moses-Mendelssohn Institut; and the problems of oral history. Groups of students conducted and filmed interviews at the Offene Kanal, a communal TV station. The interviews, with seven well-known individuals, covered their life stories before, during, and after the Nazi period in order to show how they were affected by this caesura in their lives. *RW*

0779 Maegerle, Anton: Braune Einstiegsdroge: Rechtsextreme Musik in der Bundesrepublik Deutschland. *Tribuene* 159 (2001) 107-126.

Surveys the sources and the distibution in Germany of neo-Nazi xenophobic and antisemitic rock music, which attracts many young people to the radical right. Quotes several grossly antisemitic songs. *RW*

0780 Rohloff, Joachim: *Ich bin das Volk: Martin Walser, Auschwitz und die Berliner Republik*. Hamburg: KVV Konkret, 1999. 139 pp.

Satirizes Walser's speech upon his receiving the Peace Prize of the German Book Trade on 11 October 1998, a speech in which Ignatz Bubis saw "mental arson," and the subsequent reactions of intellectuals and publicists supporting Walser and attacking Bubis. The speech and the reactions expressed the nationalism of the new Germany; the resentment at the demand that Germans continue to face their guilt for the Holocaust; the supposed exploitation of that guilt ("Auschwitzkeule"); and opposition to the plans for the central Holocaust memorial in Berlin. Some accused Bubis of insensitivity: was he sure that if he had been a German and not a Jew, he would have behaved better? Pp. 129-139, "Die Haare sind mehr geworden," contain an interview with Bubis by Hermann L. Gremliza, first published in "konkret" 2 (1999). Bubis asserts that Jews can still live well in Germany, but there are now "more hairs in the soup." *RW*

0781 Schirrmacher, Frank, ed.: *Die Walser-Bubis-Debatte: Eine Dokumentation*. Frankfurt a.M.: Suhrkamp, 1999. 687 pp.

Presents Walser's speech on receiving the Peace Prize of the German Book Trade on 11 October 1998, in which he declared his revolt against the constant confrontation with guilt for the Holocaust and against the ritualization and instrumentalization of its memory; Bubis's sharp condemnation of the speech ("mental arson") in statements to the press and later in his speech on the 60th anniversary of the "Kristallnacht" pogrom; and the flood of reactions in public speeches, in the press, and in open letters to the two opponents. The initial speech and the numerous responses brought here cover the period October 1998-July 1999. *RW*

0782 Watts, Meredith W.: Aggressive Youth Cultures and Hate Crime: Skinheads and Xenophobic Youth in Germany. *American Behavioral Scientist* 45, 4 (Dec 2001) 600-615.

Examines right-wing violence in Germany, which increased dramatically after unification and remained at a relatively high level during the 1990s. Ca. 60% of the violent incidents were directed against foreigners. Antisemitic attacks, including desecration of graves and memorial sites, accounted for ca. 2% of the violent incidents. Given that the number of Jews in Germany is ca. one-hundredth that of foreigners, the rate of anti-Jewish attacks per capita is high. Skinheads and neo-Nazis have played a significant role in the violence, but at least one-third of the violent incidents were perpetrated by informal groups of unaffiliated young males. Contemporary xenophobia is not only linked to aggressive elements of youth culture, but appears to be increasingly connected to local and international ideological networks. Focuses on developments in the Skinhead culture. *DR*

0783 Wittenberg, Reinhard: *Antisemitische Einstellungen in Deutschland zwischen 1994 und 1998: Messprobleme und Ergebnisse*. Nuernberg: Wirtschafts- und Sozialwissenschaftliche Fakultaet, Friedrich-Alexander-Universitaet Erlangen-Nuernberg, 1999. 28, [5] pp.

Presents the results of three separate polls undertaken in 1994, 1996, and 1998. Altogether, 6,671 persons participated, 2,526 from former East Germany and 4,145 from West Germany. The interpretation of the data posed methodological problems due to unequal methods and terminology used. The findings for all of Germany show that in 1994 a majority of the respondents held antisemitic attitudes, while in 1998 the percentage was inverse. As to the average findings for East and West Germany between 1994-98, 8.7 % in East Germany and 11.7 % in the West had antisemitic attitudes. 23.2 % in the East and 21.9% in the West had no antisemitic attitudes. The lower rate of antisemites in East Germany is attributed to the virtual non-existence of the Jewish topic there before 1990 and the fact that antisemitism was disguised as anti-Zionism. The results show that higher education and professional employment decreased antisemitism; lower income augmented antisemitism in the East, while the factor of gender was significant only in the West: women had less antisemitic attitudes. *ID*

0784

צוקרמן, משה: האנדרטה לזכר השואה בברלין. זמנים 76 (סתיו 2001) 48-58.

[Zuckermann, Moshe: The Shoah Memorial in Berlin. *Zmanim* 76 (Fall 2001) 48-58.]

Discusses the debate over the Holocaust memorial in Berlin, focusing on a suggestion made in 1999 by Richard Schroeder, a German professor of theology, that the memorial should consist of the Hebrew words "lo tirzah" ("do not kill"). Argues that the memorial should carry a message to Germans, not to Jews. Although Schroeder's suggestion was not accepted, it reveals a problem in the German perception of the difference between Denkmal (a monument for remembrance) and Mahnmal (a monument which serves as a warning). Discusses the planning stages of the memorial, the architectural contest, political aspects of the debate, and universalization of the Holocaust. *RE*

Great Britain

0785 Cesarani, David: Does the Singularity of the Holocaust Make It Incomparable and Inoperative for Commemorating, Studying and Preventing Genocide? Britain's Holocaust Memorial Day as a Case Study. *The Journal of Holocaust Education* 10, 2 (Fall 2001) 40-56.

A response to the controversy concerning the first national Holocaust Memorial Day (HMD) in Britain, held on 27 January 2001. Stresses that HMD is not just about Jews and not just about the past. In response to the claim by Peter Novick and others that Holocaust commemoration and education trivialize lesser crimes, concludes that these processes are crucial to commemorating, studying, and preventing genocide. While focusing on racism and intolerance, makes the point that antisemitism does exist at present in Britain and that Holocaust study should help people understand that Nazi antisemitism, which is of the past, is not the only variety. *YC*

0786 Cowan, Paula: *The Holocaust: A Teaching Pack for Primary Schools.* Dundee: Learning and Teaching Scotland; [Edinburgh]: Scottish Executive, Education Department, 2000. vi, 50 pp.

Part of a teaching pack intended for use on Holocaust Memorial Day (January 27). The pack consists of the teacher's manual for Primary 6 and

7 pupils, accompanied by a videocassette in which Reverend Ernest Levy tells about his Holocaust experiences, and a disk which contains Fred Rendell's "Into Hiding: A Topic Study on Anne Frank." *SSC*

0787 Petersen, Judith: How British Television Inserted the Holocaust into Britain's War Memory in 1995. *Historical Journal of Film, Radio and Television* 21, 3 (Aug 2001) 255-272.

Discusses the place and role allotted to the Holocaust by four British television channels – BBC1, BBC2, ITV, and Channel 4 – in their programs commemorating fifty years since the end of World War II. Of the four channels, BBC2 devoted relatively more time to Jewish wartime experiences. BBC2 and C4 also showed several documentaries on the liberation of Nazi camps. These programs either stressed the role of Britain as a liberator of the camps or elegiacally focused on Jewish losses. Much less attention was devoted to the wartime British response to the plight of European Jews and the failure to help them effectively; only "Bringing the Holocaust Home," aired by BBC2 in January, dealt with this issue, while such documentaries as "Victory," shown by C4, perpetuated the myth that Britain's main war aim was to save Europe's Jews from annihilation. Thus, while the TV time devoted to the Holocaust in Britain rose, the 1995 programs failed to break the monolithic character of Britain's official war memory. *DR*

Hungary

0788

פרוימוביץ, קינגה: השואה בתודעה ובזהות של אינטלקטואלים יהודים ולא יהודים בהונגריה בין השנים 1994-1999. גשר 144 (חורף 2001) 58-72.

[Frojimovics, Kinga: The Holocaust in Jewish and Hungarian Intellectuals' Consciousness and Identity in 1994-1999. *Gesher* 144 (Win 2001) 58-72.].

Examines essays by non-Jewish and Jewish Hungarian writers, comparing their attitudes toward the Holocaust. The non-Jewish writers (Árpád Göncz, Szabolcs Szita, Ferenc Glatz, and Géza Komoróczy) view the Holocaust in Hungary as part of Hungarian history, and the suffering of the Jews as part of Hungarian suffering during the war. The Jewish writers (Iván Beer, Áron Hoffmann, Imre Kertész, and György Konrád), on the other hand, see it as a distinctly Jewish experience and a basis for Jewish identity. Discusses, also, reactions to István Szabó's film "Sunshine" (2000), which indicate that there is a large gap between Jews and non-Jews in Hungary, particularly in their views on the Holocaust. *RE*

0789 Kőbányai, János: *Hagyományszakadás után* [After the Rending of Tradition]. Budapest: Múlt és Jövő, 1999. 282 pp.

A collection of essays; most of them appeared in the journal "Múlt és Jövő" between 1985-1998. Describes the efforts of Jews to integrate and assimilate in Hungarian society since the emancipation in the 19th century. Attempts to answer the question what it means to be a Jew in contemporary Hungarian society, and whether it is possible for Jewish tradition to strike roots in Hungary after the Holocaust. Pp. 72-119 discuss the Holocaust in the artistic works of Imre Ámos and the poetry of Miklós Radnóti. Pp. 224-282 contain reactions to antisemitism in Hungary and to antisemites like István Csurka. *MA*

0790 Verő, Gábor; Nuber, István: A Magyar Auschwitz Alapítvány – Holocaust Dokumentációs Központ működése 1990 óta [The Activities of the Hungarian Auschwitz Foundation – Holocaust Documentation Center since 1990]. *Holocaust Füzetek* 14 (2000) 11-33.

Describes the goals of the Foundation and its activities. The Foundation plans enlargement of its collection, which consists of books, documents, articles, photos, tapes and video-tapes about the anti-Jewish measures, deportations, labor service, concentration camps, labor camps, and Holocaust remembrance. The Zionist movement's actions to rescue Jews are also documented. Fields of research at the Foundation include neo-Nazism and Holocaust revisionism. The Foundation organizes conferences, establishes memorials, conducts exhibitions, and produces publications. *AV*

Poland

0791 Friszke, Andrzej: Żydzi w szkolnych podręcznikach [Jews in School Textbooks]. *Więź* 42, 7 [489] (1999) 23-36.

Criticizes Polish history textbooks published in the 1990s. Most of them do not include sufficient information about the presence of Jews in Poland throughout the centuries. There is no information given about the multinational character of Poland before World War II. The exposition of Endecja ideology is found lacking, which makes it difficult to understand later developments. The subject of antisemitism as a social problem is marginalized. The Holocaust is presented as part of Nazi policy towards the Polish population, and the Nazis' different treatment of Poles and Jews is not explained. Discusses some of the textbooks and their contents, and recommends the best of them. *HV*

0792 Mich, Włodzimierz: Antysemityzm na łamach prasy polskiej po 1989 roku [Antisemitism in the Polish Press after 1989]. *Annales Universitatis Mariae Curie-Skłodowska. Sectio F: Historia* 54-55 (1999-2000) 283-293.

Examines themes of antisemitic discourse in Poland after the fall of communism and the liquidation of censorship, between 1989-1996. Concentrates on two nationalistic journals, both of which have small circulation. As in the interwar period, the antisemites perceive the Jews as a dangerous element which desires to transform Poland into Judeo-Polonia. They negate the authenticity of the official number of Holocaust victims, and they argue that Poles suffered more than Jews during World War II. They argue that Poles are not friendly toward Jews because of the anti-Polonism of Jews, and that the Jews are responsible for communism. The antisemitic press in Poland has reprinted some articles which appeared in the prewar press, considering them still valid today. *HV*

0793 Oppenheim, Bohdan W., ed.: *Kościół katolicki w Polsce a walka z antysemityzmem: Wymiana doświadczeń amerykańskich i polskich* [The Catholic Church in Poland and Struggle with Antisemitism: An Exchange of American and Polish Experiences]. Kraków: Fundacja Judaica-Centrum Kultury Żydowskiej, 2000. 146, 45 pp.

A collection of addresses and discussions held at a conference convened in Kraków in June 2000. The participants discussed antisemitism in contemporary Poland and criticized the attitudes of Catholic Church authorities toward this phenomenon, as well as ways of fighting antisemitism through educational programs. The American teaching experience was presented as a model for education against antisemitism. The conference was followed by a panel discussion. The participants were: Bohdan W. Oppenheim, Stanisław Obirek, Martin I. Bresler, Michał Czajkowski, Konstanty Gebert, Stanisław Musiał, Andrzej Korbonski, John Pawlikowski and others. Pp. 147-193 contain a reprint of an antisemitic pamphlet by Michał Morawski (a priest), entitled "Stanowisko Kościoła wobec niebezpieczeństwa żydowskiego w dawnej Polsce" ["The Attitude of the Church toward the Jewish Danger in Old Poland"], published in 1938. *HV*

0794 Oppenheim, Bohdan W., ed.: *Rachunek sumienia: Kościół polski wobec antysemityzmu, 1989-1999* [Examination of Conscience: The Polish Church Confronts Antisemitism, 1989-1999]. Kraków: Wydawn. WAM, 1999. 244 pp.

A collection of papers (and related documents) from a conference which took place at Loyola Marymount University, Los Angeles, in January 1999. The conference was organized by Oppenheim, a Polish immigrant to the U.S., an engineer and professor at Loyola Marymount, who is active in the field of Polish-Jewish relations. The conference, even before it took place, provoked many protests by Polish conservative circles, both in Poland and in the U.S., which were opposed to the very idea that the Church needed to examine its conscience. *HV*

0795 Tych, Feliks: The Image of the Holocaust in Polish Historical Consciousness. *Polin* 14 (2001) 315-325. Appeared in Polish in "Biuletyn Żydowskiego Instytutu Historycznego" 189 (1999).

0796

וויינבאום, לורנס: רוחות הרפאים מידוובנה. גשר 144 (חורף 2001) 37-57.

[Weinbaum, Laurence: The Ghosts of Jedwabne. *Gesher* 144 (Win 2001) 37-57.]

Discusses reactions in Poland to Jan Tomasz Gross's "Neighbors" (2000), which exposes the role of Poles in the massacre of Jedwabne's Jews. The reactions range from denial and accusations of Jewish anti-Polonism to sincere regret and calls for repentance. Postwar communist authorities had whitewashed the role of collaborators in the Holocaust, insisting that all Poles were "victims." There actually were less collaborators in Poland than in other countries, because the Nazis preferred the help of ethnic Germans there. Due to traditional antisemitism, many Poles welcomed the Nazi intention of making the country "judenrein," but most were bystanders, not collaborators. The Jews continue to view all Poles as collaborators because of traditional Polish antisemitism; this view has been strengthened by Gross's book, although there were Polish rescuers even in Jedwabne. States that Poles should accept the events at Jedwabne and express regret, and Jews should overcome their negative feelings about Poles. *RE*

0797 Wollaston, Isabel Louise: *Auschwitz and the Politics of Commemoration*. London: Holocaust Educational Trust, 2000. 24 pp. (Holocaust Educational Trust Research Papers, 1, 5).

Discusses the Christianization of the Holocaust in relation to the Carmelite convent and the erection of crosses at Auschwitz. Concludes that this was not a systematic, centrally-inspired policy but rather a mixture of the unacceptable (the rewriting of history) and the insensitive (Christian failure to realize that the Carmelite presence or crosses would offend Jews). Accepts the genuineness of Christian attempts to confront the significance of the Holocaust with their own symbols. The issue of a Christian presence at Auschwitz since the 1980s is related to Polish Catholic inclusivism. The crosses at Auschwitz may reflect either Christian triumphalism or an attempt to associate the location with Polish suffering during the war. The conflict has been aggravated by the media and by the failure to differentiate between the geography and the history of the Auschwitz camp. One solution would be to have no religious symbols there, although many Catholics do want a Jewish religious presence. The basic problem relates to the cross as a symbol that is offensive to Jews due to their long history of persecution by Christians. *YC*

0798 *Znak* 52, 6 (2000).

This issue, entitled "Shoah: Pamięć zagrożona?" ("Shoah: Memory in Danger?"), contains discussion about the memory of the Holocaust among Poles and Jews in response to the lead article by Zdzisław Krasnodębski. *HV*

Russia

0799 Butkevich, Nickolai, ed.: ***Antisemitism, Xenophobia and Religious Persecution in Russia's Regions, 1999-2000.*** Washington, DC: Union of Councils for Soviet Jews, 2001. xii, 180 pp.

Presents results of annual monitoring of antisemitism in 73 of Russia's 89 administrative regions. Reporting on each region, discusses such topics as official antisemitism, the activities of extremist groups (e.g. Russkoye Natsionalnoye Yedinstvo, the Kuban Cossacks) and the Communist Party of the Russian Federation, antisemitic incidents and publications, and the kidnapping of Jews. Notes that, after an increase in 1998-99, antisemitism decreased in Russia but remained a danger due to two factors: alliances between provincial officials and antisemitic groups and the growing power, under President Putin, of the traditionally antisemitic secret services. Notes antisemitism also in the Russian military and the Russian Orthodox Church. Includes two appendices: "Antisemitism and Xenophobia in the December 1999 Russian Parliamentary Elections" (pp. 167-173) and "Antisemitic Literature Sold in the State Duma" (pp. 175-180). *YC*

0800 Butkevich, Nickolai, ed.: ***Antisemitizm, ksenofobiya i religioznye presledovaniya v rossiiskikh regionakh 1999-2000.*** Washington, DC: Union of Councils for Jews in the Former Soviet Union, 2001. 168 pp. Appeared in English as "Antisemitism, Xenophobia and Religious Persecution in Russia's Regions, 1999-2000" (Washington, DC: Union of Councils for Jews in the Former Soviet Union, 2001).

0801 Gudkov, Lev D.: Antisemitizm v postsovetskoi Rossii [Antisemitism in Post-Soviet Russia]. *Neterpimost v Rossii: Starye i novye fobii*

[Intolerance in Russia: Old and New Phobias], eds. Galina S. Vitkovskaya, Alexei V. Malashenko. Moskva: Moskovskii Tsentr Karnegi, 1999. Pp. 44-98. On back cover: Carnegie Endowment for International Peace.

A study based on results from more than 20 surveys on inter-ethnic relations carried out in Russia between 1988-97; focuses on the surveys relating to antisemitism in 1990, 1992, and 1997. The results show that there is a group of ideological and aggressive antisemites (ca. 6-9% of the respondents) whose antisemitism is mostly tribalist and compensatory, and two groups which share moderate or latent forms of antisemitism. For many respondents, moderate antisemitism serves to enhance their positive ethnic self-identity. A large number of respondents oppose the appointment of Jews to key positions (in the state, army, etc.). Nevertheless, on average, the attitude toward Jews became more tolerant in the 1990s, and the attitude toward the State of Israel and the emigration of Jews from Russia improved greatly during this period. But the fact that the ethnic consolidation of Russian society preserves ethnic negativism toward the "Other" is disturbing. *DR*

0802 Rock, Stella: Russian Revisionism: Holocaust Denial and the New Nationalist Historiography. *Patterns of Prejudice* 35, 4 (2001) 64-76.

Based on three months of research in Russia and on the Internet, contends that Holocaust denial, which has appeared only recently in Russia and has attracted little attention both academically and from monitoring organizations, is marginal and mostly an import from the West. Despite the high level of antisemitism in Russia, post-Soviet skepticism concerning historiography, and a lack of Holocaust education, anti-Nazi feeling resulting from World War II militates against Holocaust denial. Concludes that other antisemitic revisions of Russian history, such as blaming Jews for the "genocide" of Russians after the Revolution, are more likely to encourage antisemitism in post-Soviet Russia. *YC*

Slovakia

0803 Špitzer, Juraj: Scepticism and Hope. *Scepticism and Hope: Sixteen Contemporary Slovak Essays,* ed. Miro Kollar. Bratislava: Kalligram, 1999. Pp. 27-57. Translated from Slovak – source not given.

Špitzer (1919-1995) was a Slovak writer, and twice served as editor of the non-conformist monthly "Kultúrny Život" (1956-58 and 1966-68). He was ostracized and forced to leave the Institute for Slovak Literature; most of his writings were published only in the 1990s. In this address (apparently to a conference audience), he discusses the phenomenon of antisemitism –

his skepticism as to whether it can be eradicated, and his hopes that it might be. Interspersed with personal reflections, discusses manifestations of antisemitism in Slovakia during and after World War II. Questions whether he is a Slovakian Jew or a Jewish Slovak; in the face of past and continuing antisemitism, he opts for the former. He pities Jews who internalize society's demonization of them. With the future of Slovak Jewry doubtful, Spitzer feels an obligation to preserve monuments of its past. Concludes that while the end of Hitler and Stalin did not mean the end of antisemitism, one can hope that tolerance will prevail. *YC*

Sweden

0804 *Knowledge and Remembrance of the Holocaust in Sweden: A Public-Opinion Survey Conducted for the American Jewish Committee by Sifo, Stockholm, December 13-16, 1999.* New York: American Jewish Committee, 2000. 31 pp. Cover-title. On cover also: Prepared for the Stockholm International Forum on the Holocaust.

Interviews were conducted with 1,000 Swedes to determine their knowledge about the Holocaust, feelings about Holocaust remembrance, attitudes toward Jews, and awareness of Sweden's "Living History" campaign, whereby free distribution of the book "Tell Ye Your Children" is intended to encourage parents to learn about and transmit to their children information about the Holocaust. The survey found that almost all Swedes favor keeping alive the memory of the Holocaust; Swedes have strong knowledge about some aspects of the Holocaust; almost all reject Holocaust denial, although a very large number have encountered such denial; slightly less than half know about the "Living History" campaign, and 16% belong to families that ordered the book. Swedes' attitudes toward Jews are very positive; they are at the middle-to-high end of the scale of knowledge about the Holocaust when compared with other nationalities surveyed, and are first in terms of advocating Holocaust remembrance. *YC*

Switzerland

0805 Benz, Wolfgang: Antisemitismus in der Schweiz. *Judaica* 56, 1 (Mar 2000) 4-18.

Surveys the manifestations of antisemitism that swept Switzerland in the late 1990s in reaction to Jewish demands for the restitution of unclaimed bank accounts and insurance policies of Holocaust victims and revelations of Swiss cooperation with the Nazis and the closing of the borders to Jewish refugees. The monetary demands were perceived as blackmail and revived latent stereotypes of Jewish cupidity and world conspiracy; the historical revelations seemed to impugn the Swiss people's heroic stand during the Second World War. The Swiss felt themselves attacked from all sides and developed a siege mentality, in which minorities are always seen as the enemy. Emphasizes that antisemitism was not confined to the radical right but appeared in carnival floats applauded by the public, in meetings, and in letters to the press. *RW*

0806 Buffat, Françoise; Cohen, Sylvie: *Suisses et Juifs: Portraits et témoignages*. Lausanne: Favre, 1998. 240 pp.

Presents brief biographies of 33 Swiss Jews, based on interviews conducted in 1997-98. Dwells on their reactions to antisemitic polemics which arose in 1996-97 following accusations by Jewish organizations that Swiss banks had confiscated funds from the dormant bank accounts of Holocaust victims. Examines the diverse reactions of Swiss Jews to an interview with Jean-Pascal Delamuraz, a member of the Swiss Parliament, who accused Jews of disloyalty to their government. Although he later apologized, many Swiss Jews were surprised by the sudden resurgence of antisemitism. Most of those interviewed recalled antisemitic events in their past, comparing those to recent accusations in the Swiss media that Jews were guilty of inciting an international scandal. *TS*

0807 Dreyer, Philipp: *Zwischen Davidstern und Schweizerpass: 24 Portraets juedischer Jugendlicher*. Zuerich: Orell Fuessli, 1999. 192 pp.

Contains interviews with 24 Jewish adolescents living in Switzerland in the 1990s; they deal with questions of identity and conflicts concerning the non-Jewish environment, religion, the status of Jewish women, etc. Some of the testimonies discuss experiences of antisemitism in school, antisemitism related to the investigations of "Nazi gold" and the dormant accounts of Jews in Swiss banks, and anti-Jewish stereotypes. *ID*

0808 Dreyfus, Madeleine: Entschuldigung und Rechtfertigung: Zum Rezeptionsmuster der antisemitischen Fluechtlingspolitik der Schweiz im Zweiten Weltkrieg. *Judaica* 56, 1 (Mar 2000) 19-31.

Attributes the Swiss' blindness in regard to their guilt in the Second World War to their mythological self-image as a nation that stood fast heroically in the face of great danger. The policy of borders closed to desperate Jewish refugees aroused unease at the time among part of the population, but was justified by reasons of state and the priority of self-defense. This was also the response to publications on the subject until the mid-1990s. The root of the policy in endemic Swiss antisemitism was ignored. The self-image was marred by Jewish-American demands for restitution and for admission of guilt. These forced a revision of the myth, but also triggered resentment of outside interference in Switzerland's cherished independence, and the projection of guilt onto the "persecutors" – the Jews. *RW*

0809 Gisler, Andreas: *"Die Juden sind unser Unglueck": Briefe an Sigi Feigel 1997-1998*. Zuerich: Epoca, 1999. 239 pp.

Discusses various forms of verbal antisemitism expressed in Switzerland following the scandal evoked by the issue of the dormant accounts of Holocaust victims in Swiss banks. Analyzes letters addressed to Sigi Feigel, honorary president of the Juedische Cultusgemeinde Zuerich, between 1997-98. Feigel received 1,073 letters, among them 340 that were strongly antisemitic or had antisemitic allusions. Examines the "anatomy" of an antisemitic letter: usually anonymous, handwritten, full of orthographic errors and stereotypical images. Antisemitic letters came not only from extreme right-wingers or non-educated persons, but from wide sections of the population *ID*

Yugoslavia

0810 Byford, Jovan; Billig, Michael: The Emergence of Antisemitic Conspiracy Theories in Yugoslavia during the War with NATO. *Patterns of Prejudice* 35, 4 (2001) 50-63.

In the wake of the NATO bombings in Yugoslavia in spring 1999, conspiracy theories emerged in the country, some of them disseminated through mainstream media like the daily "Politika." Antisemitism, in the form of a Jewish conspiracy theory, is firmly embedded in the conspiratorial culture and has also appeared in Yugoslavian publications. The main exponent of the Jewish conspiracy theory is Ratibor Đurđević, whose mystical interpretations of current events and world history are based on U.S. conspiracy literature and on the teachings of the Serbian Orthodox theologian Nikolaj Velimirović. Distinguishing between expressions of the conspiracy theory in a moderate and quasi-academic form (e.g. by Smilja Avramov) and by paranoidal extremists like Đurđević or Gordana Knežević, notes that the borderline between them is not clear and that the dissemination of these ideas by the mainstream press makes the extreme versions of the conspiracy theory more credible. *DR*

1991-2001: Far East

Indonesia

0811 Siegel, James T.: "Kiblat" and the Mediatic Jew. *Indonesia* 69 (Apr 2000) 9-40.

Discusses antisemitism without Jews in Indonesia, where there were never any significant number of recognizable Jews. The Jews are, thus, not the other, as the local Chinese are, but a mysterious force from elsewhere that threatens Indonesians in their lack of clear identity. Focuses on the "Zionist" (actually having nothing to do with Israel) threat to Islam warned about by "Media Dakwah," the organ of former Indonesian modernists. The latter criticize the views of Nurcholish Madjid, a Muslim who studied in the U.S., for "mistranslating" Islam under the influence of "Jewish Orientalists," who are one variety of the distant, dangerous, and amorphous Jew. *YC*

Japan

0812 Kowner, Rotem: Tokyo Recognizes Auschwitz: The Rise and Fall of Holocaust Denial in Japan, 1989-1999. *Journal of Genocide Research* 3, 1 (2001) 257-272.

Notes that Japanese antisemitism is sui generis: it appeared in a society without Jews and has almost exclusively been confined to writing rather than actions. The Christian pastor Masami Uno has been a leading antisemite. In the 1980s he espoused Holocaust denial, partly under the influence of the U.S.-based Institute for Historical Review. In 1995 an article in "Marco Polo" claimed that there were no Nazi gas chambers. A combination of international Jewish protest and the threat of economic sanctions led to the closing of the magazine. In the same year a Japanese terror group identified some Japanese leaders as secret Jews. In 1999 a Japanese court rejected the claims of a Holocaust denier, confirming the basic facts of the Holocaust. Concludes that while few Japanese are interested in Jews, sympathy for Holocaust denial is related to the denial of Japan's own war crime (the 1937 "rape of Nanjing") and a feeling of victimization (after Hiroshima and Nagasaki) about their own "Holocaust." Japanese antisemitism basically reflects domestic problems rather than concerns about Jews or Israel. *YC*

0813 Makino, Uwe: Im Osten nichts Neues? Zur Leugnung des Holocaust in Japan. *Jahrbuch fuer Antisemitismusforschung* 10 (2001) 204-222.

States that the Holocaust deniers in Japan are not of the radical right, which uses the facts of the Holocaust in order to trivialize, by comparison, Japan's own alleged war crimes. But Holocaust denial and theories of a Jewish world conspiracy are popular in Japan. They were preceded in the 19th century by a theory of a Christian conspiracy – a secret religion out to dominate the world. When the "Protocols of the Elders of Zion" were published in Japan, they fit this theory perfectly; "Christian" only had to be replaced by "Jewish." The theory was reinforced by the Russian Revolution, seen as a Jewish plot. The chief contemporary disseminators of Holocaust denial are Masami Uno, whose books are bestsellers, and who asserts that the "fascist" Zionists manipulate the Holocaust for the sake of power; and Masanori Nishioka and Aiji Kimura. The latter base their arguments especially on the "Leuchter Report" and on Rudolf Hoess. *RW*

1991-2001: Islamic World

General

0814 *L'Arche* 523 (Sept 2001).

Pp. 58-95, "Dossier: La montée de l'antisémitisme dans le monde arabe," contain an introduction by Meïr Waintrater describing the spread of antisemitism, not just anti-Zionism, in the Arab world today. He then presents numerous extracts from the press and media demonstrating the prevalence of this antisemitism. The material is divided into four sections: religious antisemitism, accusations of ritual murder and poisoning, from the "Protocols" to Hitler, negation of the Shoah. Pp. 90-95 contain an article by Raphael Israeli, "Une image démoniaque du Juif," relating to the demoniacal image of the Jew in Palestinian and other Arab propaganda sources. *SSC*

0815 Nirenstein, Fiamma: How Suicide Bombers Are Made. *Commentary* 112, 2 (Sept 2001) 53-55.

Gives examples of antisemitic propaganda circulating in the Arab world and in Iran. Characterizes the methodology of this propaganda as "the big lie," i.e. the spread of false information about Israel and the Jews. Among its motifs are Holocaust denial, statements that Israel or the Jews have outdone the Nazis in brutality, and the necessity for brutal "retaliatory" measures. It is a mistake to regard this propaganda as a means in the political struggle of the Arabs against Israel: antisemitism in the Middle East is an end in itself. *DR*

0816 Nordbruch, Goetz: Leugnungen des Holocaust in arabischen Medien: Reaktionen auf "Die Gruendungsmythen der israelischen Politik" von Roger Garaudy. *Jahrbuch fuer Antisemitismusforschung* 10 (2001) 184-203. An English version appeared as a monograph in the series "Analysis of Current Trends in Antisemitism" 17 (2001).

Garaudy's work (published 1995/6) was widely and favorably discussed in the Arab press. Treats four elements of the discussion: 1) the attribution of European antisemitism to Jewish particularism and the perceived Jewish threat; 2) the Nazi-Zionist collaboration theory, on the basis of the affinity between the two racist ideologies and a common interest in Jewish emigration from Germany, and the view that Israeli policy continues to be Nazi; 3) Holocaust denial or relativization, based on Western sources such as the "Leuchter Report"; 4) condemnation of the suppression of Holocaust denial by European countries under Jewish-Zionist pressure, and of exploitation of the Holocaust by Jews to justify Zionism and to extort reparations payments. Notes that in the Arab press there is no differentiation between Jews and Zionists, who are all accused of lust for power, control of the media, and world conspiracy. The arguments of Islamists and leftists are much the same, though with differences in emphasis. *RW*

0817 Nordbruch, Goetz: *The Socio-Historical Background of Holocaust Denial in Arab Countries: Reactions to Roger Garaudy's "The Founding Myths of Israeli Politics."* Jerusalem: Hebrew University, Vidal Sassoon International Center for the Study of Antisemitism, 2001. 27 pp. (Analysis of Current Trends in Antisemitism, 17). A German version appeared in "Jahrbuch fuer Antisemitismusforschung" 10 (2001).

The reaction to Roger Garaudy's "The Founding Myths of Israeli Politics" (1996) in Arab countries highlighted the main themes of Holocaust denial and historical revisionism which are widespread in the Arab world. These include blaming the Jewish victims themselves for the Holocaust, attempts to present both Nazism and Zionism as manifestations of the same destructive tendency in Western civilization, and outright Holocaust denial. Contends that Arab Holocaust denial is not merely a tool used to delegitimize the State of Israel. Antisemitism, rather than anti-Zionism, prevails in contemporary Arab nationalist and Islamist rhetoric. The image of the Jew as an antagonist of the Arab and the Muslim has been constructed by Arab nationalists as a means for promoting an Arab national and religious "awakening." Holocaust denial attracts Arab nationalists as the most contemporary expression of antisemitic thought. *DR*

0818 Pattir, Dan: Arab Demonization of Israel and the Jews. *Justice* 27 (Spr 2001) 21-25.

Presents 15 cartoons from the state-controlled Arab press (six from Egypt, five from Jordan, two from the Palestinian Authority, one from Syria, and one from Dubai) that show how antisemitic and anti-Israel stereotypes and images (including swastikas) have been used since 1993 to demonize Jews and delegitimize the State of Israel. Some of the caricatures resemble those of the Nazi "Der Stuermer." Stresses that such cartoons particularly affect illiterate Muslims and contradict – in the case of Egypt, Jordan, and the Palestinian Authority – peace agreements with Israel that obligate the parties to refrain from incitement and propaganda. *YC*

Syria

0819 Groiss, Arnon; Manor, Yohanan, eds.: *Jews, Zionism and Israel in Syrian School Textbooks*. New York: Center for Monitoring the Impact of Peace, 2001. 78 pp.

The Center (established in 1998) examined 68 Syrian textbooks on various subjects, all of which were part of the curriculum in 2000. The survey (containing extensive excerpts from the textbooks) shows that the Arab-Israeli conflict is a major theme in the curriculum. The essence of the conflict is presented as the question of Israel's right to exist. Israel is portrayed as an alien and artificial entity, not as a legitimate state; it serves imperialism, against the Arabs, and is a cause of the lack of Arab unity and of Arab backwardness. The textbooks are meant to educate the children in a spirit of hatred toward the enemy, who is defined alternatively as Zionist or Jew. The Jews are denied the characteristics of a nation and are said to have no connection with the land of Palestine. The Jewish religion is described as racist, and the Jews themselves are blamed for antisemitism. The textbooks justify the Holocaust, and they claim that its magnitude is exaggerated by the Jews. One passage was found that calls for the elimination of the Jews. *DR*

0820 Wurmser, Meyrav: *The Schools of Ba'athism: A Study of Syrian Schoolbooks*. Washington, DC: Middle East Media Research Institute, 2000. xiii, 122 pp.

Surveys 40 contemporary Syrian textbooks for the 4th-11th grades, concluding that antisemitism is an integral part of the totalitarian ideology that the Ba'athist regime is promulgating. The hostility toward Israel reflects a pan-Arab unity that has no place for Israel in the Middle East, and a belief that conflict with Israel and the West is permanent. Islam calls for "jihad" to eliminate Israel and the Jews in general, who are demonized in images that the secular regime borrows from the Islamic tradition. Holocaust denial, along with other aspects of antisemitism, is introduced into the history, national (patriotic) education, geography, literature, and social and Islamic education curricula. *YC*

1991-2001: Israel

0821 Caplan, Kimmy: Have "Many Lies Accumulated in History Books"? The Holocaust in Ashkenazi "Haredi" Historical Consciousness in Israel. *Yad Vashem Studies* 29 (2001) 321-375. Appeared in Hebrew in "Yad Vashem" 29 (2001).

Surveys ultra-Orthodox historiography on the Holocaust and textbooks on this topic (aimed largely at Orthodox womens' educational institutions) written mainly in the 1980s-90s. Focuses on a substantial number of these writings which maintain a complex relationship with academic research on the subject. Orthodox historiography was conceived in part as a counterhistory to "Zionist" historiography; it contrasts the behavior of Orthodox Jews during the Holocaust (heroism, rescue efforts, etc.) with that of Zionist leaders. Many Orthodox texts are remarkable for their ahistorical approach to the Holocaust, but there is a gradual rapprochement taking place of Orthodox historiography with academic research. Focuses on writings by Ruth Lichtenstein, Dina Spiegel, and Esther Farbstein as most typical of recent Orthodox Holocaust historiography. *DR*

0822

קפלן, קימי: האומנם "נצברו הרבה שקרים בספרי התולדות"? השואה בהיסטוריוגרפיה ובתודעה ההיסטורית החרדית האשכנזית בישראל. יד ושם 29 (2001) 249-292.

[Caplan, Kimmy: *Have "Many Lies Accumulated in History Books"? The Holocaust in Ashkenazi "Haredi" Historiography and Historical Consciousness in Israel. Yad Vashem* 29 (2001) 249-292.] Appeared in English in "Yad Vashem Studies" 29 (2001).

0823 Gur-Ze'ev, Ilan: The Production of Self and the Destruction of the Other's Memory and Identity in Israeli/Palestinian Education on the Holocaust/"Nakbah." *Studies in Philosophy and Education* 20, 3 (May 2001) 255-266.

The Israeli educational system instrumentalized the memory of the Holocaust for the aims of building the nation. The resulting hegemonic and normalizing education is inevitably ethnocentric and tends to neglect the other, in this case, the Palestinian Arabs, their sufferings and collective memory, and thus impedes Israeli-Palestinian dialogue. On the other hand, the Palestinian educational system tends to neglect the

Holocaust; the Arab establishment and intellectuals have sided with the deniers of the Holocaust or, at least, minimized its dimensions. A new trend of acknowledging the Holocaust in its full dimensions and universal moral implications is discernible among Palestinian intellectuals. However, its main exponents, Edward Said and Azmi Bishara, regard the "Nakbah," the uprooting of Palestinians following the establishment of the State of Israel, as the inevitable outcome of the Holocaust, and the Palestinians as the ultimate victims of the Holocaust. Calls for humanistic-oriented counter-education for Israeli Jews and Arabs, based on mutual acknowledgment and open dialogue. *DR*

0824 Kressel, Gideon M.: Let Palestine Return to Khaybar. *Israel as Center Stage: A Setting for Social and Religious Enactments,* eds. A. Paul Hare, Gideon M. Kressel. Westport, CT: Bergin & Garvey, 2001. Pp. 165-187.

Khaybar was a Jewish rural community in the Arabian Peninsula (near Mecca) which, according to Muslim tradition, was exterminated by the Prophet Muhammad. In the 1990s the story of Khaybar gained a new popularity among Israeli-Palestinian Arabs. Presents excerpts from the transcripts of three conversations conducted by Kressel in 1992, 1994, and 1995 with Israeli Arabs from various walks of life, and gives examples of use of the Khaybar story by Arabs. For many of them, the story symbolizes their rejection of an undesirable reality – the existence of a Muslim region where Jews are not subject to Muslims as "dhimmis" – and expresses their aspirations to reestablish what they consider the proper order of things. For opponents of the peace process, Khaybar is a metaphor for the line that should be adhered to in relations with Jews. Khaybar has become a slogan of fundamentalist groups like Hamas. *DR*

0825 Lentin, Ronit: *Israel and the Daughters of the Shoah: Reoccupying the Territories of Silence.* New York: Berghahn Books, 2000. xiv, 256 pp.

Contends that the unavoidable inclusion of the Holocaust in the Israeli social myth was accompanied by the stigmatization of Holocaust survivors and the silencing of their memories. Israeli memorialization of the Holocaust became a tool in the negation of the diaspora; the Zionist discourse "masculinized" the "New Hebrew" brought up in the Land of Israel, and "feminized" the Holocaust and its survivors. Moreover, this type of Holocaust memory serves to justify the Israeli occupation of Palestinian lands. Israeli women writers and filmmakers of the second generation have challenged the Zionist version of Holocaust memory. In 1992 Lentin interviewed nine such writers and filmmakers, "daughters of the Shoah." Her book contains excerpts from the interviews (and in one case the full text) and offers a gendered analysis of Israeli society and the Holocaust. *DR*

0826

שצקר, חיים: בעיות בהוראת השואה בימינו. חינוך והיסטוריה: הקשרים תרבותיים ופוליטיים, עורכים רבקה פלדחי, עמנואל אטקס. ירושלים: מרכז זלמן שזר לתולדות ישראל, 1999. עמ' 447-455.

[Schatzker, Chaim: Problems in Contemporary Holocaust Teaching. *Hinukh ve-historia: Heksherim tarbutiim u-politiim* [Education and History: Cultural and Political Contexts], eds. Rivka Feldhay, Immanuel Etkes. Jerusalem: Zalman Shazar Center for Jewish History, 1999. Pp. 447-455.]

Discusses conflicts in Holocaust education in Israel, including emotional commemoration vs. rational study of the Holocaust, and universalization vs. stressing the uniqueness of the Holocaust. The first stage of Holocaust education, which began in the 1960s, was emotional; the second stage was instrumental, stressing universal aspects of the Holocaust. In the 1990s, Holocaust education became existentialist, with the aim of having students identify with the victims. Schatzker questions the usefulness of this approach. In light of historical developments, suggests that more attention be paid to ethnic struggles and their role in the Holocaust, the fate of Jews in the USSR, and the psychological effects of terror. *RE*

1991-2001: Latin America

Argentina

0827 Braylan, Marisa; Feierstein, Daniel; Galante, Miguel; Jmelnizky, Adrián: *Report on Anti-Semitism in Argentina – 1998.* Buenos Aires: Social Research Center of DAIA, 2000. 45 pp. Appeared in Spanish in "Indice" 20 (2000).

Presents a chart (pp. 20-30) listing antisemitic incidents in Argentina during 1998, based on accounts in the press and on the files of the DAIA legal issues department. The events include threats to individuals, antisemitic verbal expressions, damage and threats to Jewish institutions, physical aggression, desecration of cemeteries, and discrimination in schools and in the judiciary. States that these expressions of antisemitism lack support among the majority of Argentinians, especially after the bombing of the Israeli Embassy and the AMIA building in 1992 and 1994 respectively. Asserts that although antisemitism is still present in Argentinian society, it is no longer politically useful. Pp. 7-19 summarize the history of antisemitism in Argentina from 1880 onward. *MZ*

0828 Kollmann, Raúl: *Sombras de Hitler: La vida secreta de las bandas neonazis argentinas*. Buenos Aires: Editorial Sudamericana, 2001. 251 pp.

A study of neo-Nazi groups in Argentina. Examines their structure, ideology, and activities. In the 1990s, neo-Nazi groups, in conjunction with the police and other security organs, carried out several attacks on Jewish cemeteries. Denounces the hesitation of state institutions in investigating and undertaking legal measures against these groups. *MZ*

Brazil

0829 *Knowledge and Remembrance of the Holocaust in Brazil: A Public-Opinion Survey Conducted for the American Jewish Committee by IBOPE, São Paulo, March 22-25, 2001*. New York: American Jewish Committee, 2001. 26 pp.

Presents the results of interviews with 1,000 Brazilians, carried out by IBOPE, a leading Brazilian opinion-research organization. The poll shows that Brazilians lack basic factual knowledge about the Holocaust (ranking lowest compared with other countries surveyed); 77% do not know the term "Holocaust," yet a majority favor keeping the memory of the Holocaust alive. There is a degree of openness to Holocaust denial; 11% believe that the Holocaust may not have happened, while 12% do not know if it did. Brazil has the highest proportion of people (37%) who prefer not to have Jewish neighbors. In regard to antisemitism in the country, 50% believe that it is a "very serious problem" or "somewhat of a problem"; 59% believe there are groups in the country that have a Nazi ideology, but there is a split regarding whether these groups are influential. Education is the key variable in the survey; e.g. 86% of those with college education vs. 57% with primary education want to continue keeping the memory of the Holocaust strong. *YC*

Uruguay

0830 Aronovitz, Alberto M.: Uruguay: The Example of a Small Country Fighting against Racism and Anti-Semitism. *Justice* 27 (Spr 2001) 26-30.

Describes the little-known success of Uruguay in detecting and convicting members of two racist groups, one of which was operating on the Internet. Uruguay has had anti-racist laws since 1989. In 1999 the Orgullo Skinhead racist web site was discovered and neutralized; the three youths who ran it were tried and convicted. In 2000 the head and several members of the racist group Frente Nacional Revolucionario del Uruguay, who also targeted local "Zionists," were arrested. This group used more traditional means, e.g. graffiti, for spreading hatred. To stop cyber-racists, recommends holding web access providers and web hosts responsible if they refuse to neutralize racist sites. Advocates making Holocaust denial illegal also when spread electronically, and supporting anti-racist groups, including those who monitor web sites. *YC*

1991-2001: North America

General

0831 Lauckner, Nancy A.; Jokiniemi, Miriam, eds.: *Shedding Light on the Darkness: A Guide to Teaching the Holocaust*. New York: Berghahn Books, 2000. xxiii, 261 pp.

A collection of essays by professors of German Studies from various colleges and universities in the USA and Canada. The essays relate their experiences of teaching Holocaust studies and literature, or Holocaust-related topics in the framework of their courses in German Studies, mostly in the late 1980s-90s. All of them used literature and films in their lessons. Most of the essays also present the curriculum used. *DR*

0832 Levin, Brian: History as a Weapon: How Extremists Deny the Holocaust in North America. *American Behavioral Scientist* 44, 6 (Feb 2001) 1001-1031.

Examines the evolution of the Holocaust denial movement in the USA and Canada from the immediate postwar period to the present. Many of its leading figures have been antisemitic extremists. The whole movement, despite the presence of some academics in it, is a part of the extreme-right world in North America. Relates the activities of Willis Carto, David Irving (during a stay in the USA), and other personalities. Discusses periodicals that propagate the views of deniers, and some organizations, e.g. the Institute for Historical Review, that support denial. Discusses, also, the Zundel affair in Canada in the 1990s. *DR*

USA

0833 *Dimensions* 15, 1 (May 2001).

The section entitled "An American Perspective: The Holocaust in Religion, Politics, and Culture" contains four essays; a study guide for teachers accompanies each essay. Partial contents: Boyer, Paul: Rapturous

Tidings: The Holocaust, Bible Prophecy Belief, and Conservative America [Reflects on the evolution of Holocaust awareness in the U.S.; focuses on how conservative Protestant Church groups interpret prophetic and apocalyptic portions of the Christian Bible and how this interpretation shapes their understanding of Jewish history, including the Holocaust.] (3-8); Petropoulos, Jonathan: The Roller Coaster of Restitution: The United States Government's Involvement with Holocaust Victims' Assets (9-17); Berenbaum, Michael: Consciousness of the Holocaust: Promises and Perils [Examines manifestations of the present high level of interest in the Holocaust in the U.S.] (25-30); Study Guide Supplement (31-48). *SSC*

0834 Frolick, David A.: An Experiment in Learning: A Cross-Disciplinary Approach in Teaching the Holocaust. *Shofar* 17, 4 (Sum 1999) 38-65.

Describes the preparations for and running of a first-time program of teaching about the Holocaust at North Central College, Illinois. The experiment was interdisciplinary (involving the fields of political science, speech and communications/theater, sociology/anthropology, and French). The core seminar, taught by Frolick, dealt with "unique and universal aspects of the Holocaust." Use was made of outside speakers (including survivors) and outside exhibitions, as well as an on-campus dramatic production. Many problems of preparation and presentation are described in an effort to encourage other colleges to offer similar programs. Includes 20 pp. of syllabi from the courses developed. *YC*

0835 Gallant, Mary J.; Hartman, Harriet: Holocaust Education for the New Millennium: Assessing Our Progress. *The Journal of Holocaust Education* 10, 2 (Fall 2001) 1-28.

Calls for an assessment of Holocaust education, focusing on the situation in the U.S. today. Discusses objectives, goals, and curricular strategies now in use. However, educators do not know what effect their teaching has had on the students. Stresses that an assessment of Holocaust education must deal with cognitive, affective, and actional predispositions, i.e. changes in behavior. Advocates a systematic multi-dimensional approach to assessment and comparison of strategies in order to select the right programs. *YC*

0836 Hartman, Geoffrey H.: Memory.com: Tele-Suffering and Testimony in the Dot Com Era. *Raritan* 19, 3 (Win 2000) 1-18.

The genre of video testimony on the Holocaust poses a danger of banalization (packaged imagery) and raises problems regarding empathy. The contemporary dominance of audio-visual media affects both morality and mentality. Yale University's Video Archive for Holocaust Testimonies attempts to counter some of these problems; it tries to maintain the autonomy of the person who is testifying and prevent the "derealization" that film and video tend to cause. The Yale project focuses on the witness, leaving the interviewer unseen. The Yale project aims at creating a bond between testifiers and viewers so that traumatic knowledge gained by the latter leads to empathy rather than guilt or overidentification. *YC*

0837 Heilman, Uriel: Murder on the Brooklyn Bridge. *Middle East Quarterly* 8, 3 (Sum 2001) 29-37.

On the morning of 1 March 1994, a Lebanese cab driver named Rashid Baz fired on a van of hasidic Jewish boys on the Brooklyn Bridge, killing one (16-year-old Ari Halberstam) and wounding several others. The next day Baz turned himself in to the police. He was convicted and sentenced to 141 years in prison. Discusses here the motive for the crime. The police and FBI considered it as a homicide due to "auto rage" (Baz's defense). But others – Halberstam's mother, the prosecutor, the mayor of New York, senators and congressmen – called the shooting an act of terrorism. The shooting occurred one week after the Hebron massacre in which Baruch Goldstein murdered 29 Arabs and injured many more. Given that Baz was agitated about that event, and that he was carrying three lethal weapons in his taxi cab on March 1st, posits that he set out to kill Jews intentionally. In December 2000, federal authorities changed their minds and announced that Baz's actions were "the crimes of a terrorist." States that the authorities have been slow to recognize the threat to domestic security posed by militant Islamic groups in the USA. *SSC*

0838 Jackson Dowd, Susan Emily: *Jews and Judaism in Protestant Religious Education Materials: Educating without Antisemitism.* Diss. – Spalding University, 2000. 152 pp. Unseen.

This study on the mainline Protestant portrayal of Jews and Judaism in religious education materials is a "next step" in the fifty-year tradition of content analysis. Examines resources since completion of the latest dissertation in 1992 and discusses interviews with ten experts engaged in religious education, interfaith relations, and other related fields. Because religious education materials historically have been a source for anti-Judaism and antisemitism, and current research shows that the problem persists, this study does not provide a new content analysis. Instead, it focuses on results from past content analyses and presents a practical model for use by creators of religious education materials and teachers. This abbreviated model provides denominational policy – the Presbyterian Church (U.S.A.) is used as an example, and a word list of themes and concepts that have been problematic in religious education, from pulpit to classroom. The model also contains sample anti-bias guidelines and information on current print, electronic, and human resources.

0839 Sutherland, John: Mainstreaming Neo-Nazism. *Dimensions* 13, 1 (1999) 15-17.

The wide expansion of the Internet has facilitated the dissemination of racist materials. This is illustrated by the fate of "The Turner Diaries," a racist and antisemitic novel by Andrew Macdonald (pen-name of William L. Pierce), on the Internet. The book first appeared in 1978 and is marketed through the website of the Amazon company. Amazon even invited "online book reviews" from purchasers and received dozens of positive reviews. *DR*

0840 Toll, Nelly: *Integrating Holocaust Art and Aesthetics into the Curriculum.* Diss. – University of Pennsylvania, 2000. 148 pp. Unseen.

A study of the impact of Holocaust visual imagery and aesthetics on the curriculum based on an examination of elementary and high school students' responses to learning about the Holocaust through Holocaust art and aesthetics in two New Jersey public schools. The teachers were provided with age-appropriate pictures, poetry, literature, and historical texts. Means of heightening students' awareness to the ramifications of prejudice, racism, and stereotyping, and a focus on the importance of tolerance were discussed with the teachers. The study found that students responded by drawing and writing their interpretations as well as constructing meanings significant to their own interests and concerns. The interconnection between personal relevancy, aesthetics, and cognition provided the students with a heightened awareness and critical understanding of the moral implications of the Holocaust.

0841 Weitzman, Mark; Eaton, Rick, eds.: *The New Lexicon of Hate: The Changing Tactics, Language and Symbols of America's Extremists: A Simon Wiesenthal Center, Snider Social Action Institute Report.* 2nd, rev. ed. Los Angeles: Simon Wiesenthal Center, 2001. 36 pp.

Provides brief information (e.g. history, geographical location, names of leaders) on American extremist groups which are basically racist and antisemitic. They include white supremacists, the Ku Klux Klan, neo-Nazi organizations, Skinheads, Peckerwoods (drug-financed Skinheads, many of whom have served time in prison), militia-patriot-conspiracy groups (which aim their hatred at the U.S. government, called ZOG), Christian Identity, and youth and women's activities of various extremist groups. Includes six pages of racist tattoos, as well as other information such as hate acronyms, the names of racist bands that produce albums of hate music, instructions for lone racist terrorists, an Aryan declaration of war, and selected Internet hate websites, including neo-Nazi ones. *YC*

ANTISEMITISM IN LITERATURE AND IN THE ARTS

0842 Band, Arnold J. (Avraham): Refractions of the Blood Libel in Modern Literature. *Memory and History in Christianity and Judaism,* ed. Michael A. Signer. Notre Dame, IN: University of Notre Dame Press, 2001. Pp. 113-133.

A paper delivered at the third Crown-Minow Conference, University of Notre Dame. Follows Alan Dundes' model of projective inversion, construing the blood libel as displacement on the Jews of Christian guilt for consuming the host. The blood libel cases illustrate vividly Christianity's difficulty in accepting Jewish otherness. Explores modern literary responses to the blood libel. Dwells on the influence of blood libel trials, especially the Beilis trial, on Kafka's works. While Kafka and several other writers have a limited view of the blood libel (e.g. as a threat to the Jews), Lamed Shapiro in his "Di vaise hale" and Agnon in his "Ha-adonit ve-ha-rokhel" capture the irony involved in the Christian accusation against Jews and understand it as a projection of Christian guilt and desires. Pp. 134-143 contain a response by Krzysztof Ziarek, "Trials of Modernity: Responses to Blood Libel." Contends that the blood libel expresses Christianity's inability to respect the otherness of the Jews. Kafka's texts, rather than having a limited view of the blood libel, universalize it and turn it into allegory. Shows that for Adorno, the blood libel trials constitute a testimony to the persistence of myth in the legacy of the Enlightenment. *DR*

0843 Barker, Andrew W.: Race, Sex and Character in Schnitzler's "Fraeulein Else." *German Life and Letters* 54, 1 (Jan 2001) 1-9.

Discloses Jewish aspects of Arthur Schnitzler's story "Fraeulein Else" (1924), set in 1896. The heroine's prototype was a young Jewish female friend of Schnitzler who committed suicide. Else's milieu was largely Jewish, and she is characterized by a self-hating Jewishness. Anti-Jewish stereotypes in the story link Jews with money and sexuality. The linking of Jewish women, particularly, with sex and prostitution (Else agrees to sell herself to a rich Ostjude to save her father from financial ruin) recalls the combination of Jewish self-hatred and misogyny of Otto Weininger. Schnitzler is viewed as having been tainted by a similar perspective, as well as having been prejudiced against Ostjuden. *YC*

0844 Budziak, Anna: Ideological Vulnerability of Imagery: T.S. Eliot's Jewish Representations. *Anglica Wratislaviensia* 34 (1999) 27-40.

Discusses the offensive antisemitic imagery in Eliot's poetry as representing his underlying political ideology. The images reveal the poet's personal and social anxieties. It is not the Jew in general (e.g. not the Orthodox or Sephardic Jew) but the rich, cosmopolitan, free-thinking Jew who is a threat to Eliot's mythic order which is based on nationalism, Christianity, and Empire, and which represent for him barriers to crisis and chaos. Eliot did not appear to be conscious of why or the extent to which he scapegoated Jews. *YC*

0845 Chernenko, Miron M.: *Krasnaya zvezda, zhyoltaya zvezda: Kinematograficheskaya istoriya yevreistva v Rossii 1919-1999* [Red Star, Yellow Star; A Cinematographic History of Jewry in Russia 1919-1999]. Vinnitsa: Globus-Press, 2001. 430 pp.

A study of Jewish themes and images in Russian films. The main trend in Soviet films was gradual de-Judaization of the Jews. In the 1920s-30s, issues of antisemitism and pogroms were used by Soviet filmmakers to discredit non-Soviet regimes (e.g. those of pre-revolutionary Russia, Poland, and Nazi Germany) and to distance themselves from those regimes. The Soviet film industry was the first to depict the Holocaust on the screen (in M. Donskoi's "Unconquered," 1945). Discusses a protest letter against antisemitism in the film industry sent by Mikhail Romm to Stalin in 1943. After the war, film producers who tried to project a positive Jewish image on the screen encountered obstacles; their films were not shown. Notes that the cinema was less involved in the anti-Zionist campaign than was literature. In Brezhnev's era, Jewish topics and images disappeared from film and returned only with "perestroika." Concludes that the Russian film industry has dealt extensively with Jewish topics, including antisemitism in the past and present, the Holocaust, and the rise of right-wing extremism in Russia. *DR*

0846 Cheyette, Bryan H.: Between Repulsion and Attraction: George Steiner's Post-Holocaust Fiction. *Jewish Social Studies* 5, 3 (1999) 67-81.

Steiner is split between interest in Western culture and suspicion of it for its complicity in mass murder. His collection of stories "Anno Domini" includes characters who are attracted and repelled by the horrors of the war they had escaped. The protagonist of "Cake" opposed and feared the Nazis but also had antisemitic attitudes. The story's dialogue encapsulates Steiner's philosophy of antisemitism, which is elaborated in "In Bluebeard's Castle." In the late 1970s Steiner published "The Portage to San Cristobal of A.H.," which describes a search for Hitler in South America. Mentions Steiner's preoccupation with Heidegger and his sympathy, expressed in his autobiography "Errata: An Examined Life," with apologists for antisemitism and Nazism. *YC*

0847 Coppel-Batsch, Marthe: Georges Perec, romancier de la psychanalyse. *Les Temps Modernes* 604 (May-July 1999) 190-203.

Analyzes the novel "W ou le souvenir de l'enfance" by Perec, an allegory based on the author's childhood memories of the Holocaust period. Compares the fictional account of life on an island to his own recollections of the death of his father in 1940, his departure from Paris to safety in the countryside, and the disappearance of his mother in 1942. Contends that the novel's significance lies in its use of psychoanalysis as a tool by Perec to aid survivors in dealing with the painful memories of the Holocaust period. *TS*

0848 Crossan, John Dominic: Troubling History at Oberammergau. *Queen's Quarterly* 108, 2 (Sum 2001) 223-231.

Reflects on the Oberammergau passion play, which the author saw in 1960 and again in 2000. Focuses on the scene of the crowd standing before Pilate, shouting "crucify him, crucify him." In 1960, Crossan was struck by the fact that the same crowd that had earlier enthusiastically embraced Jesus was now calling for his death. Nothing in the plot explained why. In the revised version of 2000, there are three crowds shouting against Jesus and one crowd for him. Contends that the noise is so great that it is doubtful whether the audience can grasp the point that not "all of the Jews" were against Jesus. Holds that the description of this scene in the Gospels was first written by Mark, and then copied by Matthew, Luke, and John. Each one changed and embellished the description so that what in Mark was "the crowd, " became in Matthew "the crowds" and "all the people," in Luke "the chief priests, the leaders, and the people," and in John "the Jews." Although most of the antisemitic statements have been removed from the text of the passion play, the real problem lies in the text of the New Testament which requires examination as to its historicity and the meaning of gospel truth. *SSC*

0849 Dayan-Rosenman, Anny: Vercors et le statut des Juifs sous l'Occupation: Une révolte militante. *Les Temps Modernes* 604 (May-July 1999) 131-151.

Analyzes the major works of Vercors (1902-1991) which describe the Vichy era. Born Jean Bruller, son of a naturalized Hungarian Jewish father, he was a pacifist who became a writer after he joined the French resistance. He utilized his stories as a vehicle to condemn the persecution of French Jewry. Portrays the personality of Thomas Muritz in "La marche à l'Etoile" as the epitome of the assimilated French Jew, whose Jewish identity is exclusively defined by antisemites. Focuses on Vercors' story "Le songe" (1943), which reflects his anxiety at the possible deportation of family members and friends. Vercors was the first French writer to denounce the persecution of Jews in France during the Vichy era; he felt that it was his duty both as a Frenchman and a Jew to demonstrate how precarious the Jews' situation was during the Occupation years. *TS*

0850 *Dialectical Anthropology* 24, 3-4 (Dec 1999).

Entitled "Poetry and Ethics after the Holocaust – a Millennium Memorial," this issue contains articles explicitly or implicitly responding to Adorno's dictum about the impossibility of writing poetry after Auschwitz. Partial contents: Kovel, Joel: Poetry after the Holocaust (239-253); Bernstein, Charles: The Second War and Postmodern Memory [Emphasizes the necessity of poetry after Auschwitz and discusses the influence of the Holocaust on American poets in the 1960s-70s.] (255-278); Rothenberg, Jerome: "Khurbn" and Holocaust: Poetry "after" Auschwitz [Focuses on his own poetry.] (279-291); Kama'ila, Linda LeValley: Poetry and History: Reflections on Stanley Diamond's "Arbeit Macht Frei" [Discusses Diamond's poem (printed on pp. 305-322) as poetry of history that illuminates the Holocaust.] (293-304); Mandel, Naomi: Speaking Corpses and Spectral Spaces: Representing Testimony after the Holocaust [Discusses the problems of testimony in literature and film about the Holocaust, focusing on Elie Wiesel's "La nuit," Claude Lanzmann's "Shoah," and Art Spiegelman's "Maus."] (357-376); Lubich, Frederick A.: A German Jewish Double Bind: Anja Lundholm's Novel "Ein ehrenhafter Buerger" or How a Jewish Daughter (Barely) Survives Her Nazi Father [The novel was first published in 1972 as "Der Gruene."] (377-385); Eskin, Michael: A Survivor's Ethics: Levinas's Challenge to Philosophy (407-450). *YC*

0851 Freedman, Jonathan: *The Temple of Culture: Assimilation and Anti-Semitism in Literary Anglo-America.* Oxford: Oxford University Press, 2000. vi, 264 pp.

Discusses the prominence of the stereotyped image of the Jew as artist, sexual pervert, and source of social degeneration in late 19th-century English and American culture. Artists used this image to deflect attention from their own deviance and, at the same time, to promote high culture. One of the main images was that of Svengali in Du Maurier's immensely popular novel "Trilby." Focuses on the novelist Henry James, due to his ambivalent attitude toward Jews which included aspects of antisemitism. Concerning the entrenched view that it was not appropriate for Jews, as "outsiders," to teach English literature, notes the rise in the postwar period of Jewish teachers and critics who became guardians, if not priests, of high English culture; paradoxically, they largely ignored James' antisemitism. *YC*

0852 Fuchs, Anne: *A Space of Anxiety: Dislocation and Abjection in Modern German-Jewish Literature*. Amsterdam: Rodopi, 1999. vi, 200 pp.

Analyzes the works of a number of Jewish writers (Freud, Kafka, Joseph Roth, Albert Drach, and Edgar Hilsenrath) which were composed just before, during, or after the Holocaust. Finds a common thread in their use of the "travel paradigm" to express homelessness. Reacting to their rejection by German society, these "borderliners" expressed alienation from Jewishness (Freud and Roth) and from the self (Kafka). They responded to a social order which had no place for the Jew by rejecting aspects of that order that might have provided them with possible sources of identity. They also internalized some of the hostility, as can be seen in their projection of negative aspects of "the Jew" onto women. In the works of these writers, assimilation is revealed as both failing to satisfy antisemites and leading to identity confusion among Jews. *YC*

0853 Galchinsky, Michael: "Permanently Blacked": Julia Frankau's Jewish Race. *Victorian Literature and Culture* 27, 1 (1999) 171-183.

Contends that the Anglo-Jewish writer Julia Frankau (1864-1916) expressed Jewish self-hatred in her novels "Dr. Phillips: A Maida Vale Idyll" (1887) and "Pigs in Clover" (1903), which she legitimated by recourse to an idiosyncratic form of "scientific" racism. Early and mid-Victorian novelists regarded the Jews as a religious group rather than a racial one, but in the 1880s-90s, with the influx of Eastern European Jews and the rise of nativist racism, the Jews were classified as "non-whites." Frankau shared this racist view, and felt that biologically inherited traits explained the moral degeneracy of Jews. Although she projected negative Jewish traits on Eastern European Jews, and thus divided the Jewish community into "good" and "bad" factions, it does not mean that she approved of "West End" Jews. The only way to improve Jewry in Frankau's view was the kind of mixed marriage in which a Jewish woman married a Christian man, because only the latter was able to "whiten" Jewry. Contends that Frankau's racism was a leitmotif in Anglo-Jewish thinking. *DR*

0854 *Le Genre Humain* 36 (Dec 2001).

This issue, entitled "L'art et la mémoire des camps: Représenter exterminer," edited by Jean-Luc Nancy, was published under the auspices of La Maison d'Izieu. Partial contents: Nancy, Jean-Luc: La représentation interdite (13-39); Loraux, Patrice: Les disparus (41-57); Benslama, Fethi: La représentation et l'impossible [On the behavior of individuals in extreme situations in the camps as interpreted by Primo Levi, Bruno Bettelheim, and Giorgio Agamben.] (59-80); Rancière, Jacques: S'il y a de l'irreprésentable (81-102). *HV*

0855 Genzeleva, Rita: *Puti yevreiskovo samosoznaniya: Vasilii Grossman, Izrail Metter, Boris Yampolskii, Ruf Zernova* [Paths of Jewish Self-Consciousness: Vasilii Grossman, Izrail Metter, Boris Yampolskii, Ruth Zernova]. Moskva: Mosty Kultury; Ierusalim: Gesharim, 1999. 318 pp.

The works of these Russian Jewish writers show that, under the influence of the Holocaust and of postwar antisemitism, their Jewish consciousness became stronger. In particular, the representation of and attitude toward antisemitism changed in the works of Grossman and Metter (1909-1996). Grossman was the first of the Soviet writers who, breaking with the stereotypical depiction of the "friendship of peoples" during the Soviet-German war, showed the wide collaboration of non-Jews with the Nazis. Contrary to Ehrenburg, for whom antisemitism was an atavism, Grossman showed it as an actual phenomenon in the USSR. One of Grossman's themes was the negative effect of antisemitism on the Jews themselves. Metter, who had depicted antisemitism as a phenomenon of little importance which does not prevent the Jew from being a Russian patriot, realized its dangers and changed his mind about Zionism, adopting a positive view. Postwar antisemitism also affected the works of Yampolskii and Zernova. *DR*

0856 Gill, Jonathan Peter: *The Law of the Modernist Letter: Anti-Semitism and the Technology of Representation in the Poetry of Ezra Pound.* Diss. – Columbia University, New York. 1999. 366 pp. Unseen.

Pound's conception of literary form derives its motivation from the threats that Jews supposedly present to the production and circulation of value and meaning in America and to the nation's cultural and social order. His antisemitism is about representation itself, and is therefore intimately related to his ideas about literature, shaping the "Cantos" and exemplifying the broader cultural roots underlying the complex relationship between ethics and literature in 20th-century America. In the 1930s-40s, Pound became interested in the radio, a new technology of representation (invented by Pound's cousin, Mahlon Loomis). Radio inspired in Pound an explosive anti-Jewish rhetoric. Investigates his final years, with their continuing fear of "Jewish" textuality and the Talmud in particular. Pound's last writings are less a pastoral recantation of antisemitism than an attempt to shore up a poetics of "antisemitism" by coming to terms with his ultimate literary ancestor, the reputed author of the Talmud, the biblical Ezra.

0857 Gilman, Sander L.: "We're Not Jews": Representing "Jews" in Contemporary Multicultural Literature. *Arcadia* 36, 1 (2001) 2-20.

Discusses the image of "the Jew" in contemporary multicultural literature. The Jew is treated in terms of physical stereotypes (e.g. body, nose, and tongue), reflecting perceptions of Jews as different. These attributes are associated with differences in language, culture, and values – all perceived as foreign. Jews are usually characterized as victims (especially of the Holocaust) or as examples of assimilation. Even when they are sometimes identified with power, Jews are stereotyped. Discusses four works: the short story "We're Not Jews" (1997) by Anglo-Pakistani writer Hanif Kureishi, the novella "Kafka's Curse" (1997) by South African writer Achmat Dangor, British-born Zadie Smith's "White Teeth" (2000), and the Turkish-German Zafer Senocak's "Gefaehrliche Verwandtschaft" (1998). *YC*

0858 Goffman, Ethan: *Imagining Each Other: Blacks and Jews in Contemporary American Literature*. Albany: State University of New York Press, 2000. xiii, 262 pp.

Ch. 2 (pp. 25-48), "Black (E)Masculinity and Anti-Semitism," discusses Black stereotypes of Jews, who were largely identified with Whites, as oppressors. States that antisemitism is linked to Black self-hatred. Notes that stereotypes of Jews in Black literature were less acceptable than those of Blacks in Jewish literature. Ch. 5 (pp. 91-110), "Burning Bridges: Black Nationalism and Anti-Semitism," shows how, in the late 1960s, questioning of liberal assumptions about U.S. society led to Black nationalists' viewing Jews as paternalistic, self-interested, and largely responsible for oppression of Blacks. Black nationalist literature, e.g. that of LeRoi Jones (Imamu Amiri Baraka), was rife with antisemitic stereotypes. After the 1960s a hybrid African-American literature developed, more differentiated in regard to Jews. Ch. 6 (pp. 111-141), "Jewish Backlash," presents Saul Bellow's "Mr. Sammler's Planet" (1970) and Bernard Malamud's "The Tenants" (1971) as Jewish reactions to threats raised by the Black Arts Movement and Black nationalists. *YC*

0859 Gold, Helmut; Heuberger, Georg, eds.: *Abgestempelt: Judenfeindliche Postkarten*. Heidelberg: Umschau Braus, 1999. 380 pp. On title page also: Auf der Grundlage der Sammlung Wolfgang Haney. Eine Publikation der Museumsstiftung Post und Telekommunikation und des Juedischen Museums Frankfurt am Main [anlaesslich der Ausstellung... (14.4.1999-1.8.1999)].

A collection of articles, accompanied by numerous small black-and-white reproductions of the postcards as well as 23 larger ones in color. *RW*

0860 Gruenzweig, Walter: Leben mit dem Holocaust: Die "Erinnerungsbuecher" Fred Wanders. *Das Juedische Echo* 50 (Oct 2001) 302-309.

Discusses the Holocaust theme in the works of Wander (b. 1917), an Austrian Jewish writer published and known primarily in East Germany, where he lived for 30 years. He is the survivor of a long series of concentration and labor camps and death marches. The striking element in his work is his curiosity and passion for exact observation, which was a lifesaver in the camps. It was missing in German bystanders and is still missing in their descendants. *RW*

0861 Heidelberger-Leonard, Irene: Klaerung oder Verklaerung? Ueberlegungen zu Ilse Aichingers Roman "Die groessere Hoffnung." *Verschwiegenes Wortspiel: Kommentare zu den Werken Ilse Aichingers: Akten des Internationalen Colloquiums vom 27. bis 28. April 1998 an der Vrije Universiteit Brussel,* ed. Heidy Margrit Mueller. Bielefeld: Aisthesis, 1999. Pp. 157-168.

Criticizes Aichinger's novel (1948) for aestheticizing the Holocaust. Her heroine, Ellen, sees the yellow star, which, as she knows, means death, as a "higher hope," and acceptance of the Jewish fate as salvation. Thus, Aichinger did not jar the consciences of her colleagues in Gruppe 47, who at that time were writing about the sacrifice of German soldiers, not of the Jews. Contrasts Aichinger's laudation of hope with Tadeusz Borowski's description of its paralyzing effect on the concentration camp victims and Jurek Becker's ambivalent treatment in "Jacob the Liar." *RW*

0862 Herring, Scott: "Her Brothers Dead in Riverside or Russia": "Kaddish" and the Holocaust. *Contemporary Literature* 42, 3 (Fall 2001) 535-556.

Discusses Allen Ginsberg's poem "Kaddish," an elegy for his mother written in 1961 and relating to her insanity, in terms of the Holocaust. The poem is an elegy not only for one woman but for all Holocaust victims. Naomi Livergant Ginsberg immigrated to the U.S. in 1905 from Nevel, where Nazi Einsatzgruppen later wiped out the Jewish community; she is presented as an unheeded prophet of disaster. Her paranoia reflects both irrational fears and the Nazi enemies of everything she was – a Jew, a communist, a teacher of retarded children, and a mother of a Jewish son. The figure of Hitler, prominent in the poem, epitomizes the Holocaust. Naomi's madness contrasts with the "madness" of the indifferent leaders of the world who knew about the Holocaust but did not act. Also linking the mother's mental illness and the Holocaust are her fears of poison gas and medical experiments. The poem also reflects its time of writing, the 1950s; the poet composed his "Kaddish" for the victims of the Holocaust, whom the world was then aware of but cared little about. *YC*

0863 Huntemann, Willi: Zwischen Dokument und Fiktion: Zur Erzaehlpoetik von Holocaust-Texten. *Arcadia* 36, 1 (2001) 21-45.

The scandal aroused by the discovery that Binjamin Wilkomirski's "Bruchstuecke" was an invention shows that in Holocaust memoirs the authenticity of the authorship counts. What would have been accepted as a literary work had it dealt with any other topic is not acceptable when dealing with the Holocaust. Wolfgang Koeppen, in "Jakob Littners Aufzeichnungen aus einem Erdloch," transmits as a sort of ghost writer the real story of a Holocaust survivor. States that the primary function of a Holocaust memoir is to testify to the facts; secondly, it expresses subjective memory; and thirdly, it represents all the victims in a call for public memory. A literary memoir must go beyond a documentary report; yet it cannot represent the inhumanity of the Holocaust aesthetically, or indeed present it adequately at all. Also, the usual demand from an autobiography, that it give a personal significance to the writer's life, is out of place where he is a passive sufferer of a collective fate. Analyzes the ways Ruth Klueger, in her literary memoir "weiter leben," and Imre Kertész, in his autobiographical novel "Fateless," deal with these problems. *RW*

0864 Inglot, Mieczysław: *Postać Żyda w literaturze polskiej lat 1822-1864* [The Figure of the Jew in Polish Literature in the Years 1822-1864]. Wrocław: Wydawn. Uniwersytetu Wrocławskiego, 1999. 296 pp. With an English summary (pp. 286-288).

Distinguishes two main types of representation of the Jew in Polish prose and poetry. The first is a romantic figure, modeled on images from the Hebrew Bible – a fighter for the independence of his suppressed nation, such as Moses or the Maccabees. The second presents the Jew in a stereotypical negative light, as a usurer or banker (sometimes a convert to Christianity) or as a participant in an international plot. During the period covered, a Jew is only rarely the hero of the novel, poem, or play. Notes antisemitic accents in Zygmunt Krasiński's "Non-Divine Comedy." Relates to connections between literature and politics, e.g. the positive image of the Jew in the period of the January insurrection (1861-63), when the Jews were perceived as allies in the struggle against the occupiers of Poland. Pp. 193-280 present an anthology of Polish poetry from the period studied. *HV*

0865 Jacobs, Steven Leonard: Binjamin Wilkomirski ("Fragments") and Donald Watt ("Stoker"): When the Holocaust Story Is a Lie. *CCAR Journal* 47, 2 (Spr 2001) 56-70.

Summarizes the cases of two "Holocaust memoirs" published in 1995 – Wilkomirski's "Fragments: Memories of a Wartime Childhood" and a less-known book by the Australian, Donald Watt, "Stoker: The Story of an Australian Soldier Who Survived Auschwitz-Birkenau" – which have been proved to be fictional creations. The identity of Wilkomirski, born in Biel in 1941 to a Swiss Protestant woman and adopted by the Doessekker family, has been ascertained. Watt's book has been partially discredited; it is too fraught with historical errors. States that the publication of the two books, the good reception they received, and the scandals they caused afterwards damaged Holocaust research and brought into question the veracity of all other Holocaust memoirs. *DR*

0866 Janicka-Świderska, Irena; Jarniewicz, Jerzy; Sumera, Adam, eds.: *Jewish Themes in English and Polish Culture*. Łódź: Wydawn. Uniwersytetu Łódzkiego, 2000. 196, [21] pp.

This collection of articles is based on a four-year (1995-1999) scholarly link between the University of York and the University of Łódź, at which papers were delivered by visiting lecturers. Partial contents: Berthoud, Jacques: The Figure of the Jew in English Renaissance Drama (11-45); Drain, Richard: "The Shark in the Harbour": Anti-Semitism and the Plays of Harold Pinter (46-69); Edelson, Maria: The Holocaust and the Language of Silence in "Isolation" by Jerzy Peterkiewicz [On the 1959 novel about a Holocaust survivor; Peterkiewicz left Poland during the war and settled in England.] (70-77); Jarniewicz, Jerzy: Stephen Romer's Ghetto Poems [On two poems he wrote in 1989-90 about the ghetto of Łódź.] (89-96); Luckhurst, Mary: The Case of "Theresa": Guernsey, the Holocaust and Theatre Censorship in the 1990s [On the play by Julia Pascal (1990) about an Austrian Jewish woman who was deported from Guernsey to Auschwitz in 1942 and murdered there.] (138-148); Moody, A. David: Ezra Pound with Two-Pronged Fork of Terror and Cajolery: The Construction of His Anti-Semitism (up to 1939) [A paper presented at the 17th international Ezra Pound Conference, 1997.] (149-169); Wicher, Andrzej: "The Most Unkindest Cut of All": Models of Prejudice in Chaucer's "The Prioress's Tale" and Shakespeare's "The Merchant of Venice" [On the association of Jews with "murderous cutting," including circumcision, as a literary stereotype.] (182-196). *DR*

0867 Klugherz, Dan: "Schindler's List" and "Schindler": The Movie and the Documentary. *Television Quarterly* 30, 2 (Fall 1999) 74-78.

Compares Spielberg's film "Schindler's List" with the British television documentary "Schindler," produced, written, and directed by Jon Blair in 1983. Klugherz, himself a documentary filmmaker, concludes that the Hollywood film is important because it has already been seen by millions, but "Schindler," the more authentic documentary, has the ring of truth throughout, including in terms of the survivors' conveying what it was like to be a victim of the Nazis. Summarizes the reaction to the two films in terms of audience interest, stating that audiences are not much concerned with historical truth, and that it is the fate of "Schindler," as it is of most documentaries, to be seen by a few. *YC*

0868 Kőbányai, János: *Az apokalipszis aggadája: Ámos Imre és kora* [The Haggadah of the Apocalypse: Imre Amos and His Times]. Budapest: Múlt és Jövő, 1999. 64; [50] pp. Published also in English, French, and German. The last 50 pp. comprise a notebook of drawings by Ámos exhibited in the graphic art collection of the János Damjanich Museum in Szolnok.

Describes the life and works of the Jewish painter and graphic artist Imre Ámos (né Unger, 1907-1944). Ámos is compared with the poet Miklós Radnóti, who was also of Jewish origin but converted to Catholicism. In reaction to the Hungarian anti-Jewish laws in the 1930s, Ámos took pride in his Jewishness. Subsequently he continued to create between and during his stints at forced labor service, starting in 1940. The Szolnok Sketchbook, appended here in a facsimile edition, contains his last work (he died in a concentration camp in Germany), replete with apocalyptic symbolism and focusing on victims of the Holocaust. The "Haggadah" presents Ámos as an artist-prophet, whose creativity represented the survival of the soul despite disaster. Contrasts the posthumous fame of Radnóti with the fate of Ámos, whom the Hungarians have failed to include in their cultural pantheon. *YC*

0869 Koerte, Mona: *Die Uneinholbarkeit des Verfolgten: Der Ewige Jude in der literarischen Phantastik*. Frankfurt a.M.: Campus, 2000. 347 pp. Based on the author's diss. – Technische Universitaet Berlin, 1999.

Describes the many aspects and functions of the figure of the eternal wanderer in late 18th-20th century literature. In "Ahasver, der Jude" (pp. 27-68) states that after the secularization of the Christian legend, the Eternal Jew represented the Jewish people, especially its negative qualities as perceived by antisemites, who claimed that because of his timelessness, the Jew can adapt himself to all times while essentially remaining himself, stubborn and loveless, the qualities for which he was condemned. He is anachronistic – a living corpse that inspires fear. His salvation can come only from complete assimilation or "Untergang." Also surveys Jewish writers' images of the Eternal Jew. The Nazis presented the Eternal Jew in literature, and in an exhibition and a movie by that name, which showed him constantly changing his masks in order to achieve world domination. *RW*

0870 Kowal, Kristopher: Krzysztof Kieślowski's "Dekalog 8": Narrating Jewish-Polish Reconciliation. *The European Legacy* 4, 4 (Aug 1999) 58-76.

Analyzes the eighth film in the "Dekalog" series by the Polish film director, which tells the story of two women: Zofia, a Polish woman and a professor of ethics at Warsaw University, and Elżbieta, a Holocaust survivor and Holocaust researcher who came to Poland from the USA to engage in research. But the two women had met in Warsaw before, in 1943, when Elżbieta was a child and Zofia and her husband were asked to become her godparents for a baptism which could save her life. They refused on the grounds that it would require them to disobey the commandment "Thou shalt not bear false witness against thy neighbor." Elżbieta was then helped by another couple. This story is brought by Elżbieta to the ethics class of Zofia and is discussed by the students. The film deals with the delicate relationship between the two women, the moral implications of Zofia's refusal, and their reconciliation. Concludes that the film deals with forgiveness and reconciliation as an individual experience and cannot be interpreted as a film dealing with Polish-Jewish relations in the 20th century. *HV*

0871 Krall, Hanna, ed.: *Es war der letzte Augenblick: Leben im Warschauer Getto.* Trans.: Roswitha Matwin-Buschmann. Stuttgart: Deutsche Verlags-Anstalt, 2000. 119 pp. The texts in the book were translated from the Polish.

Pp. 11-35 present excerpts from the official journal "Gazeta Żydowska," which appeared in Kraków but was distributed to all the ghettos between July 1940-August 1942. Pp. 37-69 comprise watercolor drawings of life in the ghetto by Teofila Reich-Ranicki (b. 1920), who was interned in the Warsaw ghetto from November 1940. In July 1942 she married Marcel Reich; in February 1943 they escaped to the "Aryan side" and survived in hiding. Pp. 71-89 present an essay by Hanna Krall on the drawings. Pp. 91-115 contain drawings of opera heroines by Reich-Ranicki. *ID*

0872 Lachman, Jeffrey Bryan: *The "Raquel" Plays: Theatrical Representation of Political Intrigue and Antisemitism in Seventeenth- and Eighteenth-Century Spain.* Diss. – Georgetown University, 1999. 245 pp. Unseen.

The reputed affair between King Alfonso VIII and a Jewish woman became the subject of many literary works. Three plays are analyzed here: "Las paces de los reyes y Judía de Toledo" by Lope de Vega Carpio (1617), "La desgraciada Raquel" by Antonio Mira de Amescua (1635), and "Raquel" by Vicente García de la Huerta (1778). The works differ in their focus and exposition of political intrigue and antisemitism. Examines the reasons for the legend's continuing popularity, how contemporary political and social events influenced audience reception, and the lingering antisemitism shown in the plays long after the expulsion of 1492. The royal iconography always serves to exalt the dynasty, while the Christian iconography of Jews portrays them as weak and inferior. Antisemitism persisted in Spain, affecting the portrayal of Raquel as the representative of the Jewish people. Jewish stereotypes endure, even though theatrical paradigms change.

0873 Lafferty, Maura K.: 'Limping Jacob': The Image of the Jew in Walter of Châtillon's Lyrics. *Mittellateinisches Jahrbuch* 33, 1 (1998) 133-147.

The use by Walter (fl. 1170-1180) of anti-Jewish typologies in his lyrics, particularly the Christmas lyrics, is generally seen today as reflecting an anti-Judaic mentality. Points out that such imagery was common in the 12th century, and was not out of place in the Christmas season. Contends that the anti-Judaic elements in his poems (written in Latin) were meant to serve as a vehicle for comparing Christians and Jews and exploring the problematic nature of textual interpretation. *REK*

0874 Lange, Sigrid: *Authentisches Medium: Faschismus und Holocaust in aesthetischen Darstellungen der Gegenwart*. Bielefeld: Aisthesis, 1999. 256 pp.

In addition to analyses of Nazi films, and of postwar films and literature on fascism, discusses film and literature on the Holocaust, including Lanzmann's "Shoah," Spielberg's "Schindler's List," and the writings of Ka-Tzetnik, Imre Kertész, Edgar Hilsenrath, and Peter Weiss. *RW*

0875 Lathey, Gillian: *The Impossible Legacy: Identity and Purpose in Autobiographical Children's Literature Set in the Third Reich and the Second World War*. Bern: Peter Lang, 1999. 258 pp.

Contends that although autobiographical children's literature (fictionalized or straight accounts) about World War II may ostensibly be directed toward juvenile readers, they are primarily written for the authors themselves, to help them establish their identity or to serve a therapeutic purpose. Compares works written in 1970-95 by Germans, Britons, and Jews (mainly refugees from Nazism). Understandably, their perspectives are quite different. The Jews write about persecution and exile. Germans who want to write about their childhood in the Third Reich are faced with an "impossible legacy": how to depict their childhood experiences (sometimes including membership in the Hitler youth movement) and, at the same time, express rejection of Nazi values. Concludes that this type of literature has not yet established its true audience and that further research might study children's responses and the contrast in ideological perspectives between works written in West and East Germany. *YC*

0876 Latzer, Robert: Orwell's Jews. *Midstream* 46, 2 (Feb-Mar 2000) 36-41.

Analyzing stereotypical images of the Jew in Orwell's writings before and after World War II, concludes that he retained the antisemitic prejudices of the English society in which he grew up. Orwell's views were altered somewhat by the Holocaust, which sensitized him to the murderous danger of prejudice and racism. For example, in an article in 1945, he noted that English antisemitism "does not at present lead to open persecution but it has the effect of making people callous to the sufferings of Jews in other countries." Nevertheless, Orwell continued to overgeneralize about "the Jew," whom he still viewed as an alien. Orwell was suspicious of "-isms," and he spoke against antisemitism and communism, but also against Zionism. *YC*

0877 Lauer, Gerhard: Erinnerungsverhandlungen: Kollektives Gedaechtnis und Literatur fuenfzig Jahre nach der Vernichtung der europaeischen Juden. *Deutsche Vierteljahrsschrift fuer Literaturwissenschaft und Geistesgeschichte* 73 [Sonderheft] (1999) 214-245.

States that all Holocaust literature faces the dilemma of the need for authenticity in communicating the Holocaust experience and its incommunicability. Until the mid-1980s, while memory of the Holocaust was still based on living witnesses, literature such as the documentary dramas of Rolf Hochhuth and Peter Weiss put the emphasis on straightforward communication and authenticity; these did not seem open to question. Since then, with memory becoming collective and based on writing, works by authors such as Ruth Klueger, Imre Kertész, Louis Begley, Edgar Hilsenrath, or Art Spiegelman problematize communicability and deal with the Holocaust by literary devices that subvert the consensual narrative and frustrate the wish for meaning. *RW*

0878 Lazarus, Joyce Block: *Strangers and Sojourners: Jewish Identity in Contemporary Francophone Fiction*. New York: Peter Lang, 1999. 141 pp.

Discusses six autobiographical novels by French-language Jewish writers dealing with Jewish identity in societies that were characterized by antisemitism and/or affected by the Holocaust. Albert Memmi's "La statue de sel" (1953) is set in Tunisia, where there was racism, pogroms, religious prejudice, and, finally, Nazi occupation. Elie Wiesel's "Les portes de la forêt" (1964) is about racism and Christian antisemitism, and bystanders as well as perpetrators of genocide. André Schwarz-Bart's "Le dernier des justes" (1959) discusses centuries of antisemitism, culminating in Auschwitz. Georges Perec's "W ou le souvenir d'enfance" (1975) describes his childhood during the war and raises questions about the French under Vichy. Patrick Modiano's "La Place de l'Etoile" (1968) explodes myths about French resistance to the Nazi occupation and discusses collaboration with Vichy antisemitism. Paula Jacques' "L'héritage de tante Carlotta" (1987) describes the author's feelings of exile in her native Egypt (and later in France), and refers to the pogroms in 1948 and 1956-57, the expulsion of Egyptian Jews in the 1950s, and the antisemitism faced by Jews who remained. *YC*

0879 Leak, Andrew N.; Paizis, George, eds.: *The Holocaust and the Text: Speaking the Unspeakable*. Basingstoke, Hants: Macmillan Press, in association with the Institute for Romance Studies and the Institute for English Studies, School of Advanced Study, the University of London, and the Wiener Library, 2000. ix, 196 pp. Appeared in the USA in New York: St. Martin's Press, 1999.

A collection of essays on representation of the Holocaust in literature, testimony, and philosophy. Contents: Lang, Berel: Holocaust Genres and the Turn to History [Another version appeared in his "Holocaust Representation" (2000).] (17-31); Gordon, Robert S.C.: Holocaust Writing in Context: Italy 1945-47 [Discusses a dozen texts by or about Jewish deportees.] (32-50); Hardman, Anna: Representations of the Holocaust in Women's Testimony (51-66); Cheyette, Bryan H.: Between Repulsion and Attraction: George Steiner's Post-Holocaust Fiction [Appeared in "Jewish Social Studies" 5 (1999).] (67-82); Reiter, Andrea Ilse Maria: The Holocaust as Seen through the Eyes of Children [Discusses works by Gmeyner, Aichinger, Wiesel, Klueger, Wilkomirski, and Kertész.] (83-96); Eaglestone, Robert: From behind the Bars of Quotation Marks: Emmanuel Levinas's (Non)-Representation of the Holocaust (97-108); Parry, Ann: Idioms for the Unrepresentable: Postwar Fiction and the Shoah [Appeared in the "Journal of European Studies" 27 (1997).] (109-124); Vice, Sue: The Demidenko Affair and Contemporary Holocaust Fiction (125-141); Yudkin, Leon I.: Is Aharon Appelfeld a Holocaust Writer? (142-158); Khalifa, Samuel: The Mirror of Memory: Patrick Modiano's "La Place de l'Etoile" and "Dora Bruder" (159-173); Crowley, Martin: 'Il n'y a qu'une espèce humaine': Between Duras and Antelme (174-192). *YC*

0880 Levitina, Viktoria B.: *Yevreiskii vopros i sovetskii teatr* [The Jewish Question and Soviet Theater]. Ierusalim: [The Author], 2001. 308 pp.

Examines the stage image of the Jew in the Soviet and post-Soviet theater and the treatment of the "Jewish question" in theatrical works. Antisemitism, which was opposed by the regime in the early Soviet period, returned to the stage in the years of the New Economic Policy (1921-28). The image of the affluent and tricky Jewish "nepman" replaced the image of the Jewish revolutionary. In the 1930s, a positive image of the "denationalized" Soviet Jew appeared on the stage. The Western Jew as a victim of fascism also appeared for a short time. During World War II and in the postwar period, the Jew disappeared from the Soviet stage; however, negative heroes with "Jewish-sounding names" were portrayed on the stage in the 1950s-70s. Discusses the anti-cosmopolitan campaign in the late 1940s and the purges of Jews from the theaters. With the "perestroika" of the late 1980s, it was possible to discuss the "Jewish question" and the Holocaust on the stage. Appendix II (pp. 228-251),

"Sluchai Kugelya," discusses the Russian-Jewish theater critic Aleksandr (Avraham) Kugel (1864-1928); he was inconsistent in fighting antisemitism on the Russian stage. *DR*

0881 Lorenz-Lindemann, Karin: Die Tore von Frage-und-Antwort sind verschlossen: Klage im Werk von Paul Celan, Dan Pagis und Tuvia Ruebner. *Jahrbuch fuer Biblische Theologie* 16 (2001) 233-266.

Examines similar ways in which the three poets, all of them Holocaust survivors, dealt with their despair over the breakdown of language, cultural tradition, and faith as a result of the Holocaust. *RW*

0882 Lowy, Vincent: Les têtes parlantes: Analyse croisée de la pratique de l'interview dans les films de Marcel Ophuls et de Claude Lanzmann. *Cahier International sur le Témoignage Audiovisuel* 7 (Sept 2001) 53-86.

Analyzes the role of the interview in Marcel Ophuls' and Claude Lanzmann's films, which have shaped Holocaust awareness and contributed to filmed interviews becoming the preferred form of Holocaust testimony since the 1980s. Compares their interviewing techniques, choice of interviewees, use or rejection of archival material – all of them expressions of different views of the past. For Ophuls, history unfolds horizontally, storylike, while Lanzmann's "psychoanalytical" interviews reflect a vertical view of history as sacred. Despite their ideological, methodological, and temperamental differences, Ophuls and Lanzmann have influenced each other, making each other's work possible. Concludes that reconciling their worldviews is invigorating and fruitful for both historians and film lovers, opening up a dynamic field of exploration into the Shoah. *EO*

0883 Maechler, Stefan: *Der Fall Wilkomirski: Ueber die Wahrheit einer Biographie*. Zuerich: Pendo, 2000. 367 pp.

Analyzes the phenomenon of Wilkomirski's "false" memoirs, tracing the author's real biography, his transformation from Bruno Grosjean to Binjamin Wilkomirski, and the creation of his book "Fragments." Examination of documents and conversations with witnesses clearly call into question Wilkomirski's story. States that this case goes beyond Grosjean's tragic life story, and indicates how individual traumatic experiences, even when imagined, can be used to manipulate the public. *ID*

0884 Maechler, Stefan: *The Wilkomirski Affair: A Study in Biographical Truth*. Trans.: John E. Woods. New York: Schocken, 2001. ix, 496 pp. Unseen. Originally published as "Der Fall Wilkomirski: Ueber die Wahrheit einer Biographie" (Zuerich: Pendo, 2000). The English edition includes the text of Wilkomirski's "Fragments."

0885 Makarova, Elena: *Friedl Dicker-Brandeis: Ein Leben fuer Kunst und Lehre: Wien – Weimar – Prag – Hronov – Theresienstadt – Auschwitz.* Wien: Verlag Christian Brandstaetter, 2000. 240 pp.

An album published on the occasion of the eponymous exhibition held at the Palais Harrach in Vienna, October-November 1999. Pp. 11-39 present a biography of the Jewish artist (Vienna 1890-Auschwitz 1944), who lived in Czechoslovakia from 1934. She was deported to Theresienstadt in 1942, and to Auschwitz in 1944. In Theresienstadt she taught art to the children, thereby helping them to cope with their Holocaust experiences. Pp. 155-235 deal with the period in Theresienstadt, presenting Dicker-Brandeis' drawings, drawings of the children she taught, and memoirs of pupils about her. Ca. 4,000 drawings by children were recovered when the camp was liberated in 1945. *SSC*

0886 Mandell, Sherri Lederman: *Writers of the Holocaust.* New York: Facts on File, 1999. xxvi, 132 pp.

Presents brief biographies and discussions of the work of ten Jewish writers of the Holocaust, three of whom did not survive. Their testimonies, in the form of diaries, fiction, and poetry, present various perspectives on Jewish suffering under the Nazis. The writers are: Yitzhak Katzenelson, Abba Kovner, Primo Levi, Hannah Szenes, Ida Fink, Arnošt Lustig, Elie Wiesel, Anne Frank, Aharon Appelfeld, and Jerzy Kosinski. Mentions events in many places, including small towns in Poland, the Vilna and Warsaw ghettos, occupied Hungary, Transnistria, Ukraine, Amsterdam, Theresienstadt, and the camps of Auschwitz and Buchenwald. *YC*

0887 Mannes, Stefan: *Antisemitismus im nationalsozialistischen Propagandafilm: "Jud Suess" und "Der ewige Jude."* Koeln: Teiresias Verlag, 1999. 122 pp.

Discusses the role of films as propaganda tools during the Third Reich. Although the first productions of antisemitic films were far below the usual standard of German movies, with the mise-en-scène of "Jud Suess" and "Der ewige Jude" antisemitic films achieved a higher quality. "Jud Suess" falsified historical facts and presented the character of Joseph Suess Oppenheimer as the embodiment of corruption. Goebbels controlled the production of the film at every stage. The première took place on 5 September 1940 in Venezia, with great success. In the documentary film "Der ewige Jude" too, history was distorted by traditional anti-Jewish charges and stereotypes. The message that "German equals good, Jewish equals bad" was relayed through technical manipulation, such as camera perspective, cutting, simplified texts, and emotionalized music. *ID*

0888 Massey, Irving: *Philo-Semitism in Nineteenth-Century German Literature.* Tuebingen: Max Niemeyer, 2000. vi, 199 pp.

Argues that most of the non-Jewish 19th-century German authors considered here expressed both antisemitic and philosemitic views. Focuses on Leopold von Sacher-Masoch, but also discusses works by Karl Gutzkow, Bettine von Arnim, Annette von Droste-Huelshoff, Friedrich Hebbel, Gustav Freytag, Wilhelm Raabe, Theodor Fontane, Marie von Ebner-Eschenbach, Ludwig Anzengruber, Ferdinand von Saar, and by the Alsatian Alexandre Chatrian. Ascribes philosemitism to the authors' guilt feelings about Jews, which are compared to Jewish self-hatred. A short chapter on women and philosemitism touches on complex, often antagonistic attitudes of non-Jews toward Jewish women. Stereotypes play a large role in the presentation of Jewish figures in these literary works. *YC*

0889 McGee, Daniel T.: Dada Da Da: Sounding the Jew in Modernism. *ELH – English Literary History* 68, 2 (Sum 2001) 501-527.

Argues that T.S. Eliot's antisemitism is implicit in the image of the rat in "The Waste Land," which reflects the poet's view of the Jew as an anti-poetic barbarian. Claims that Eliot was influenced by Charles Maurras and Richard Wagner. Finds a common thread of antisemitism in the modernisms of Eliot and of the Black American Marxist poet Langston Hughes, despite considerable differences in their views. *YC*

0890 Mellinkoff, Ruth: *Antisemitic Hate Signs in Hebrew Illuminated Manuscripts from Medieval Germany.* Jerusalem: Center for Jewish Art, Hebrew University, 1999. 158 pp.

Rejects the widely-held theory that medieval artists distorted the human faces in Hebrew illuminated manuscripts due to anti-iconic attitudes of German Jewry. Presents the hypothesis that the illustrations were done by Christian artists under the supervision of Jewish scribes and for Jewish patrons. Concludes that artistic devices such as profile portrayal and animal features were antisemitic signs, whereby the artists expressed hatred and denigration. The animals chosen had negative connotations (e.g. asses and pigs). Another feature, closed eyes, suggested the "blindness" of the Jews. Further, human features were often distorted according to stereotypes (beaked noses) or deformed to the point of grotesque or demonic, suggesting that Jews were less than human. To the question of how the Jewish scribes and patrons overlooked such signs, responds that the "looking but not seeing phenomenon" operated in the past as it does in the present. *YC*

0891 Morahg, Gilead: Israel's New Literature of the Holocaust: The Case of David Grossman's "See Under: Love." *MFS – Modern Fiction Studies* 45, 2 (Sum 1999) 457-479.

Until the mid-1980s it was taboo in Israel for a non-survivor to write about the Holocaust, especially from an imaginative perspective. This changed with Grossman's novel "See Under: Love" (1986). This work employs elements of the fantastic to break down the strongest barrier separating the world of Holocaust victims from that of their Israeli descendants, the latter's inability to understand or accept the victims' going "like sheep to the slaughter." In the novel, fantasy enables the grandchildren to enter the concentrationary world and gain an empathy that allows the possibility of dialogue with important voices that have long been suppressed. *YC*

0892 Nader, Andrés José: *Emergency Poetry: Lyric Production in the Concentration Camps.* Diss. – Cornell University, 1999. 206 pp. Unseen.

Analyzes poems written by inmates in the ghettos and concentration camps, focusing on German-language texts by authors such as Hasso Grabner, Alfred Kittner, Ruth Klueger, Edgar Kupfer-Koberwitz, Fritz Loehner-Beda, Karl Schnog, Georg von Boris, and Ilse Weber. Traces the reception and publication history of their writings. Argues that poems from the concentration camps evince multiple modes of responding to the events. In terms of aesthetic ideology, these poems show that poetry was not only possible in the camps but also perhaps "useful" to some inmates. They demonstrate the enduring power of "old forms" in critical situations and suggest that classical meter, conventional rhythm, and rhyme provided the needed sense of order and stability. Poems from the camps present a stark contrast to the stylistic experimentation, ellipsis, and fragmentation of post-Holocaust poetry of the kind Paul Celan eventually wrote. Concludes that traditional poetic forms provided the necessary structure for expressing the "inexpressible" in the camps.

0893 O'Riley, Michael F.: The Recoding of "La Place de l'Etoile" in Modiano, Simon, and Le Clézio. *Romance Notes* 40, 1 (Fall 1999) 93-102.

Discusses four works of postwar French fiction: Patrick Modiano's "La Place de l'Etoile" (1968) and "Dora Bruder" (1997), Claude Simon's "L'Acacia" (1989), and J.M.G. Le Clézio's "Etoile errante" (1992). Focuses on the use of the symbol of the star to contrast two aspects of historical memory, as represented by the Place de l'Etoile in Paris, with the national emblem of the Arc de Triomphe at its center, and the Jewish star, with its dual resonance as an emblem of Jewish continuity and of Jewish fate under the Nazis. Modiano's "Dora Bruder" recounts the search for a Jewish woman who disappeared during the Holocaust. Her absence represents the deaths of those Jews whom French society did not protect during World War II. *YC*

0894 Ozick, Cynthia: The Rights of History and the Rights of Imagination. *Commentary* 107, 3 (Mar 1999) 22-27.

Discusses the relationship between fiction and history in regard to the Holocaust. Notes the problem with fiction which claims to be true, when imposture "escapes the bounds of fiction and invades life"; it is then defined as a hoax. In this case, the imaginative creation generally permitted the author of fiction is immoral; "the argument for fictional autonomy collapses and the rights of history can begin to urge their own force." Also faults fictional works about the Holocaust, like "Sophie's Choice" by William Styron and "The Reader" by Bernhard Schlink, which (intentionally or not) deflect readers' attention from the destruction of the Jews of Europe and those responsible for it. *YC*

0895 Paasch-Beeck, Rainer: Zwischen "Boykott" und "Pogrom": Die Verdraengung und Ermordung der juedischen Bevoelkerung Mecklenburgs im Spiegel der "Jahrestage." *Text + Kritik* 65-66 [2. Aufl., Neufassung] (May 2001) 119-134.

Discusses the centrality of the Holocaust in Uwe Johnson's novel and the authenticity of specific incidents, including several that actually took place in Mecklenburg, one of the two scenes of the novel "Jahrestage" (the other is New York, where Gesine, the central character, meets Holocaust survivors and is confronted with the guilt of her parents' generation). States that a much misunderstood matter is the supposed Jewishness of Dr. Semig, a victim of the Nazis in Gesine's hometown; Johnson emphasizes that he is not a Jew but a third-generation Christian, with one grandfather who was a baptized Jew. Thus, he points to the perversity of Nazi racist policy, but many of his critics, contrary to his intention, call Semig a Jew and thus follow Nazi categorization. *RW*

0896 Pape, Matthias: "Ich moecht' Jerusalem gesehen haben, eh' ich sterbe": Antisemitismus und Zionismus im Spiegel von Arthur Schnitzlers Roman "Der Weg ins Freie" (1908). *Jahrbuch des Freien Deutschen Hochstifts* (2001) 198-236.

Based on a lecture delivered at the University of Karlsruhe, October 1998. After describing the spread of antisemitism in Austria in the late 19th century (triggered by legal emancipation and by the influx of culturally alien East European Jews) and its espousal in differing degrees by all the political parties except the Social Democrats, as well as Jewish reactions ranging from assimilation to Zionism, shows how Schnitzler embodies these reactions in his Jewish characters. *RW*

0897 Parry, Ann: The Caesura of the Holocaust in Martin Amis's "Time's Arrow" and Bernhard Schlink's "The Reader." *Journal of European Studies* 29, 3 (Sept 1999) 249-267.

This study takes its departure from the philosophies of Theodor Adorno, Jean-François Lyotard, and Philippe Lacoue-Labarthe and their views about the Holocaust as a caesura or radical break in values. Both Amis's "Time's Arrow" (1991) and Schlink's "The Reader" (1997) deal with the Holocaust indirectly, with Jewish victims basically "offstage." Nevertheless, these two novels express the caesura of our times. Amis's book presents a disorientation of space, time, and understanding that reflects the radical evil of Auschwitz. This is symbolized by a Nazi doctor whose experiments in Auschwitz are succeeded by his treatment of people as subhuman after the war. Schlink's "The Reader" also deals with a female perpetrator who was responsible for many deaths at Auschwitz. But in this book, the impact of the caesura is relativized and loses its status as a rupture. In both novels society is depicted as relativizing and normalizing radical evil, which continues even after the Holocaust. *YC*

0898 Patterson, Lee: "The Living Witnesses of Our Redemption": Martyrdom and Imitation in Chaucer's "Prioress's Tale." *The Journal of Medieval and Early Modern Studies* 31, 3 (Fall 2001) 507-560.

Argues that in "The Prioress's Tale" Chaucer indirectly but definitely undercuts the antisemitism expressed by the prioress. Notes her intolerance and ignorance, as well as the primitive nature of her Marian Christianity. While she presents herself as an innocent for whom history is irrelevant, Patterson shows the key relevance of historical context, both European and specifically English, in terms of Christian massacres of Jews during the Crusades and Jewish reactions to them. "Kiddush Hashem," Jewish martyrdom in order to avoid forced conversion, was influenced by Christianity but was intended as a riposte to it. Demonstrates that the prioress represents a Christianity that "misread" the painful sacrifice by Jews of their children and believed that the Jews killed Christian children. Aside from the German, French, and English connections with the Crusades, stressed in medieval Hebrew literature, notes that Chaucer probably knew about Christian persecution of Jews, including pogroms, from his experience in Spain. *YC*

0899 *Perspectives* 6 (1999).

This issue is entitled "Ecrire la Shoah." Partial contents: Lévi-Valensi, Jacqueline: Ecriture et mémoire de la Shoah (9-19); Koren, Roselyne: "Unicité" versus "identité": Quelques aspects polémiques des qualifications de la Shoah (21-34); Mole, Gary D.: Pierre Créange: "Un Juif à ne pas rayer" (37-54); Bellos, David: Ce que Momo veut dire: La mémoire de la Shoah dans "La Vie devant soi" de Romain Gary (55-66); Jaron, Steven: Georges Perec ou "La vérité de la littérature" (67-80); Malachy, Thérèse: La mise en scène de la Shoah (81-86); Aslanov, Cyril: Tropes narratifs dans l'évocation de la Shoah: Métonymie et aposiopèse dans "L'oublié" de Wiesel et dans "Le temps des prodiges" d'Aaron Appelfeld (87-100); Wardi, Charlotte: Une représentation israélienne de la Shoah: "Voir ci-dessous: Amour" de David Grossman (101-115); Atlan, Liliane: Rendre la vérité vivante [An expanded version of a paper delivered at a colloquium in Naples, published in "La Sho'ah tra interpretazione e memoria" (1999).] (127-139). *HV*

0900 Pfanner, Helmut F.: Jewish Outsiders in Two Dramas by Felix Mitterer. *Modern Austrian Literature* 34, 1-2 (2001) 89-101.

Two plays by Austrian writer Felix Mitterer deal with Jews and the Third Reich; they are based on real events and people. In "Kein schoener Land" (1989), a converted Tyrolian Jew is deported to a camp and beaten by a young Nazi. The Nazi, recognizing the Jew as his father, kills him and then commits suicide. At the end of the play, the mayor of the converted Jew's village encourages his fellow Austrians to forget the past. The play premiered at the time of the Waldheim affair. "In der Loewengrube" (1998) is about a German theater company performing "The Merchant of Venice" in 1933-34; it highlights German antisemitism, with references to Lessing's "Nathan der Weise" providing a contrasting message of tolerance. Mitterer creates sympathy for his Jewish protagonists, who are victimized by Nazis and their sympathizers. Both plays include images of perpetrators who have survived the war but remained the same opportunists as before. *YC*

0901 Poniž, Denis: Antisemitizem v drami "Herman Celjski" Antona Novačana [Antisemitism in the Drama "Herman Celjski" by Anton Novačan]. *Jews and Slavs, Vol. 6: Jerusalem in Slavic Culture,* eds. Wolf Moskovich, Oto Luthar, Samuel Schwarzband. Jerusalem: Center for Slavic Languages and Literatures, Hebrew University; Ljubljana: Scientific Research Center of the Slovenian Academy of Sciences and Arts, 1999. Pp. 145-156.

Examines the historical drama "Herman Celjski" (1928), by Slovenian poet, playwright, and politician Anton Novačan (1887-1951). Novačan introduces the character of a Jew, Aron Salobir, to whom he ascribes "typical Jewish traits": he is greedy, perfidious, scheming, and part of the "world Jewish conspiracy." Novačan based the image of the Jew on negative stereotypes spread by antisemitic propagandist literature. *DR*

0902 Pressler, Mirjam: "If You Prick Us, Do We Not Bleed?" Der Jude Shylock als Theaterfigur, als Mensch, als Vater. *Shakespeare Jahrbuch* 137 (2001) 11-22.

A lecture delivered at the conclusion of the Shakespeare-Tage in Bochum, April 2000. Asserts that Shakespeare's portrayal of Shylock has been more baneful in its antisemitic influence than even the images of ritual murder in the churches. Just because Shylock is a differentiated human being, he leaves a deeper impression than if he were pure stereotype. Accuses Shakespeare of antisemitism: he allows Portia and the other Venetian aristocrats to humiliate Shylock and trample on his rights; he doesn't bother to investigate the social circumstances, the restrictions, and the insecurity that made Shylock what he was; he caters to the prejudices of the masses, which Shakespeare himself must have been too intelligent to share. Describes her own research on the life of the Jews in the 16th-century Venetian ghetto, which serves as background to her novel "Shylocks Tochter" (1999) in which she tries to rehabilitate Shylock and his relationship to Jessica. *RW*

0903 Priel, Beatriz: Negation in Borges's 'The Secret Miracle': Writing the Shoah. *International Journal of Psychoanalysis* 82, 4 (Aug 2001) 785-794.

After discussing Freud's concept of negation and recent approaches to symbolization and functions of fiction, proceeds to discuss Borges's short story "The Secret Miracle" (1943), in which a Jewish poet is murdered by the Nazis in Prague in 1939. The nightmare reality is accentuated and undermined by dreams, and time is confused. Human freedom and rationality are questioned, as is human creativity in the face of lawlessness and arbitrariness that lead to death. The reader is left perplexed by Borges's subversive narrative technique, which uses negation to suggest unspeakably horrendous reality. Contends that Borges's art of continually moving between reality and fantasy is an appropriate way to write fiction about the Holocaust. *YC*

0904 Reichelt, Matthias, ed.: *Zwei Entwuerfe zum Holocaust-Denkmal in Berlin*. Nuernberg: Verlag fuer Moderne Kunst, 2001. 160 pp.

A collection of essays commenting on two projects submitted jointly by Rudolf Herz and Reinhard Matz to competitions for the design of memorials for the Jewish victims of the Holocaust. The essays discuss their provocative proposal entitled "Ueberschrieben," which involved dedication of a kilometer of the superhighway near Kassel to the commemoration of the murdered Jews. Since the proposal was to pave the highway with stones, there was little chance that it would be adopted. Nevertheless, it presents a conceptual reorientation of commemoration. Their other proposal, "Leerstelle," was also meant as an anti-memorial. The plan was to create a huge cavity with walls of cement, expressing emptiness and negation. *ID*

0905 Reinbold, Wolfgang: Der Text der Oberammergauer Passionsspiele 2000: Ein Produkt des christlich-juedischen Dialoges und ein Testfall fuer dessen gegenwaertigen Stand. *Zeitschrift fuer Theologie und Kirche* 98, 2 (June 2001) 131-160.

The text of the Oberammergau passion play was based until 1990 on Joseph Alois Daisenberger's text for the performances of 1860 and 1870. This text was criticized by both Jews and Christians for its extreme anti-Judaism, which went far beyond the Gospel story: the Jews were depicted as full of demonic hatred for Jesus and as solely guilty of his death. Since the 1970s, the Oberammergau community, which is responsible for the play, has engaged in dialogue with American Jewish organizations; these resulted in minor changes in 1980 and 1990 and a thorough revision by Christian Stueckl and Otto Huber in 2000. Jews are now presented in a much more differentiated way, and Jesus' Jewish identity is emphasized. This is great progress, reflecting the progress in Christian-Jewish relations in Germany; but it does not go far enough because it still adheres to the Gospel story, which is itself anti-Jewish. Suggests that the play could permit itself more freedom on the basis of new theological and historical understandings of the New Testament. *RW*

0906 Rothberg, Michael: *Traumatic Realism: The Demands of Holocaust Representation*. Minneapolis: University of Minnesota Press, 2000. xii, 324 pp.

Presents a post-Holocaust view of contemporary culture. Examines, in particular, the question of realism as one of the central problematics that the Holocaust forces back into view. Pt. 1 (pp. 17-96), "Modernism 'After Auschwitz'," discusses the philosophers Theodor Adorno and Maurice Blanchot. Pt. 2 (pp. 97-177), "Realism in 'the Concentrationary Universe'," deals with the literary works of Ruth Klueger and Charlotte Delbo. Pt. 3 (pp. 179-273), "Postmodernism, or 'the Year of the Holocaust'," deals with Philip Roth, Art Spiegelman, and Americanizing the Holocaust. Derives from Holocaust testimonies the concept of traumatic realism as a way of superseding the realist vs. anti-realist dichotomy. Stresses the relation between present and past, and a shift from events to their transmission, including in mass culture, via such forms as comic books, feature films, and museum exhibits. *YC*

0907 Rudtke, Tanja: "Eine kuriose Geschichte": Die Pikaro-Perspektive im Holocaustroman am Beispiel von Imre Kertész' "Roman eines Schicksallosen." *Arcadia* 36, 1 (2001) 46-57.

Suggests that Kertész's novel, which has repelled some readers because of the protagonist's strange amorality and the lack of emotion with which he looks at the horrors of the Holocaust, can best be understood as a picaresque tale. In such a tale, the hero is a rogue (Schelm) characterized by naïveté and ignorance, by his readiness to accept the words, actions, and dictates of others, and by a distorted, dissociated, and innocuous view of events. The use of this genre enables Kertész to break through the conventions of Holocaust literature and, in his own words, "to say only the sayable and to be confident that the completed work will say more about the unsayable than if I tried to do so directly." At the end of Kertész's book, however, the protagonist is no longer naïve but possesses a knowledge that estranges him no less from others. Compares Kertész's work with two other, but very different, picaresque novels on the Holocaust – Becker's "Jakob der Luegner" and Hilsenrath's "Der Nazi & der Friseur." *RW*

0908 Sagarra, Eda: Grillparzer, the Catholics and the Jews: A Reading of "Die Juedin von Toledo" (1851). *Leo Baeck Institute Year Book* 46 (2001) 67-79.

Contends that one of the goals of Grillparzer's play was to criticize traditional Catholic attitudes toward Jews. "Die Juedin von Toledo" was written in the wake of the Revolution of 1848, but it reflects views regarding the relationship between the Jews and the surrounding society in Metternich's Austria. This society still retained many prejudices against Jews; idealizing the pre-Napoleonic and even medieval past, it opposed Jewish civil rights and extolled Christian society, a society in which Jews cannot depend on protection before the law. In the play, the society which calls itself "Christian" murders the Jewess Rahel, and the king is unable to prevent the murder of his beloved. Remarkably, Grillparzer modified the plot taken from Lope de Vega by placing Rahel, instead of the king, at the heart of the play. The play ends with a call to both Christians and Jews for mercy and reconciliation. *DR*

0909 Schaefer, Julia: Verzeichnet: Ueber "Judenbilder" in der Karikatur als historische Quelle. *Jahrbuch fuer Antisemitismusforschung* 10 (2001) 138-155.

Analyzes caricatures of Jews collected in Eduard Fuchs's "Die Juden in der Karikatur" (Muenchen, 1921). Whereas Fuchs sees in caricatures a way of penetrating to the real essence of the figures portrayed, they are in fact a compilation of stereotypical features. A fat man with cigar, diamond, top hat, and a long hooked nose is automatically decoded as representing a Jewish capitalist; conversely, all these signs are ascribed to Jewish capitalists in complete disregard of individual reality. The society projects its perception of the cultural degeneration of the age onto the Jews in the form of physical degeneration and disease. Fuchs's book, intended by him primarily as entertainment, contributed to the distorted stereotype of the Jew. *RW*

0910 Scheit, Gerhard: *Verborgener Staat, lebendiges Geld: Zur Dramaturgie des Antisemitismus*. Freiburg i.Br.: ça ira, 1999. 587 pp.

Using psychoanalytic and Marxist theories, analyzes antisemitism in drama, including medieval German passion plays, Elizabethan drama (especially "The Merchant of Venice"), late 18th-19th-century German farce, Wagner's operas and essays, German pre-Nazi and Nazi films, and postwar German drama, as well as Jewish treatment of the theme. Shows that in these works Jews stand for money (usury), and money for blood – the blood of the sacrificed Christ. *RW*

0911 Schlant, Ernestine: Die Sprache des Schweigens: Die westdeutsche Literatur und der Holocaust. *Tribuene* 160 (2001) 194-204. An English version appeared in "Dimensions" 14, 2 (2000).

Based on a lecture delivered in Berlin, September 2001. Analyzes four phases in the treatment of the Holocaust by German writers. Between 1945-ca. 1960, young writers like Boell wrote on the evil of the Nazi regime and its continuation into the present, but hardly touched on the Holocaust, even when describing parallel events such as the killing of non-Jews. Between 1960 to the mid-1970s, under the impression of the war crimes trials, there appeared documentary dramas and novels, but at the same time postmodern questioning of the validity of documents. 1975-1990 saw the rise of a new generation whose members considered themselves free of guilt for the Holocaust, of which they accused their parents; they saw themselves as victims of their parents' generation, ignoring the real victims, the Jews. In the 1990s at last, with a growing Jewish presence in Germany, though many writers continue in the old vein, a few have succeeded in portraying sympathetically the fate of individual Jews. *RW*

0912 Schnell, Ralf: "Holocaust-Literatur" als Generationen-Problem. *Zeitschrift fuer Literaturwissenschaft und Linguistik* 120 (Dec 2000) 108-129.

Asks how memory of the Holocaust is transmitted to later generations, and analyzes the relationship between the experiencing and the writing self – the self who gives the experience literary form – in the memoirs of Jean Améry, Cordelia Edvardson, Imre Kertész, and Ruth Klueger. Literary writing on the Holocaust requires some sort of distancing and objectification. Améry wrote from a distance of twenty years, the others of forty and more. Each chose different methods: Améry, narrative broken by reflection; Kertész, a simple-minded young narrator and his naïve observation and acceptance of horrible situations; Edvardson the use of the third person and of montage; and Klueger, reflections on the time of writing as the context of her narrative. *RW*

0913 Schreckenberg, Heinz: *Christliche Adversus-Judaeos-Bilder: Das Alte und Neue Testament im Spiegel der christlichen Kunst*. Frankfurt a.M.: Peter Lang, 1999. 469 pp.

Presents 250 mainly medieval and early modern illustrations taken from manuscripts and church buildings, intended to accompany the author's three-volume collection "Die christlichen Adversus-Judaeos Texte" (1988-94). Like the texts, the images deal mainly with biblical and New Testament themes. They are arranged in chronological order, and are accompanied by explanatory notes and references. The introduction notes that the Old Testament was considered important only for its foreshadowing of the New, and explains the most common typological themes found in the illustrations. Many show Jews as enemies of Christianity; the few good Jews are usually depicted without Jewish characteristics, while the others are distinguished by their Jewish attire and often by ugly "Jewish" physiognomies. Other pictures show the supersession of the Synagogue by the Church. The glossary (pp. 401-429) contains not only definitions but also broader descriptions of the conditions under which Jews lived among the Christians. *RW*

0914 Schumacher, Manfred: *Barabas' Enkel: Juden-Bilder in der englischen Literatur bis zur Moderne*. Frankfurt a.M.: Peter Lang, 2000. 209 pp.

Against the background of the mentality of the times, the social, political, and religious issues, and the attitudes toward the Other, analyzes images of the Jew in English literature: as criminal (Marlowe's Barabas, Dickens' Fagin); as usurer obstructing modern enterprise (Shylock); as virtuous benefactor (some Victorian novels); as precariously assimilated (Joyce's Bloom). Diverse as these are, all of them are outsiders. Discusses stereotypes of the Jew, and the Jew as symbol and as an object of fear – whether medieval fear of the infidel or modern fear of the agent of social change. *RW*

0915 Schwarz, Daniel R.: *Imagining the Holocaust*. New York: St. Martin's Press, 1999. 353 pp.

Deals with the ethics of imagining the Holocaust – representation, responsibility, and reading – as illuminated by works of literature in four categories: memoirs; realism; myth, parable, and fable; and fantasy. Ch. 1 (pp. 45-73), "The Ethics of Reading Elie Wiesel's 'Night'," was first published in "Style" 32 (1998). Other authors or works discussed are Primo Levi, "The Diary of Anne Frank," Sophie Goetzel-Leviathan's "The War from Within," Tadeusz Borowski's "This Way for the Gas, Ladies and Gentlemen," John Hersey's "The Wall," Gerald Green's "Holocaust: A Novel of Survival and Triumph," Jerzy Kosinski's "The Painted Bird," William Styron's "Sophie's Choice," Keneally's and Spielberg's

"Schindler's List," André Schwarz-Bart's "The Last of the Just," Aharon Appelfeld, Leslie Epstein's "King of the Jews," Art Spiegelman's "Maus," Cynthia Ozick's "The Shawl" and "Rosa," Bruno Schulz's "The Street of Crocodiles" and "Sanatorium under the Sign of the Hourglass" (as prophecies of the Holocaust), and Cynthia Ozick's "The Messiah of Stockholm" (a response to Schulz). *YC*

0916 Seng, Joachim: Damit der Schrei der Opfer nicht verstummt: Paul Celan und der Dokumentarfilm "Nacht und Nebel." *Neue Rundschau* 112, 3 (2001) 166-172.

Celan prepared the German translation of Jean Cayrol's filmscript (1956). He accepted the task because he was witness to a resurgence of antisemitism in Germany, and felt that "after Auschwitz no-one has the right to be apolitical." The exclusion of the film from the Cannes film festival, following the intervention of the German ambassador to France, who feared the public would not distinguish between the criminal Nazi regime and present-day Germany, angered many in both countries, and certainly Celan. Celan also detected antisemitism in the campaign slandering him as a plagiarist (the Goll affair). Analyzes changes Celan made in the French original: where Cayrol described life in the concentration camps in the past tense, Celan used the present; for him, it was not over. He added details relevant to his own fate and that of his parents, and omitted names seemingly identifying individuals as non-Jews (Cayrol and Resnais ignore the fact that most of the victims were Jews). The film influenced several of Celan's poems. *RW*

0917

שפי, נעמה: היסטוריה של שנאה: היהודי זיס בעין המצלמה הנאצית. זמנים 72 (סתיו 2000) 89-98.

[Sheffi, Na`ama: A History of Hate: Jud Suess in the Eye of the Nazi Camera. *Zmanim* 72 (Fall 2000) 89-98.]
The film "Jud Suess" was directed by Veit Harlan and produced in 1940 by Universum Film, on the orders of Joseph Goebbels. It was intended to convince the German and European public that the Jews must be destroyed. The Nazis realized that film was an effective way to reach a large percentage of the population, including those who were uneducated. "Jud Suess" was one of the 30 most popular German films in 1940-42. Joseph Suess Oppenheim was chosen by Goebbels as the film's subject because he epitomized the "evil Jew" – he was involved in finance, accused of treason and fraud, and executed in 1738 for having relations with a Christian woman. After the war, Harlan escaped punishment by insisting that had been forced to make the film. *RE*

0918

שיינטוך, יחיאל: א.דמ.ע.: לבירור מושג מפתח בכתבי ק. צטניק. חוליות 5 (חורף 1999) 290-275.

[Szeintuch, Yechiel: A.dm.e: Clarification of an Essential Concept in Ka-Tzetnik's Writings. *Chulyot* 5 (Win 1999) 275-290.]

An expanded version of a lecture presented at the 12th World Congress of Jewish Studies, Jerusalem, 1997. Deals with the abbreviation "a.dm.e." [alef.daled/mem.ayin] – four Hebrew letters inscribed as a motto in all of Ka-Tzetnik's writings in Yiddish and in Hebrew, as well as in all of the editions of his translated writings. In "Tsofen: adme" (1987), which relates in detail to the abbreviation, Ka-Tzetnik explained that during the Holocaust the symbol was perceived by him as an incantation invoking God to save him. He used to scream it silently every time he encountered life-threatening experiences. Traces the origin of the abbreviation to Jewish incantations, spells, and amulets, based on a prayer first mentioned in a talmudic legend relating to Rabbi Meir, who had been victimized by the Emperor Hadrian's restrictive anti-Jewish laws. Points to similarities in the experiences of the Jewish people during the time of Hadrian and in the Holocaust period, as well as in Rabbi Meir's experiences and those of Ka-Tzetnik. *LFo*

0919

טרסי, אניטה, עורכת: אמנות ורפואה בגטו טרזיינשטאט (טרזין): רישומים מהשנים 1944-1942. חיפה: הפקולטה לרפואה ע"ש ברוך רפפורט, הטכניון; לוחמי הגטאות: בית לוחמי הגטאות, 2001. 56 עמ'.

[Tarsi, Anita, ed.: *Art and Medicine in Ghetto Theresienstadt (Terezín): Drawings from the Years 1942-1944*. Haifa: Bruce Rappaport Faculty of Medicine, The Technion; Lohamei Haghetaot: Ghetto Fighters' House, 2001. 56 pp.] In English and Hebrew.

The catalogue of an exhibition first shown on Holocaust Day 2000 at the Bruce Rappaport Faculty of Medicine, Haifa. Presents 44 drawings by 11 artists (four of whom are physicians and one is a nurse) who drew physicians and scenes illustrating different aspects of health care in the ghetto. Pp. 50-51 contain a list (in English) of physicians who gave lectures in Theresienstadt. Includes a brief introduction and biographical information on each artist, in English and in Hebrew. *SSC*

0920 *Teofila Reich-Ranicki: Bilder aus dem Warschauer Ghetto*. Frankfurt a.M.: Juedisches Museum, [1999]. [50] pp. Opposite title page: Beiheft zur Ausstellung "Teofila Reich-Ranicki: Bilder und Erinnerungstuecke aus dem Warschauer Ghetto" in der Boernegalerie im Museum Judengasse vom 9. Nov. 1999 bis 9. Jan. 2000.

The catalogue of an exhibition of paintings by Teofila Reich-Ranicki (née Langnas). She was born in 1920 in Łódź and fled with her parents to Warsaw in 1939. Her father, who had been a successful businessman in Łódź, committed suicide in January 1940. Teofila married Marcel Reich in July 1942 in the Warsaw ghetto; in February 1943 they escaped and hid on the "Aryan side." Most of these works were painted in 1942 in the ghetto and smuggled out before their escape. They represent the terror, starvation, struggle for life, and deportations. The brochure also contains excerpts from Marcel's memoirs. *ID*

0921 Vordermayer, Martina: *Antisemitismus und Judentum bei Clemens Brentano*. Frankfurt a.M.: Peter Lang, 1999. 300 pp. Based on the author's diss. – Universitaet Muenchen, 1997.

Brentano (1778-1842) incorporated folkloristic material, including traditional Jewish literary types and anti-Jewish themes, into his writings. For instance, in "Das Maerchen von Gockel, Hinkel, und Gackeleia" (1815-16), he satirizes the Rothschilds in the form of three Jewish swindlers. Toward the end of his life, he published Catholic tracts, among them a record (with his comments) of the mystical visions of a nun, including a vivid account of the Passion and the guilt of the Jews. He transmitted medieval beliefs, such as ritual murder, but suggested that even if such cases did occur they would have been the act of a fanatical Jewish fringe group. Asserts that Brentano had nothing in common with the racist antisemitism that arose in his time. He valued ancient Judaism as the root of Christianity and took a great interest in Jewish tradition; his relationships with (mainly converted) Jews were based on personal inclinations and literary rivalries that had little to do with their Jewishness. But he was influenced by the circles in which he moved: first, the Romantic-national movement and his friend Achim von Arnim; then, the Catholic Church and Joseph Goerres. *RW*

0922 Witthoefft, Cornelis, ed.: *Komponisten in Theresienstadt*. 2nd, rev. ed. Hamburg: Initiative Hans Krása, 2001. 96 pp. First published in 1999.

A collection of essays on the activities and fate of six Jewish composers in Theresienstadt. Contents: Eckstein, Pavel: Pavel Haas (1899-1944) (8-19); Slavický, Milan: Gideon Klein (1919-1945) (20-31); Witthoefft, Cornelis: Hans Krása (1899-1944) (32-51); Witthoefft, Cornelis: Karel Reiner (1910-1979) (52-63); Schultz, Ingo: Siegmund Schul (1916-1944) (64-73); Schultz, Ingo: Viktor Ullmann (1898-1944) (74-87). *ID*

0923 Wright, Melanie J.: 'Don't Touch My Holocaust': Responding to "Life Is Beautiful." *The Journal of Holocaust Education* 9, 1 [Sum 2000] (2001) 19-32.

Reflects on Roberto Benigni's "Life Is Beautiful" (1999) and on the ability of the cinema to handle the Holocaust. Suggests that part of the criticism of this film is inappropriate; humor may be used to approach a tragedy, and Benigni is not the first to make a humorous film dealing with fascism and the Holocaust. However, and above all, cinema is unable to represent the Holocaust authentically; thus, Benigni's language is not that of documentary, but is closer to that of fable or opera. Notes that, in the words of Gillian Rose, "fascism and representation are inseparable," and many Holocaust films share with the Nazi camps the willingness to combine detachment and brutality, distance and cruelty, pleasure and indifference. Benigni does not avoid this pitfall. Besides, his film neglects the Jewish aspect of the Holocaust. *DR*

0924 Young, James Edward: *At Memory's Edge: After-Images of the Holocaust in Contemporary Art and Architecture*. New Haven: Yale University Press, 2000. viii, 248 pp.

Presents seven essays (most of them published previously) dealing with postwar artists and architects whose "memory" of the Holocaust is mediated and vicarious. They reject any redemptory function for art. Chs. 1-3 (pp. 12 -89) deal with the works of Art Spiegelman, David Levinthal, and Shimon Attie. Chs. 4-6 (pp. 90-183) deal with monuments or "counter-monuments" in Germany which need to relate to the absence of the Jews who were killed. The artists discussed are Horst Hoheisel, Micha Ullman, Rachel Whiteread, Renata Stih and Frieder Schnock, and Jochen Gerz, as well as Daniel Libeskind and his Jewish Museum in Berlin. Ch. 7 (pp. 184-223) deals both with the dilemmas of Germany memorializing its victims and the author's own role in this process as he moved from being a skeptical member of the German Commission to one who tried to make constructive suggestions for a way for Germany to memorialize its Jews. *YC*

0925 Zelizer, Barbie, ed.: *Visual Culture and the Holocaust*. New Brunswick, NJ: Rutgers University Press, 2001. vii, 364 pp.

This collection of essays explores representation of the Holocaust in film, photography, comic books, art, museums and museum artifacts, the human body, television and video, and the Internet. *YC*

AUTHOR INDEX

SUBJECT INDEX

Bei Fragen zur Produktsicherheit wenden Sie sich bitte an:
If you have any questions regarding product safety,
please contact:

Walter de Gruyter GmbH
Genthiner Straße 13
10785 Berlin
productsafety@degruyterbrill.com